Economic Geography

Exploring Geography

Series Editor: David H. Kaplan

David H. Kaplan's series, "Exploring Geography," offers a set of cohesive, concise overviews of key subjects in geography. Books in the series serve both as core texts for upper-division courses and as state-of-the-art synopses of a particular subfield that point the way toward future trends and research. More compact and affordable than most standard textbooks, each is written by a recognized authority and covers the major concepts and debates in the field.

Titles Include

An Introduction to the Geography of Tourism, Third Edition
By Velvet Nelson

An Introduction to Transportation Geography: Transport, Mobility, and Place
By Julie Cidell

Food Geographies: Social, Political, and Ecological Connections
By Pascale Joassart-Marcelli

Introduction to Geographic Information Systems
By Falguni Mukherjee

Biogeography in a Human-Influenced World
By Kenneth R. Young

Political Geography
By Corey Johnson

An Introduction to Health and Medical Geography
By Sue C. Grady

Economic Geography

Capitalism, Globalization, and Landscapes

Barney Warf

BLOOMSBURY ACADEMIC
NEW YORK • LONDON • OXFORD • NEW DELHI • SYDNEY

BLOOMSBURY ACADEMIC
Bloomsbury Publishing Inc, 1359 Broadway, New York, NY 10018, USA
Bloomsbury Publishing Plc, 50 Bedford Square, London, WC1B 3DP, UK
Bloomsbury Publishing Ireland, 29 Earlsfort Terrace, Dublin 2, D02 AY28, Ireland

BLOOMSBURY, BLOOMSBURY ACADEMIC and the Diana logo are
trademarks of Bloomsbury Publishing Plc

First published in the United States of America 2026

Series Design by Aneeka Makwana
Cover image: Aerial image of illuminated buildings in Manhattan, New York City © Michael H via Getty Image

Library of Congress Cataloging-in-Publication Data

Names: Warf, Barney, 1956- author
Title: Economic geography : capitalism, globalization, and landscapes / Barney Warf, University of Kansas.
Description: New York : Bloomsbury Academic, 2027. | Series: Exploring geography | Includes bibliographical references and index. | Summary: "Economic Geography offers a concise overview of how "place" impacts economic outcomes and processes. It balances U.S. and international examples with a unique chapter on consumption, seldom treated in-depth in competing texts. In addition, end-of-chapter study questions will help students develop critical thinking skills"– Provided by publisher.
Identifiers: LCCN 2025053287 (print) | LCCN 2025053288 (ebook) | ISBN 9781538168318 hardback | ISBN 9781538168325 paperback | ISBN 9781538168332 epub | ISBN 9798765150078 pdf
Subjects: LCSH: Economic geography
Classification: LCC HF1025 .W37 2027 (print) | LCC HF1025 (ebook) | DDC 330.9–dc23/eng/20260401
LC record available at https://lccn.loc.gov/2025053287
LC ebook record available at https://lccn.loc.gov/2025053288

ISBN: HB: 978-1-5381-6831-8
PB: 978-1-5381-6832-5
ePDF: 979-8-7651-5007-8
eBook: 978-1-5381-6833-2

Series: Exploring Geography

Typeset by Integra Software Services Pvt. Ltd.

For product safety related questions contact productsafety@bloomsbury.com.

To find out more about our authors and books visit www.bloomsbury.com
and sign up for our newsletters.

Contents

Figures

Tables

Preface

This volume is intended to give students a comprehensive overview of economic geography. Economic activities are profoundly important: they shape every aspect of our lives, from how and where we work to what we consume. But economies do not exist suspended in the sky: they are grounded and stretched over the surface of the earth. A geographic perspective offers many advantages: it affords us the opportunity to understand why activities occur in some places and not others, or why the same process (e.g., mechanization of agriculture) can have very different impacts in different places. A geographic view encourages us to contextualize our observations, showing that *where* they occur affects *how* they occur. It highlights how places come into being and the processes that tie places together, whether through trade, migration, or diffusion.

There are many ways to study economic geography. For decades, the field consisted of little more than the recitation of facts. The quantitative revolution in geography gave rise to a tradition of mathematical models, such as those based on transport costs, which is still active today.

This volume adopts a political economy perspective on economic geography, which has arguably become the core of this sub-discipline. This approach emphasizes the role of social relations in the construction of economic landscapes, viewing them as outcomes of processes that inevitably and inescapably involve class, power, gender, ethnicity, and inequality. This line of thinking invites us to think about social processes that are both material (e.g., investment, trade, consumption) and simultaneously ideological (e.g., religion, nationalism, consumerism); the world of social practices and that of ideas and knowledge are seamlessly integrated. A political economy view stresses that landscapes, such as those characterized by wealth and poverty, are made, not given. This view is at odds with many prevailing conceptions that emphasize supply and demand.

Throughout, the volume emphasizes an historical understanding of economic geography. The landscapes of the present did not appear out of thin air: they were produced by processes that often took centuries to unfold. Historicizing the social world helps to denaturalize it, to show that events and phenomena were made by people, even if unintentionally. Moreover, an historical perspective allows us to appreciate capitalism as one of several ways people have created economies, not as some eternal, unchangeable society. By the standards of world history, capitalism is relatively recent, but since its emergence in the sixteenth century, it has come to dominate almost all of the planet. Economic geographies today are essentially the landscapes of capitalism, which are characterized by uneven spatial development.

Finally, a political economy view foregrounds the causal processes of capital accumulation in the production of wealth and poverty. The book proceeds by sketching the origins and dynamics of capitalism, the creation of the world economy, and the nature of poverty and development in the developing world. For this reason, it holds off on the discussion of population growth until after these subjects have been covered; population change is seen primarily as the outcome rather than the cause of economic activities.

1 Introduction

Geography is the study of space, of how the earth's surface is used, of how societies produce places, and how human activities are stretched among different locations. In many respects, geography is the study of space in much the same way that history is the study of people in time. This conception is very different from popular stereotypes that portray geographers as a boring bunch concerned only with drawing boundaries and obsessed with memorizing the names of obscure capital cities. Essentially, the *discipline of geography examines why things are located where they are.* Simply knowing where things are located is relatively simple; anyone with an atlas can find out, say, where bananas are grown or the distribution of petroleum. Geographers are much more interested in explaining the processes that give rise to spatial distributions, not simply mapping those patterns. Much more interesting than simply describing patterns on the earth's surface is the *explanation* linking the spatial outcomes to the social and environmental processes that give rise to them. Thus, geographers examine not only where people and places are located, but how people understand those places, give them meaning, change them, and are in turn changed by them.

All social processes and problems are simultaneously spatial processes and problems, for everything social occurs somewhere, that is, in a place. More important, *where* something occurs shapes *how* it occurs. Place is not some background against which we study social issues, but it is part of the nature and understanding of those issues. Geographers ask questions related to location: Why are there skyscrapers downtown? Why is there hunger in Africa? Where are cell phones made, and why? How does international trade affect local economies? Why is China rapidly becoming a global economic superpower? How has the microelectronics revolution changed productivity and competitiveness and the global locational dynamics of this sector? What can be done about rural poverty?

To view the world geographically is to see space as socially produced, as made rather than simply given, i.e., as a product of social relations, a set of patterns and distributions that change over time. This means that geographic landscapes are social creations, in the same way that your shirt, your computer, your school, and your family are also social creations.

Geographers argue that the production of space involves processes that operate at different spatial scales, ranging from the smallest and most intimate—the body—to progressively larger areas, including neighborhoods, regions, nations, and the global economy. Individuals create a geography as they move through time and space in their daily routines. Societies and geographies are formed by the aggregate movements of people through space and time in everyday life. In local communities, neighborhoods, and cities—the next larger scale—these movements form regular patterns that reflect a society's organization, its division of **labor**, cultural preferences and traditions, and political opportunities and constraints. Geographies thus reflect economic, political, and cultural relations that give rise to class, gender, ethnicity, age, and other categories that shape our lives. Moreover, because geographies are produced historically, spatial patterns reflect the historical legacy of earlier social relations, political and economic organization of resources, the technologies of production, transportation, and communications, the cultures that inform behavior and guide it, and legal and regulatory systems. The global economy itself—an intertwined complex of **markets** and countries—involves planetary patterns of production, transportation, and consumption, with enormous implications for the standard of living and life chances of people in different areas.

Geographers study how societies and their landscapes are intertwined. To appreciate this idea, we must recognize that social processes and spatial structures shape each other in many ways. Societies involve complex networks that tie together economic relations of wealth and poverty, political relations of power, cultural relations of meanings, and environmental processes as well. Geographers examine how societies and places produce one another, including not only the ways in which people organize themselves spatially, but also how they view their worlds, how they represent space, and how they give meaning to it. Separating one dimension, say the economic, from another, such as the political, is ultimately fruitless, but to make the world intelligible, we must approach it in manageable chunks. This text centers upon only one aspect of this set of phenomena, economic landscapes.

Economic geography is a subdiscipline concerned with the spatial organization and distribution of economic activity (production, transportation, communication, and consumption), the use of the world's resources, and the geographic origins, structure, and dynamics of the **world economy**. Economic activities are often grouped into a primary sector (extraction of **raw materials**, agriculture, mining, forestry, fishing), a secondary sector (manufacturing, or the transformation of raw materials into finished goods), and a tertiary sector (finance, services, transportation, communications). Each of these creates its own distinctive distribution and landscape, and is intertwined with the other sectors. In addition, economic geography involves issues of wealth and power, such as population distributions, uneven **development**, and the role of governments in shaping economies and landscapes. Economic geographers address a wide range of topics at different spatial scales using different theories and methodologies. Trade, migration, poverty, flows of capital, **globalization**, crises, and **transnational corporations** are all objects of economic geographic inquiry.

1.1 Five Analytical Themes for Understanding Economic Geography

One means of starting a comprehensive analysis of economic geography is through five analytical themes, which will reappear in different ways throughout this book. These are broad generalizations designed to encourage you to think about economic landscapes, and include: (1) the historical nature of geographies; (2) the interconnectedness of regions, particularly since the rise of the global capitalist economy; (3) the interpenetration of human and biophysical systems; (4) the importance of culture and everyday life in the creation of social and spatial relations; and (5) the centrality of comprehending social structures and their spatial manifestations.

1. **The study of space is inseparable from the study of time.** If geography is the study of human beings in space and history as its temporal counterpart, then geography and history are inseparable. All geographies are produced historically, and all histories unfold geographically. Historical awareness undercuts the notion that the present is the "typical" or "normal" way in which human beings organize themselves, for it is history as much as geography that teaches us the full range and diversity of human behavior, cultures, and social systems.

Such a view invites us to ask *how* histories and geographies are produced, particularly through the everyday lives of ordinary individuals. Historical geography—a redundancy, for *all* geography is inescapably historical—is thus much more than simple reconstructions of past worlds, but the analysis of social systems over space and time.

Taking history seriously means acknowledging that geographies are always changing, that they are forever undergoing change, that landscapes are humanly created and therefore plastic and mutable. History is produced through the dynamics of everyday life, the routine interactions and transient encounters through which social formations are reproduced. Time is therefore synonymous with historical change and the capacity of people to make, and change, their worlds. Views that purport to represent a "snapshot in time," therefore, are more deceiving than illuminating; it is the *process* that underlies the creation of places that is central, the social dynamics at work, not their appearance at one instant in time.

2 **Every place is part of a system of places.** This theme stresses that all places, cities, regions, and countries are interconnected; that is, they never exist in isolation from one another. Places are invariably tied together to a greater or lesser extent by the biophysical environment (e.g., winds and currents, flows of pollution), flows of people (migration), capital (investment), and goods (trade), and the diffusion of information, innovations, and diseases. It follows that what happens in one place must affect events in others; the consequences of action are never purely local. For example, the outbreak of the COVID pandemic in Wuhan, China, in 2019 soon reached across the globe and changed life everywhere. Russia's invasion of Ukraine in 2022 led to food shortages in the Middle East. Decisions made in New York, Washington DC, London, and Beijing shape people's lives everywhere.

Places have always been connected to one another, even if only to a minimal extent. But this theme is especially pertinent since the rise of **capitalism** on a global basis since the sixteenth century. More broadly, the global system of nations and markets has tied places together to an unprecedented degree, including international networks of trade, investment, subcontracting, telecommunications, and worldwide markets, as any trip to the grocery store will attest.

3 **Human geographies are produced in interaction with the biophysical environment, or nature.** The biophysical environment includes the climate, topography, soils, vegetation, and mineral and water resources of a region. Nature affects everything from the length of a growing season to transport costs to energy supplies. It is important to acknowledge that these factors *affect* the construction of histories and geographies but do not determine them. But the interpenetration of people and nature is a two-way street: everywhere, nature has been changed by human beings, such as through the modification of ecosystems, annihilation of plant and animal species, deforestation, soil erosion, pollution, changed drainage patterns, agriculture, desertification, disruptions of biogeochemical cycles, and more recently, global warming. Indeed, human beings cannot live in an ecosystem without modifying it. In short, the formation of geographies is neither reducible to nature nor independent of it.

This theme illustrates how nature shapes human action and how in turn those environments are always and everywhere changed through human actions. For example, think about human modifications of the New World prior to the Columbian encounter, dispelling the myth that Indigenous peoples inhabited their world without changing it; in reality, they burned grasslands, hunted, planted crops, and modified ecosystems extensively. Agriculture has had perhaps the greatest human impact on the natural world because it is so space-extensive. The geography of the Industrial Revolution was profoundly preconditioned by the location of the large coal deposits in the northern European lowlands stretching from Wales to Silesia. More recently, geopolitical conflicts over water or petroleum illustrate the continued significance of nature in human affairs.

4 **Culture is fundamental to economic geography.** Because human beings are sentient, they have consciousness about themselves and their world. This consciousness is shaped by societies through socialization, and sets the contours of individual perceptions, cognition, identity, and language. Traditional definitions of culture view it as the sum total of learned behavior or a "way of life," including religion, language, mores, traditions, roles, etc. More recent interpretations hold that culture is what we take for granted, i.e., common sense, the ideologies that allow people to negotiate their way through their everyday worlds. In every society, culture defines what is normal and what is not, what is important and what is not, what is acceptable and what is not. People reproduce their worlds, largely unintentionally, in their everyday lives, and in turn, the world reproduces them through socialization. Put differently, individuals are both the producers, and are produced by, history and geography. Everyday thought and behavior hence do not simply mirror the world, they constitute it. In addition, culture is deeply tied to social relations of power, including divisions of class, gender, and ethnicity.

   This theme is useful in appreciating how the "economy" is not sealed off from other domains of social action; culture enters deeply into economic and political behavior. For example, the ideology of nationalism was vital to the historical emergence of the nation-state. Many industries that rely on face-to-face interaction, such as investment banking, are heavily conditioned by cultural norms of trust and behavior. Ethnicity is critical to knowing how many economies operate, such as in immigrant communities. Gender norms deeply shape labor markets, generating unequal opportunities for men and women.

5 **Social relations are fundamental to understanding societies and geographies.** Social relations include the structure of classes, power relations, gender roles, and the opportunities and constraints faced by different groups. Viewing societies in this way means that "the social" is not simply reducible to the sum of individual behaviors. Social relations are often theorized through the lens of **political economy**, including the uneven distributions of power along the lines of class, gender, ethnicity, and place. Political economy is concerned with "who gets what, when, where, and why," that is, the ways in which resources and opportunities are distributed. As economic geographers came to rely heavily on a political economy perspective, it gradually became the primary way of understanding the world, an important distinction between the contemporary disciplines of geography and economics.

Thus, to understand contemporary economic geographies, we must understand their historical development, the class structure of a society, its relations of gender and ethnicity, and how these dimensions are tied to culture and ideology. Economic landscapes are the products of changing social relations that organize and reorganize space in a diverse array of historically distinctive ways.

Power is a fundamental characteristic of all societies and, of course, takes many forms, including violence, personal charisma, status, the capacity to withhold favors, control over information, bureaucratic rank, the role of the state, etc. Power is a fundamental part of how any social system is organized, and the economy and politics cannot be divorced, for power and wealth are always closely linked. Power is always unequally distributed among and within societies, and for political economy, therefore, social and spatial inequalities figure front and center in their analysis. Any understanding of economic geography, of who is relatively rich and powerful and who is poor and holds less power, must therefore invoke some notion of economic class, as well as gender, ethnicity, and other types of social relations. In every society, wealth and power are primary axes around which individual lives are structured. Viewed this way, the economy and politics are inseparable. Access to education, jobs, housing, healthcare, and transportation are examples of how political economy shapes everyday life. Throughout this book, we will turn to political economy to explore how geographies are produced, including the restructuring of regions, uneven development, the ways new technologies are incorporated into social systems, boom and bust cycles, the service economy, and so forth.

Much as political economy highlights the centrality of class, feminists have shown how social and geographic reality is pervasively *gendered*, that is, how gender relations are woven into existing allocations of resources and modes of thought. Generally, gender relations work to the advantage of men and the disadvantage of women. Feminist scholars have studied gender relations in families, labor and housing markets, and the structure of cities, making this view an essential part of economic geography. For example, spatial patterns of work and daily life reveal profound differences between men and women in commuting patterns.

1.2 Capitalism

This book is primarily concerned with the economic geography of capitalism. Although it is sometimes viewed as a universal type of society, capitalism in fact is only one of many possible types of society. Capitalist societies are defined by a particular configuration of economic relations centered on markets, private ownership of property, and **profit**. Historically, capitalism arose in Western Europe in the fifteenth and sixteenth centuries, and, in the form of colonialism, ultimately came to be spread over most of the world today (Chapter 2).

The fundamental institution involved in the organization of factors of production in capitalist economies is the market, by which buyers and sellers interact through supply and demand using prices. The guiding imperative in capitalist economies is profit, the difference between the revenues a firm receives and its production costs. Only profitable products will be produced, based on market demand and price. Prices reflect the value of goods and services based on what consumers are able and willing to pay. How and where the goods are produced is based on labor and technology efficiency and the spatial distribution of production costs. In competitive market economies, the most efficient producers are the survivors; their production processes and locations will dictate how and where goods will be produced.

As we shall see, the economic landscapes of capitalism are characterized by uneven spatial development, or vast differences in incomes and standards of living among places. This process occurs at several spatial scales, that is, among countries, within countries, and within towns and cities. The disparities in standards of living reflect a wide variety of processes, including the legacy of colonialism, the distribution of natural resources, types of governance, trade relations, debt, and differing rates of population growth.

Although many people believe otherwise, capitalism consists of more than just markets. A commonly held view of capitalism is that it is synonymous with the "free market" with minimal government intervention, a system sometimes called laissez-faire. However, in reality, truly free markets (with zero government rules) have never really existed. Governments have always been an integral part of capitalist economies, including the provision of the infrastructure, protecting property rights, providing public services such as education and transportation, and protecting producers from foreign competition. Indeed, the argument can be made that markets could not exist without some role of the state. This means that the various forms of capitalism are mixed systems in which both markets and governments are important decision makers, including vital domains such as transportation, education, and health care. The balance between markets and governments varies widely among countries and over time, ranging from those with high levels of government intervention, such as in Scandinavia, to those with relatively little, such as the United States.

1.3 The World Economy

The focus of this book is the world economy, the various social processes, institutions, and networks that shape the planetary system of resource distribution and consumption, create vastly different levels of wealth and poverty across the globe, and contribute to the rise and fall of national powers. The global scale is only one way in which economic geography can be studied, but given the massive processes of globalization that have been at work, particularly over the last fifty years, it is the most useful way to understand what goes on in the world. The world economy links far-flung people and places so that what happens in one place shapes what happens in another through networks of interdependence (theme 2 above). The everyday lives of billions of people, including you, are deeply shaped by these processes: your clothes and cell phone may be made in China, your gasoline is likely from the Middle East, and financial decisions made in London or New York shape your credit availability and interest rates you pay. Every trip to the grocery store is a window to the global economy. In this way, the global economy and everyday life are two sides of the same coin.

The world economy constantly undergoes changes, including geopolitical, technological, and demographic forces. These changes reflect capitalism's dynamism, its incessant pressures on firms to create new products, new production techniques, and new markets. Around the world, countries have witnessed the steady growth of large transnational corporations, the rising role of markets and diminished role for the state, and declining barriers to trade. Improvements in transportation and communications have greatly reduced the friction of distance and barriers to worldwide exchange. The world's cultures have become globalized, and often homogenized, through the internet, television, and the globalization of consumption. The principal instruments of globalization of the economy are transnational corporations, which are producing new efficiencies and new geographies in production, distribution, and the use of the world's resources. The growth of international banking and finance via telecommunications systems has made industries such as currency exchange the largest in the world by volume. Across the planet, conservative forces have reduced the roles of governments through privatization and deregulation. Transportation cost and time reductions have improved exchanges of people and goods. The internet connects two-thirds of the world's population. Rising populations in the developing world, and stagnant demographic growth in the developed world, have altered the global supply, demand, and cost of labor, shaping immigration patterns. Globalization has accelerated corporate investment, trade, and tourism. It has unleashed new diseases that shock the global economy, such as COVID. Improved fishing technologies have depleted many of the world's fisheries. The rising global demand for meat has accelerated the deforestation of tropical rainforests. Emissions of gases such as carbon dioxide and methane have altered the planet's climate, leading to unprecedented warming and biological changes. These changes have enormous implications for countries, states, and regions.

The global economy does not simply consist of markets: it also involves nation-states. The capitalist world economy is a multistate system that began in Western Europe in the early sixteenth century and grew over the next four hundred years. As this system expanded, it produced a core group of wealthy countries and a periphery of poorer, less powerful countries. This division

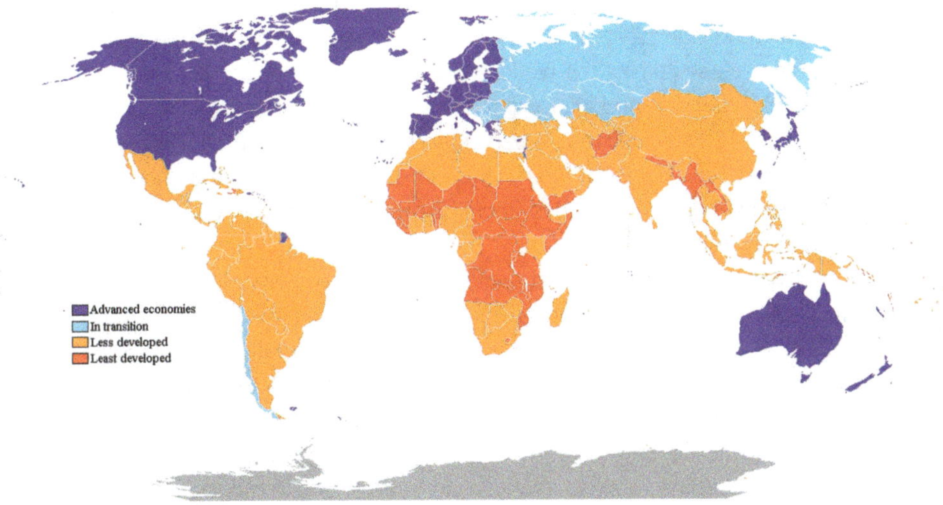

Figure 1.1 The world's developed and less developed countries.

is often framed in terms of the world's developed and developing countries. The economically developed world consists of Europe, the United States and Canada, Australia, New Zealand, and Japan (Figure 1.1). The defining feature of these countries, which comprise about one-quarter of humanity, is their relatively high standard of living, characterized by a large middle class. The *Second World* was represented by the Soviet Union and Eastern Europe, a designation that lost its meaning in the post–Cold War era (the 1990s and since), when the Second World disappeared, to be divided between the First and Third.

The poorest countries are often called developing or less developed countries, which include Latin America, Africa, the Middle East, and most of Asia. This is a broad set of societies with a diverse range of cultures, historical backgrounds, and standards of living. A few countries have climbed their way out of the developing world, such as Singapore or South Korea, to enjoy standards of living that rival those in Europe and North America. However, there are strict limits on the mobility of countries: for most, being locked into poverty offers little chance of escape.

1.4 Globalization

Globalization consists of a complex set of interrelated processes that define the world economy and make the various countries within it more interdependent. It is essentially an expansion in the scope, scale, velocity, and consequences of international transactions. Globalization is a process that shrinks the world by reducing the transport and communication times and costs among places, a process often called time-space compression. This has generated a global market in which more goods and services are widely available everywhere than ever before, a global culture dominated by the West (particularly the United States), and enormous global environmental changes and problems.

Globalization is not simply inherently beneficial or inherently negative in character. Rather, it is a mixture of both sets of qualities that varies widely by place. In some regions, social, political, and economic problems result from a tension between the processes promoting global culture, economy, and environment on the one hand, and the practice and preservation of local economic isolation, cultural tradition, and the localization of environmental problems on the other hand.

Because contemporary capitalism is so information-intensive, globalization has been greatly accelerated by a worldwide digital telecommunications network, notably the fiber optic system that forms the nervous system of the world economy. This technology enabled the internet, as well as global cable television and international news services; news travels around the world at unprecedented speeds. However, the uneven spread of fiber, the internet, and cell phones generates its own geographies. Access to communications is restricted by the uneven distribution of wealth worldwide. Even within countries, there are uneven geographies of access, particularly between urban and rural places.

This system also allowed for financial markets to become deeply intertwined, so that digital money travels at the speed of light. Globalization today involves enormous international financial flows. In the deregulated, hypermobile electronic world of international banking, telecommunications have created a single global capital market in which billions of dollars in digital funds effortlessly cross national borders. Traders can monitor and trade in national currencies, stocks, bonds, and annuities listed anywhere in the world instantaneously. Consequently, banks and corporations can react immediately to changes in the value of commodities or gold on the world market and the rate of exchange between the dollar, the euro, the yen, and other currencies.

The major actors in the globalization of the world's economy are transnational corporations (TNCs), which conduct research, operate factories, and sell products in many countries. Today, there are some 100,000 TNCs in the world controlling about 40 percent of all private sector assets and accounting for one-third of the goods produced. The sales of the largest TNCs are larger than most countries' total economies. This is important to a small country's economy, which can be affected by a decision of a global corporation. They dominate international trade. Their disparate operations are linked by globalized supply chains that integrate components made in different countries into one product, such as an automobile. Most transnational corporations maintain their headquarters in one of the core countries—North America (the United States and Canada), Western Europe (especially Germany, France, the UK, and the Netherlands), and Japan. In 2020, transnational corporations employed a hundred million workers in the core regions, with another sixty million elsewhere.

Foreign direct investment (FDI) is a term for the investments made by companies from one country in another. It is a critical source of investment capital in many countries, as well as a source of jobs. US transnational corporations are most likely to invest in Europe, Canada, China, and Latin America. Western European transnationals are most likely to invest in Eastern Europe, Russia, and African markets, as well as in North America. Japanese transnationals are most likely to invest in East Asia and in North America.

The globalization of services is also important. For example, US service exports generate one-third of the nation's foreign revenues. Business services such as law firms, advertising, marketing, public relations, and accounting provide essential inputs to transnational corporations. Many professionals, such as architects, engineers, software designers, and business consultants, sell their skills throughout the world. The globalization of services includes international tourism, visiting foreign students, and movies. Most US motion picture studios derive the bulk of their revenues from outside the United States.

Globalization affects different regions in different ways, and generally does not completely destroy their local diversity. This notion helps us to dispel simplistic views that globalization always homogenizes culture, as well as "one-size-fits-all" interpretations. Global processes, such as trade, investment, and consumption, are often tailored to individual markets and cultures. For this reason, globalization and local diversity shape each other, an idea geographers call "glocalization."

1.5 Summary and Outline of This Volume

Because capitalism is by far the dominant economic system in the world today, it is the primary focus of this volume, the theme that runs throughout subsequent chapters. Capitalism is one of many ways in which human beings have organized economies, but its spread over the last five hundred years throughout the world has led it to penetrate almost every corner of the world, leaving no place unscathed. Today, a worldwide network of firms, particularly TNCs and banks, dominates the world economy, which coexists with roughly two hundred nation-states. Contrary to the popular notion that globalization is the same everywhere, it adapts to local circumstances.

Chapter 2 outlines the historical origins and growth of capitalism, including its roots in medieval Europe and subsequent expansion. This chapter illustrates theme 1 above. It focuses on the defining features of capitalism such as markets, classes, and the ideologies that accompanied its rise. It also summarizes the Industrial Revolution, which unleashed enormous technological, social, and spatial changes. Finally, it turns to colonialism, the process by which capitalism went global and conquered the world.

The third chapter summarizes some major theoretical notions in economic geography. It starts by summarizing the factors that shape the location decisions of firms and the formation of economic landscapes, such as labor, **land**, capital, and transportation costs. Second, it discusses economic base analysis, the idea that places and countries are linked to broader economies through their exports. Third, it looks at processes of capital substitution and the formation of economies of scale and agglomeration economies, key notions to any substantive theorization. It looks at how firms grow and the product cycle. It offers an overview of how economic landscapes change through business cycles of investment and disinvestment. It concludes by pointing to the role of the state, or government, in the creation of economic geographies.

In Chapter 4, the focus is on international trade and investment. It opens by summarizing the broad contours of global trade. Then it proceeds to the theory of comparative advantage, perhaps the most famous notion in economic geography, and how that idea has been updated through the idea of competitive advantage. It looks at global capital markets, which are vital to understanding trade and the world economy, as well as the role of exchange rates. It describes the growth of the US trade deficit and the reasons that drive it. It examines FDI and its role around the world. Next, it turns to barriers to trade and investment. Following that, it turns to trade agreements and varying levels of economic integration around the world.

Chapter 5 takes up the role of resources in economic geography, an illustration of theme 3 above. It examines types of resources and their relations to population growth. It notes the growth of the world's food supply and the subsequent decline in malnutrition and hunger. It discusses nonrenewable minerals as well as energy, the input that drives the world economy. It briefly notes the environmental and ecological catastrophes that have followed in the wake of unlimited capitalist growth, and concludes by calling for sustainable development.

Because the landscapes of capitalism are typified by uneven spatial development, Chapter 6 examines this process at the global level. It discusses the notion of development, which is not a simple process. It looks at urbanization in the developing world, then at a long list of economic problems faced by the world's less fortunate countries. It summarizes development strategies and the possibilities of development in the developing world. East Asia, the most rapidly growing part of the world economy, gets particular attention. It then returns to the notion of sustainable development as the only viable solution to the problem of long-term growth.

Population growth and economic development are inextricably intertwined. Chapter 7 examines the distribution of the world's people and the dynamics of population growth. It summarizes the most popular and influential theory of population change, Malthusianism, and notes problems with this view. It points to the demographic transition as a more realistic interpretation of how and why societies change in their birth, death, and growth rates. It comments on population structures as manifested in population pyramids. Finally, it discusses international migration, which has huge implications for societies across the globe.

Agriculture is not only how human beings feed themselves, but arguably the most important way in which humans have transformed the world's surface. Chapter 8 opens by differentiating preindustrial from industrial agricultural systems. Second, it discusses the formation of a global agricultural system, which began with European plantations during the age of colonialism. It then turns to the industrialization of agriculture: like manufacturing and cities, farming was greatly affected by the Industrial Revolution. It then looks at commercial agriculture as it is shaped by the capitalist imperative of profit-maximization, leading to the rise of agribusiness and various forms of farming, dairying, and ranching. US agricultural policy, particularly subsidies, comes in for particular scrutiny. Following the theme of sustainable development, it focuses on sustainable agriculture as an alternative to the destructive forces of commercial agriculture.

Chapter 9 delves into the secondary economic sector, manufacturing. It begins by describing the distribution of manufacturing clusters across the world. Then it summarizes four major manufacturing sectors, textiles, steel, automobiles, and electronics, and how each produces a distinctive geography. Third, it discusses the epochal transformation in global manufacturing that began in the late twentieth century, the shift from Fordism to post-Fordism. The fourth part examines deindustrialization, or the decline in manufacturing in many countries, and its aftermath. Finally, it addresses the issue of sweatshops and the exploitative labor conditions found there.

Most of the world economy is engaged in the production of services of one form or another. In Chapter 10, the complexity of the tertiary sector, or services, is explored; it is no easy matter to define services and their output. The chapter discusses the forces that drive the growth of services across the globe. Next, it examines the distinctive qualities of labor markets in services, including gender roles, lack of unionization, and the demand for higher education propelled by business services. It then proceeds to discuss financial services, the core of the economy of many global cities, as well as business services that provide specialized expertise. The next section looks at the dispersal of back offices to peripheral areas. It then touches on consumer services such as retail trade, whose location is driven by the geography of population and income. Finally, it focuses on the world's largest industry by employment, or tourism.

Given the enormous role that digital technologies play in the economy and everyday life, Chapter 11 focuses on digital geographies. It begins by historicizing this issue, noting that telecommunications can be traced back to the nineteenth-century innovations of the telegraph and telephone, predecessors to the current web of fiber optic lines. Despite the rapid diffusion of technologies such as cell phones, profound social and spatial discrepancies in use remain; digital divides remain in terms of internet access. Next, it discusses the relations between telecommunications and geography, dispensing with the myth that information technologies render space meaningless. The third part examines the origins, growth, and impacts of the internet, now used by two-thirds of the world's peoples. Fourth, it summarizes e-commerce, the myriad ways in which the internet has revolutionized business and retail transactions.

The final chapter examines consumption, a topic often overlooked in economic geography. In many countries, consumption comprises the bulk of GDP (gross domestic product). It begins by historicizing consumption and consumerism, which arose on the heels of mass production in the nineteenth century. It then summarizes theoretical perspectives on consumption, including those offered by sociologists (e.g., conspicuous consumption and the role of advertising), neoclassical economists and utility-maximizing theory, and Marxists, such as commodity fetishism. The chapter then delves into global geographies of consumption, which are shaped by the distributions of income as well as cultural preferences, noting uneven patterns of consumption of commodities such as beer, wine, and meat. It touches on the spaces of the shopping mall, and then looks at consumption in light of commodity chain analysis. Finally, it explores the environmental dimensions of consumption, including landfills and ecological footprints.

Glossary

Capital: monetary funds used for investments to generate profits
Capitalism: the political and economic system that revolves around markets and private ownership of production
Development: economic growth, including rising incomes and standards of living
Economic geography: the branch of the discipline of geography concerned with landscapes and the spatial distributions of economic activities
Globalization: capitalism at a worldwide scale, which links parts of the world together in the world economy
Labor: human inputs into the production process
Land: territory on the earth used for production
Markets: institutions that allow for buyers and sellers of commodities to exchange goods and services
Political economy: the structure of a society, including how its resources are produced and distributed and networks of power
Profit: the difference between revenues and costs
Raw materials: resources extracted from the earth, including wood, mineral ores, and water
Transnational corporation: a firm that operates in two or more countries
World economy: the global network of production, distribution, and consumption

Study Questions

1. What defines the geographic perspective?
2. Define economic geography.
3. Why are geography and history inseparably linked?
4. Why can't places be studied in isolation of each other?
5. What are some ways in which nature shapes, and is shaped by, the economy?
6. How is the economy related to culture?
7. Why are social relations an important place to begin understanding economic landscapes?
8. Define the term *globalization* and list reasons why it has occurred.
9. What are four ways in which globalization is manifested?
10. What is the political economy approach to geography?

Suggested Readings

Coe, N., P. Kelly, and H. Yeung. 2007. *Economic Geography: A Contemporary Introduction*. Blackwell.
Dicken, P. 2015. *Global Shift: The Internationalization of Economic Activity*, 7th ed. Guilford.
Florida, R. 2004. *The Rise of the Creative Class*. Basic Books.

Friedman, T. 2005. *The World Is Flat: A Brief History of the Twenty-First Century*. Farrar, Straus and Giroux.

Harvey, D. 2006. *A Brief History of Neoliberalism*. Oxford University Press.

Knox, P., J. Agnew, and L. McCarthy. 2008. *The Geography of the World Economy*, 5th ed. Edward Arnold.

Shepard, E., and T. Barnes (eds.) 2002. *A Companion to Economic Geography*. Wiley.

2 The Historical Development of Capitalism

Geographies are not created overnight. The spatial distributions of people and economic activities reflect processes that often take years, centuries, even millennia to unfold. For this reason, a historical understanding is essential for understanding the contemporary economic geographies of the world. Any serious understanding of economic geography must include an appreciation of how the contemporary world was produced historically. An historical appreciation illustrates that the construction of the modern world took a long time and that the landscapes of the present are constantly changing (thus reflecting the first analytical theme in Chapter 1).

This chapter provides an historical overview of **capitalism** in several ways. First, it delves into **feudalism**, the type of society that preceded capitalism in Europe and formed the crucible in which capitalism was born. Second, it explores the characteristics of capitalism, the features that make it unique. Capitalism, today the dominant form of production and consumption around the world, is not the only way in which human beings have organized themselves but arose in the sixteenth and seventeenth centuries, primarily in Western Europe. Third, this chapter discusses the Industrial Revolution, which began in the eighteenth century and marked an exponential increase in the speed of capitalist activities, leading to large cities, the modern working class, and very high levels of productivity. Finally, it addresses **colonialism**, the process by which capitalism spilled out of Europe and expanded on a global scale, conquering the rest of the planet for the next five hundred years.

2.1 Feudalism and the Birth of Capitalism

There have existed many ways of organizing economies over historical time. For the vast bulk (95 percent or more) of human existence, people lived as hunters and gatherers, nomadic food collectors depending on nature for food and other necessities. The Neolithic agricultural revolution that began roughly ten thousand years ago saw a major transition in the ways in which people worked and lived, including settlements and the first class-based societies. Many early agricultural societies were based on slavery, which in Europe culminated in the Roman Empire that ended in the fifth century AD.

Prior to capitalism, the prevailing form of economic and social relationship was *feudalism*, sometimes called the Middle Ages or Dark Ages, which lasted from roughly the fifth to the fifteenth centuries. Feudalism was not unique to Europe, as other places had similar societies, including Japan, China, the Arab world, and India. Politically, this type of system was manifested in Europe as a changing set of empires, including the Frankish kingdoms, the Normans, the Holy Roman Empire, tsarist Russia, and the Austro-Hungarian Empire; the latter two ended only with the First World War. One of the major differences between Europe and the United States is the impact that feudalism had in Europe: in North America, capitalism emerged on a landscape that had not been shaped by more than a millennium of feudalism, as was Europe, including its land use and property systems, cities, and class and gender relations.

2.1a Characteristics of Feudalism

Feudalism was characterized by a set of interlocking characteristics that made it qualitatively different from capitalism. Compared to the dynamic, rapidly changing world in which we exist today, feudalism was a remarkably stable and conservative world that changed relatively little. Most people's lifestyles—peasants working the land in cycles of endless drudgery—would be the same from one generation to another. Tradition was the dominant shape of human experience and gave everyone a sense of where they fit into the world. Often, that which was new was widely distrusted. In this sense, feudalism actively discouraged experimentation and change. Under capitalism, in contrast, novelty is the norm, for it helps to sell goods and services in the **market** (just look at the advertising that screams "new").

However, it is erroneous to think of the feudal era as completely static and unchanging. In fact, during the later feudal period, there were significant changes: universities were established, new types of farming introduced, wetlands drained, forests cleared, plagues and diseases swept through the land, new technologies were introduced, such as the heavy plow, and wars and ideological conflicts caused enduring changes. The introduction of the longbow and guns in the fourteenth century, for example, made knights essentially obsolete.

In Europe, the Church (that is, the Catholic Church until the rise of the Protestant Reformation of the sixteenth century) was the predominant political and ideological institution. Essentially, everyone was extremely religious, and their belief in god informed every aspect of their behavior and everyday life. The population fatalistically accepted its lot in life, and the idea of progress, of change for the better, was largely unthinkable. In most towns, the cathedral was the largest and most impressive building, its size and design testimony to the wealth and power of the church. Local priests, who were often the only ones who could read and write, were important actors in the community's spiritual and social life, serving as judges, teachers, and officiating at weddings and funerals. Very little secular knowledge resembling science existed. In Rome, the pope exerted great power over kings throughout the continent. Popes were masters of politics, wealthier than anyone else, and often corrupt. The Church owned farmland and hunting estates, raised taxes, had its own armies, and the pope could excommunicate recalcitrant kings (e.g., England's Henry VIII in the sixteenth century).

The ruling class of feudalism was an aristocratic nobility. There were many tiers within this ruling class, including lords, dukes, earls, and barons. In an overwhelmingly rural society, in which the productivity of agriculture was low, the vast majority of people were peasants and farmers. Ownership of land was the basis of wealth and political power, and unlike capitalism, there was no effective division between public and private property: the property of the ruling class was the property of the state. Aristocrats typically owned vast estates of farmland and forests. In the manorial system that characterized feudalism, the extraction of surplus value occurred through the payments of rent. Tenant farmers paid tribute to their local lords, who owned the land. Often, they paid one-half or more of their output as rent in exchange for protection from invasions and robbers. Thus, rent payments were the primary form of wealth transfer from the poor to the wealthy, which occurred through the state rather than the market. The aristocracy controlled the reins of government, including the military, judiciary, and penal system, and their private interests were synonymous with those of the state. Knights and the military existed to enforce the rule of aristocratic law, suppress rebellions, and to protect local communities from brigands, robbers, and invaders.

Farming under feudalism was based on animate sources of energy, that is, living human and animal muscle power. Peasants and draft animals worked the fields, collected firewood, drew water from wells, and performed the innumerable other tasks necessary to keep their society working. Tools were very primitive. Child labor on the farms was the norm, birth rates were high (meaning women were frequently pregnant), and most people lived in large, extended families. The population lived in small hamlets and villages, many of which were self-sufficient, producing their own food, clothing, and other necessities.

Markets existed under feudalism but typically were small and poorly developed, and only the wealthy had the incomes required to buy luxury goods. Weekly fairs saw farmers and craft workers barter their goods. Larger markets consisted of seasonal fairs where itinerant merchants (often Jewish) sold metal goods, silks, or jewelry. Thus, feudalism was not a type of society in which markets were the central institution that governed the allocation of resources; rather, this function was performed primarily by the state.

A substantial share of the rural population, but not everyone, consisted of **serfs**. Serfdom was a uniquely feudal institution that differed both from slavery and from the class structure of capitalism. A serf was not a slave; that is, he or she was not owned by a master; rather, serfs were bound to the land by feudal law and custom. The standard of living for virtually everyone except the aristocrats and some merchants was very low. Serfs and other farmers (including pools of "freemen") lived a monotonous life in which each day was identical to the day before, doing the same chores, eating the same food, and seeing the same people. Most people lived very simply. Diets were usually poor, and malnutrition was common. Famines broke out every few years. Life expectancy in feudal Europe was typically under fifty years. Many women died in childbirth, and infant mortality rates were high. Water supplies were often infected by bacteria, and diseases such as cholera, plague, and tuberculosis took an enormous toll in human lives and suffering.

Agricultural work was organized around the rhythms of the seasons. Winters, for example, might be spent indoors, weaving clothes or fixing farm implements. Spring was the time for planting crops and birthing new animals. Summers involved tending to the crops and livestock. And the fall was the time of the harvest, for many, the central event of the year. A poor harvest meant hunger or famine in the winter. Most people lived in extended families, in which several generations coinhabit one dwelling or live nearby; grandparents, parents, nieces and nephews, cousins, and infants lived in crowded rooms, often without paved floors. A family sleeping together in one bed, sometimes including dogs and pigs for warmth, was not unusual. The vast majority of people were illiterate; even many kings, queens, aristocrats, and priests were unable to read and write. Peasants and farmers passed lessons to their children through poems and stories, and the populace was generally very superstitious.

Although feudal society was predominantly rural and agricultural, there were a few cities and towns. Urban areas under feudalism were very different from those of today. Because agricultural productivity rates were low, and the ability of farmers to support urbanites was therefore limited, cities were small. Most farming hamlets did not exceed a few hundred people, and cities with over ten thousand people were rare. There were a few metropolitan areas, such as Constantinople or London, but these were few and far between. Feudal cities were densely populated, with the inhabitants crowded together, often in very unsanitary conditions. There was no running water or sewer systems, and the streets were often covered with mud and animal waste. Outbreaks of disease were common. The centers of feudal cities often consisted of a walled fortress, often with a small palace located within where the local lord lived. Because land was not a commodity to be bought and sold, but allocated on the basis of power, there was little differentiation among land uses. Commercial and residential land uses were mixed together, and there was no effective distinction between the spaces of home and work.

Within the cities, feudal ***guilds***, or associations of craft workers and artisans, produced a variety of goods that most people could not make for themselves. Guilds consisted of skilled workers with years of experience, and were organized by the type of goods they produced. For example, there were guilds for blacksmiths, weavers, goldsmiths, bakers, leather workers, paper making, glass workers, and shoemakers. Young men who were chosen to work in the guild spent years as apprentices learning the trade before becoming craftsmen in their own right. The purpose of guilds was to protect local monopolies and minimize competition, and guild leaders were powerful forces in urban politics.

2.1b The End of Feudalism

The so-called Dark Ages in Europe changed relatively little from the sixth to the fourteenth centuries, testimony to feudalism's stability and slow rate of change. However, starting around the eleventh century, Europe saw a gradual agricultural revolution based on the introduction of the heavy plow, waterwheels, the horseshoe, stirrup, the three-field system of farming, and several other innovations, many of which were introduced from other, more advanced societies such as the Arabs, India, and China. Other imports included cotton, oranges, coffee, the compass, sugar, rice, silk, paper, printing, the needle, the zero, and the windmill. The Arab conquest of the Iberian Peninsula made it a primary point of entry for new ideas and technologies.

Feudal Europe was the western terminus of a much larger world system that stretched across the Mediterranean, the Middle East, the Indian Ocean, and Eastern Asia in the thirteenth and fourteenth centuries (Figure 2.1), connecting most of the Old World. The Arab world's strategic location connecting Asia, Africa, and Europe placed it at the center of the enormous trade networks that linked together places as far flung as China, Mozambique, and Belgium. Straddling the center of the feudal world system, the Arab world guaranteed safe passage between two critical worlds—the Mediterranean and the Indian Ocean—that had been separated since the collapse of Rome. Shipping across the Indian Ocean connected China, Indonesia, India, the Middle East, and the east coast of Africa.

A vital part of the Chinese Sung dynasty economy, as well as dynasties that came before and after it, was the Silk Road, the famous umbilical cord that connected Europe, the Middle East, Central Asia, South Asia, Tibet, and China with caravans carrying goods, innovations, ideas, merchants, missionaries, and armies. Few individuals traveled the entire distance of the Silk Road; rather, it was served by networks of intermediaries. Most goods were transported by a series of agents, passing through many hands before they reached their ultimate destination. For two thousand years, Silk Road caravans linked ports, trading cities, oases, and different cultures. At its height during the thirteenth and fourteenth centuries, the Mongol Empire provided protection for merchants from robbers and brigands. From China came jade, paper, the compass, gunpowder, printing, porcelain, lacquerware, silk, pearls, peaches, apricots, citrus fruits, cherries, and almonds; moving in the other direction, at various times, China acquired horses, hides, furs, dyes, amber, pistachios, saffron, castor beans, sesame, peas, onions, coriander, cucumbers, grapes, sugar beets, kohlrabi, ivory, tortoise shells, and, oddly, the chair. The Chinese established customs posts on the Silk Road to minimize smuggling and to tax goods as they passed. So too did religions flow along this highway, including Manicheism, Zoroastrianism, Buddhism, and Islam.

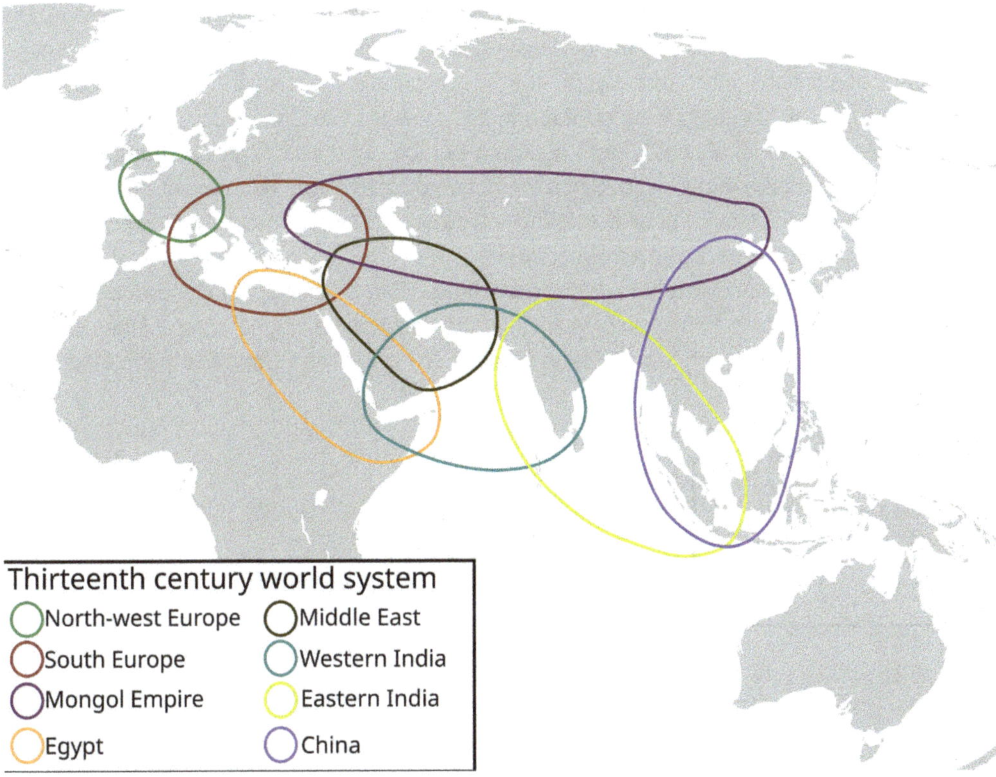

Figure 2.1 The thirteenth-century world system.

The Silk Road was therefore as important for the flows of cultures and ideas it expedited as for the movement of goods.

The introduction of innovations, which constituted a commercial revolution of sorts, improved European agricultural productivity. The heavy plow opened up the thick soils of northern Europe and expanded the amount of arable land for farming. The supply of agricultural land expanded as peasants and farmers in the late Middle Ages cut down forests and drained swamps to create new farmland. This set of circumstances led to a gradual increase in the urban populations, as fewer farmers could support more urbanites. By the fifteenth century, much of Western Europe saw the growth of a network of cities, which were called "newtowns" in Britain and *villanovas* in Spain. The increase in productivity also underscored a round of cathedral construction.

However, compared to much of the rest of the world, feudal Europe was relatively primitive. European standards of living, literacy, and rates of innovation from the fifth to the fifteenth centuries were much lower than the wealthier, more powerful, and more sophisticated societies of the Arab world, India under the Mughals, or China under the Sung and Ming dynasties. Indeed, many of the luxury goods and innovations that Europeans imported came from these wealthier societies.

One of the most important things introduced to Europe at this time was a bacterium, *yersinia pestis*, which causes a deadly disease commonly known as **bubonic plague**. In Central Asia, this disease was endemic in rodent populations in the steppes, or grasslands; plague was spread by ticks and fleas that live on them. Bubonic plague was called the Black Death for the necrotic or black tissue that it produced. In 1347, the Mongols laid siege to Caffa, a city on the Crimean peninsula, and catapulted bodies of infected people over the walls, in history's first case of biological warfare. Refugees fleeing the city on ships landed in Sicily, Genoa, Valencia, and other Mediterranean ports that were part of the expanding trade network between Europe, the Middle East, and Asia, and carried the disease with them. Within four years, at least one-quarter of Europe's population—more than twenty-five million people—was dead (some estimates are as high as one-half). The plague raced through the crowded, unsanitary cities, in places annihilating the majority of inhabitants. From southern Europe, it spread north to Germany, Britain, Scandinavia, and Russia (Figure 2.2). Since the notion of germs was nonexistent, feudal Europe had no means of understanding or controlling the plague: frequently used "remedies," such as burning Jews or witches, did not halt its spread.

The plague played a major role in destabilizing feudal Europe and opening the door to a new type of society, capitalism. Within a few years after the onset of the plague, much of the continent experienced severe labor shortages; Europe went from being land-poor and people-rich to a

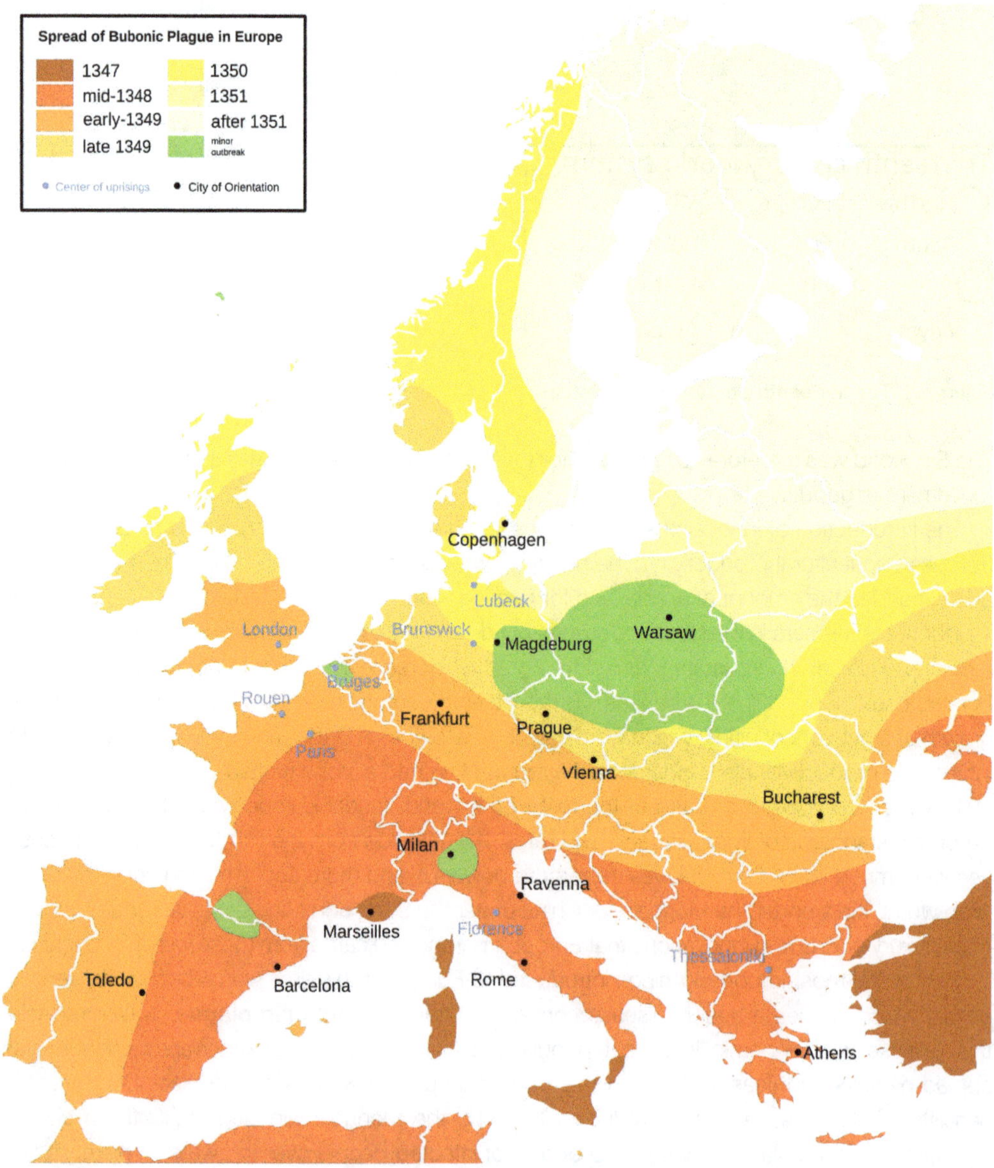

Figure 2.2 Diffusion of the bubonic plague in fourteenth-century Europe.

people-poor, land-rich group. The disintegration of legal systems allowed serfs to run away without fear of being caught and returned. In many places, peasants revolted against the aristocracy (e.g., the English Peasants' Revolt of 1381). Others have argued that feudalism was suffering from numerous problems anyway and would have collapsed even without the plague. For example, in some cities, such as Florence, Italy, capitalist social institutions such as banks were already emerging before the fourteenth century. In any case, in the face of the plague pandemic and other changes, including a mini-Ice Age that depressed harvests, the Hundred Years' War between France and England that bankrupted both empires, and the Crusades, feudalism in Europe began to crumble, and in its ashes a new economic, political, and social system emerged: capitalism.

2.2 The Emergence and Nature of Capitalism

From roughly the fifteenth to the early twentieth centuries, feudalism in Europe was gradually replaced by capitalism. Thus, feudalism served as a womb that incubated what would ultimately become the most powerful type of economic and political system in the world.

If capitalism can be said to have a birthplace, it would be northern Italy. The city-states of this peninsula, such as Florence, Venice, Pisa, and Genoa, played a key role in fomenting the new kind of society. They had large groups of wealthy merchants with active commercial ties to the Middle East, including the famous Medici family of Florence, with vast holdings in silver mines, silk production, and banking. The Medici contributed enormously to the rise of capitalism, running a wealthy bank, helping to create accounting, and creating a basis of wealth and power outside of the Catholic Church. Northern Italian city-states had flourishing trade networks: they traded with each other, with other cities around the Mediterranean, even extending into the Black Sea and the Atlantic Ocean, and traded with the Arabs in Egypt and the Middle East.

In northern Europe, a network of prosperous cities traded under the auspices of the **Hanseatic League**, which stretched from Russia and Scandinavia across northern Germany and into the North Sea. In these centers, the rising groups of ***burghers***, or merchants, accumulated wealth and power that would make them the dominant figures in the new social formation. The Hanseatic League united a disparate series of proto-capitalist urban centers, but eventually collapsed as trade increasingly shifted across the Atlantic Ocean in the sixteenth century.

Like feudalism, capitalism possesses a distinct set of characteristics that define it and give it its unique form. The major features of capitalism include the following.

2.2a Markets

Unlike feudalism and slavery, in which political power through the state is the principal way in which resources are allocated, under capitalism, the most important institution is the market. Markets consist of buyers and sellers of **commodities**, which are goods and services bought and sold for a price. Not everything under capitalism is a commodity (e.g., air); only scarce goods (i.e., those that command a price and can generate a profit for producers) can be classified as such. Capitalism excels at **commodification**, or turning everything into a commodity. As capitalism grew, the expansion of market societies saw the steady commodification of different goods and services, including food, housing, clothing, transportation, education, entertainment, medical care, and other domains. Thus, things that used to be produced on a subsistence basis or bartered came increasingly to be made for sale on a market, their value determined by the amount of money they commanded. Money displaced tradition and ancestry as the primary marker of social status.

Markets are not unique to capitalism. For example, there were markets in slave-based societies such as Rome. In feudal Europe, there were occasional markets in the form of weekly fairs or annual festivals, which brought traveling merchants and local populations together. Precapitalist markets were poorly developed, and involved relatively few luxury goods. However, markets are unique in their *importance* to capitalism: Only under capitalism are markets the major way in which resources are allocated, that is, through the private sector rather than the state. It is worth emphasizing that markets are not the *only* way in which resources are organized, for even in ostensibly "free market" societies (which do not exist in fact), the state, or government, plays a key role, including protecting property rights, building the infrastructure, providing public services, and protecting firms from foreign competition (Chapter 3).

There is a huge variety of markets based on the type of commodity being produced and consumed, as well as the amount and nature of competition. Markets range from being freewheeling and highly competitive, with many small producers who must accept the price the market dictates, to large ones dominated by a few major producers, or oligopolists, who can affect the market price. In market-based societies, private property, and the right to own it, are key requirements to production; after all, a good cannot be sold unless the seller owns it. Markets thus do not exist or function well without appropriate legal guarantees to protect property rights, such as intellectual property rights. The incentive of producers to sell goods and services is profit, the difference between gross revenues and production costs. A successful capitalist is a profit-maximizing one; those that do not maximize profits get driven out of business. Thus, markets involve production for exchange, rather than subsistence or use. In this sense, capitalism involved the triumph of the private sphere over the public one.

Because markets involve competition among different producers, they create a strong incentive to produce goods and services cheaply, efficiently, and to please consumers. Generally, the more competitive a market is, the more dynamism and innovation it exhibits. Thus, competitive markets contain a powerful incentive to innovate, which is largely responsible for making capitalism a dynamic, rapidly changing society. In this sense, capitalism differs considerably from noncapitalist societies in that it rewards innovation, change, and risk-taking.

2.2b Class Relations

Although capitalist societies are sometimes depicted as being devoid of classes (e.g., as consisting only of individual buyers and sellers), market-based systems in fact do have social classes with uneven distributions of wealth and power. In contrast to feudalism, the class system of capitalism reflected a broad-based shift from a hierarchy based on tradition to one based on money, from born rank to earned status. Historically, this process saw the rise to power of the merchant class, or what is sometimes called the capitalist class.

As the merchants and burghers of Europe gained wealth, power, and prestige in the fifteenth through seventeenth centuries, they enjoyed mounting political, economic, and cultural power. Essentially, the merchant class rose to become a capitalist class, that is, the owners of capital. This process was resisted by the feudal aristocracy, which correctly perceived the newcomers to be a lethal threat to their centuries of rule; many kings, earls, and dukes held merchants in low regard, as money-grubbers who were not motivated by ancient notions of honor. Conflicts between the merchant class and its interests on the one hand and the feudal elite on the other grew steadily in number and intensity, and revolved around issues like levels of taxation, the freedom to open markets at particular hours (e.g., on holidays), and merchants lending large amounts to kings in order to wage war, which left aristocrats deeply in debt. The demise of the feudal aristocracy came gradually in some places, such as England (e.g., during the civil wars of the seventeenth century), and suddenly in others, such as France, where the aristocracy collapsed abruptly in the revolution of 1789.

In addition to the ruling classes, capitalism changed the role and nature of the population of workers. Specifically, labor itself became a commodity, that is, bought and sold for a price (wages) in labor markets. Over several centuries, the peasants and serfs of Europe were gradually forged into a working class. Unlike feudal Europe, workers under capitalism must sell their labor power to survive rather than work for themselves. Creating the working class was not simple, but involved driving rural workers off the land and inculcating them with new norms, such as working the hours dictated to them by their new bosses, the capitalists. Time-keeping became increasingly important, as witnessed by the spread of clocks.

The capitalist class and the working class are not the only classes of capitalism, even if they are the largest. There are others, such as professionals and owners of small businesses. Farmers are often considered a separate class. So too are the homeless and indigent.

The new class structure of capitalism was central to emerging forms of identity and politics. Gradually, older forms of viewing oneself and others grounded in ancestry and religion began to decline. In many places, particularly during the Industrial Revolution, class conflict became severe as capitalists and workers struggled over wages, working hours, conditions, and benefits. It is a myth, therefore, that capitalism is classless.

2.2c Finance

The growth of capitalism saw a fundamental change in the role of money, which increasingly became the measure of all worth. In noncapitalist societies, barter plays a major role in how economic relations are organized: goods are traded for one another, or labor is traded for goods. In the context of barter-based economies, money is relegated to a relatively small role. Obviously, money existed before capitalism—the Romans, for example, had coins—but under market-based societies, money assumed a new level of importance. As the cash system replaced barter, money became standardized and ubiquitous as a measure of value. The commodifications of time, space, and labor power were thus measured in monetary terms. Wealth and power were increasingly defined along monetary lines, that is, economically rather than politically. Many traditional social roles that were defined by kinship, religion, friendship, and trust became depersonalized and formalized as they were mediated by money. Money thus has important social and political as well as economic roles.

The organization and control of money in emerging capitalism became an industry in its own right. Large, complex societies cannot function without well-established financial systems, which not only reflect the organization of production but also shape it. Banking arose primarily with the goldsmiths of feudal Europe, who stored gold for their customers and then lent it out to borrowers, for a price (interest). Modern banking arose to become a huge and complex industry linking savers and borrowers of different types and with different needs. By the seventeenth century, commercial credit became widespread, and with it, different types of banks and insurance firms. Starting with small savers who pooled their funds to purchase ships to trade with Asia, joint stock companies spread the risks of large investments over many small investors, and became the nucleus of modern stock markets. Accounting became an important profession. By the nineteenth century, financial systems were increasingly regulated by the state through central banks, which sought to control money supplies and thus interest, inflation, and exchange rates.

2.2d Territorial and Geographic Changes

As capitalism fundamentally changed the structure of Western societies, it also reshaped how they were organized geographically. Not surprisingly, the geographies of capitalism are unique to the logic of profit-maximizing societies.

Central to understanding the geography of capitalism is **uneven spatial development**, that is, varying levels of economic growth, wealth, and poverty in different regions. Uneven development is reflected in the simultaneous existence of rich and poor places, those with high and low unemployment, regions and countries with happy, prosperous residents and those with large numbers of impoverished, hungry ones. Those who believe in neoclassical economics (the mainstream type taught in the United States) often hold that uneven development is a temporary phenomenon that markets eliminate in due course through the movements of capital and labor. Others, working from a political economy view, hold that there are several mechanisms through which uneven development occurs, but the most significant is that of capital investment and disinvestment. As capital seeks out the highest rate of profit, it flows into some regions and out of others, in the process simultaneously creating prosperous places (e.g., New York City) and abandoning others to economic decline (e.g., Detroit). In this light, wealthy regions and poverty-stricken ones are intimately connected, two sides of the same coin.

The production of uneven development occurs at different spatial scales. At the global level, capitalism, through colonialism, created a worldwide system of commodity production and consumption, with Europe at the center and its colonies in the periphery, a process we will examine in more detail shortly. Essentially, Europe became wealthy, in part due to its extraction of cheap raw materials from its colonies in Latin America, Asia, and Africa. Prior to colonialism, Europe was poor and backward compared to competing powers; by the sixteenth century, when colonialism began in earnest, Europe had begun to surpass many other regions of the planet.

Within Europe, capitalism unleashed a division between a relatively prosperous northwestern part and a comparatively poorer southern and eastern part. Historically, southern Europe, including Greece, Italy, and Spain, had been the wealthiest and most powerful part of the continent. With the rise of capitalism, particularly during the Industrial Revolution in the late eighteenth and nineteenth centuries, northwestern Europe became much more advanced economically. To this day, the societies of Western Europe remain wealthier than those elsewhere in that continent (Figure 2.3).

Within the individual countries of Europe, as well as other emerging nation-states dominated by capitalism, there arose a division in wealth between cities and the countryside. Feudal Europe, with its tiny cities, had a relatively small urban-rural schism. Under capitalism, as rural areas were reshaped by waves of enclosures and the commodification of agriculture in the form of cash crops, large numbers of people migrated to the urban areas. Under feudalism, in contrast, there were relatively few differences in standards of living between urban and rural areas. Given the maritime world economy, port cities in particular thrived. Today, almost everywhere, cities have higher incomes, more jobs, and more opportunities than rural areas.

Finally, within capitalist cities, urban land became a commodity, organized through land markets. Land uses were differentiated by their capacity to earn a profit in the form of rent. This created a division between home and work, areas of production and social reproduction. As cities grew larger, particularly under the Industrial Revolution in the nineteenth century, the distances between home and work were stretched to the point where workers engaged in commuting. Thus, an intra-urban division of labor complemented the one between cities and the countryside.

It is important to see all of these scales as different versions of the same process (i.e., uneven development manifested at the global, continental, national, and local scales).

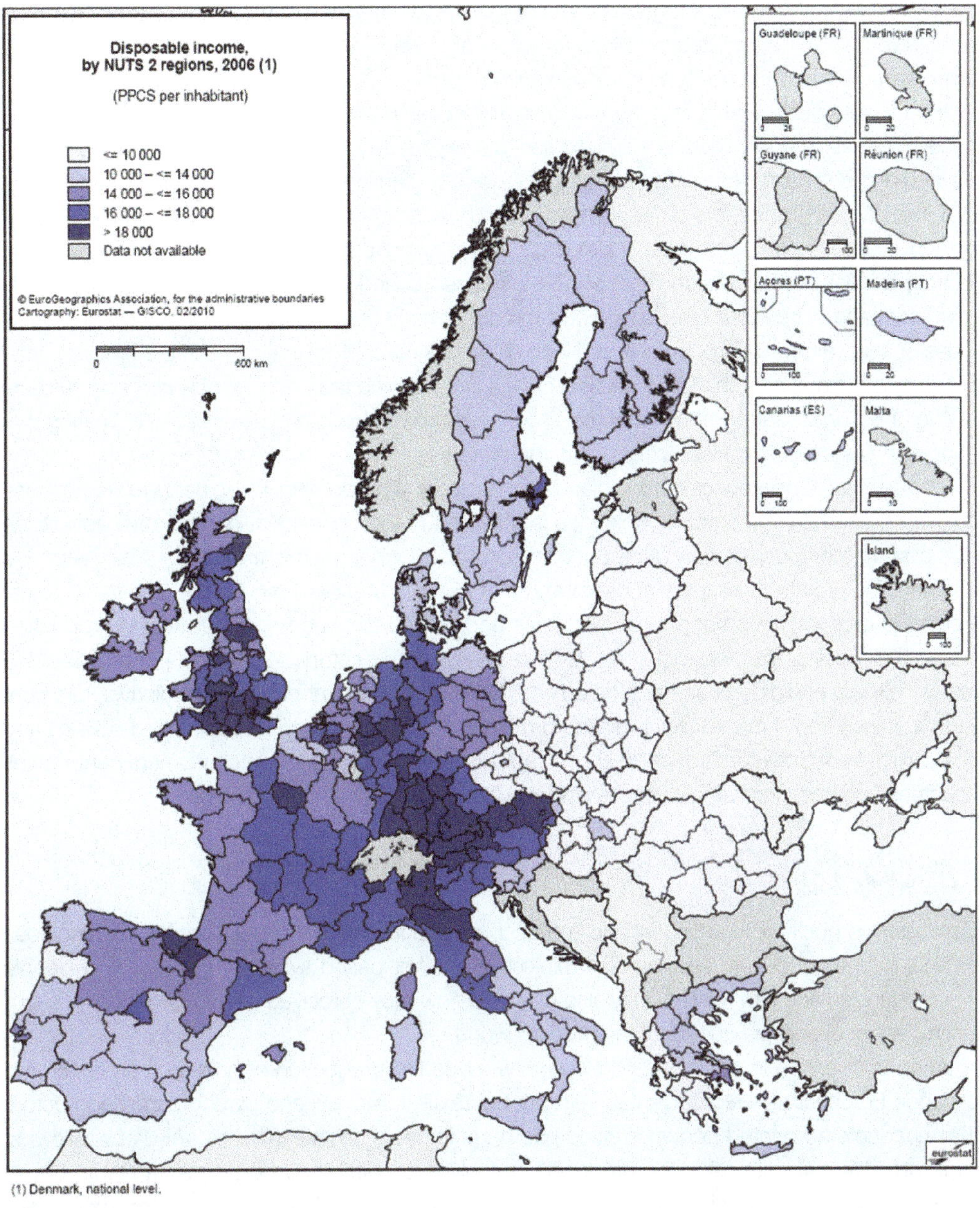

(1) Denmark, national level.

Figure 2.3 Map of European disposable incomes.

2.2e Long-Distance Trade

Buying and selling goods over long distances is an integral part of capitalism. Trade is a reflection of the geographic organization of production and consumption, linking producers and consumers who generally never see one another. As capitalism became entrenched across Europe, trade networks grew in their size and extent. There were trade routes prior to capitalism. In feudal Europe, trade with the Arab world, and through the Silk Road, with East and South Asia, brought in many goods that Europeans could not produce for themselves. On the eve of capitalism, Europe was the western end of a much larger fourteenth-century trading system that stretched across the Middle East, India, and into Eastern Asia (Figure 2.2). This form of trade primarily consisted of expensive goods such as spices, silks, porcelain, and precious metals. The consumers were the aristocrats who had the ability to purchase such luxuries.

If trade was peripheral to feudalism, it is central to capitalism. In market-based societies, trade occurs in a vast variety of goods, from luxuries to everyday items. The expansion of trade networks on land and sea routes tied different parts of Europe together and sutured the continent to Africa and Asia. Within Europe, the sixteenth, seventeenth, and eighteenth centuries saw a rapid expansion in roads, canals, and, later, railroads, which bound places together into an increasingly interdependent spatial division of labor. The increased speed with which people, capital, goods, and information circulated over space and through time is called time-space compression. New types of ships allowed Europeans to sail long distances relatively quickly, and in the process, they created new maps and charts for nautical navigation, learning the behavior of winds and currents all over the planet. Cartography emerged in conjunction with trade and colonialism, indicating that geography as a discipline was complicit in the process.

If trade reflects differences among places in the nature of production, it also helps to shape those places. In economic geography, this idea is reflected in the concept of comparative advantage (Chapter 4), the specialization of production that occurs when places trade extensively with one another. The growth of long-distance trade within Europe, and between Europe and its colonies, increased competition among local producers and helped to bring about declines in production costs, thus raising standards of living. By the seventeenth century, for example, a middle-class family in Britain could purchase salted cod from Newfoundland, timber from Scandinavia, furs from Russia, wines from France, blown glass from what is now the Czech Republic, and olive oil and citrus fruits from Spain or Greece. In short, capitalist trade relations made consumers better off as countries became increasingly interdependent upon one another.

2.2f New Ideologies

Capitalism is not simply an economic, political, or geographic set of relations; it also includes particular types of culture and ways of thinking. If market-based systems changed how people were organized and interrelated, they also changed how they perceived the world, including secularism, science, and the culture of the modern world.

After its invention in 1450, the printing press made the production of books much faster and cheaper. Europeans were acquainted with printed textiles, money, and playing cards long before they encountered printed books. Printing had huge impacts on the societies of Europe, allowing large quantities of material to be produced cheaply and distributed quickly; as a result, books declined in price dramatically in the sixteenth century. The effects of this revolution were monumental. Printing accelerated the decline of the feudal order. Literacy and printing destabilized traditional society by bringing adults—especially males, for female literacy lagged far behind—within reach of texts. Printing was the first step in the mechanization of communication, and accelerated the diffusion of information by packaging it conveniently, creating pathways for literate people by widening their access to distant people, places, and events. Printing helped to break the monopoly of information held by the church and fomented the growth of a lay intelligentsia. Ideas of many sorts began to circulate around the continent, and larger numbers of people learned to read and write.

Printing and rising rates of literacy facilitated the Italian Renaissance, the Protestant Reformation, and the rise of modern capitalism and science. In the sixteenth and seventeenth centuries, starting in Italy, Europe witnessed the explosion of artistic and scientific knowledge known as the Renaissance. Intellectuals like Leonardo da Vinci personified the rise of secular knowledge, as did

Erasmus and the Humanists in northern Europe. Although most intellectuals of the Renaissance were religious, the movement marked a dramatic shift from the god-centered view prevalent under feudalism to one that increasingly emphasized the role of human beings in the making of the world.

The sixteenth century also saw the Protestant Reformation, the second great divide in Christianity (the first was the division between the Catholic and Orthodox branches around the year 1000). Starting with Martin Luther in Germany, Protestantism offered a different view of life and god than did Catholicism. In particular, Protestantism emphasized the role of the individual and their direct relation to god, one aspect of a much broader growth of individualism in different social spheres. Bypassing priests as intermediaries between individuals and god, Protestantism spread with the growth of literacy in the aftermath of the printing press. It is also important to note the role that Calvinism, named after John Calvin, played in fostering markets by squaring Christian doctrine with the capitalist practice of charging interest, which many pastors dismissed as usury. Calvinism helped to sanctify work and commerce as part of the godly realm.

The famous sociologist Max Weber studied how Protestantism was entwined with industrial capitalism in northwest Europe, arguing that the "Protestant ethic," including delayed gratification, saving, and material success as a sign of god's grace and potential entry into heaven, was instrumental to the development of capitalism. He argued that Protestantism elevated work into a moral obligation and made profit a reward rather than a sin. This view has been challenged by those who maintain that Protestantism followed rather than caused market relations. Others hold that the origins of capitalism in Catholic Italy undermine the claim that it began in predominantly Protestant northern Europe.

Renaissance science in the sixteenth and seventeenth centuries dramatically changed how people viewed the world and their place in it. The Copernican revolution, including Galileo and the telescope, led to a heliocentric view of the universe rather than the older geocentric Aristotelian one. This shift accentuated the gradually emerging secularism of the time. Modern science was born as scientists such as Francis Bacon, Isaac Newton, Robert Boyle, Blaise Pascal, Carl Linnaeus, and Antoine Lavoisier gave rise to modern physics, mathematics, astronomy, biology, and chemistry and their applications to gravity, optics, and other fields.

By the seventeenth, eighteenth, and nineteenth centuries, Western societies underwent the Enlightenment, an explosion of science and secular political thought. In Britain, France, Germany, and elsewhere, great advances were made in geology, biology, chemistry, and physics, such as the discovery of atoms, oxygen, electromagnetism, and bacteria, which led to the germ theory of disease. The Darwinian revolution of the nineteenth century revolutionized the understanding of evolution and ecosystems. These discoveries demystified nature, revealing it to be the product of scientific laws. The process was accelerated by the proliferation of universities and institutionalized learning through academic societies. A gradual, widespread secularization of culture was underway. In political thought, thinkers such as John Locke, David Hume, and Adam Smith developed a worldview that stressed secularism, individualism, rationality, progress, and democracy. Indeed, the last, articulated by Thomas Jefferson among others, may be seen as a largely American contribution to the Enlightenment.

2.2g The Nation-State

Capitalism is not simply an economic system for organizing the production and consumption of goods through markets and trade; it is also inherently a political one that involves the state, that is, governments. The rise of capitalism witnessed a series of political changes; particularly important was the rise of the nation-state.

By definition, a nation is a group of people who share a common culture, language, history, territory, and identity. Feudal empires were culturally diverse and had many nations within their borders; the Holy Roman Empire and the Austro-Hungarian Empire, for example, contained many different ethnicities. Individual and collective identity was largely defined in religious terms, ancestry, or as subjects of a given sovereign. By the eighteenth century, however, as these empires began disintegrating, nations increasingly became the primary source of identity for many people. Nationalism describes the attachment of a person to a particular nation and the political activities to achieve statehood (or national self-determination). As the nation-state became the primary political unit in the eighteenth and nineteenth centuries, it became important for ruling elites to offer a narrative that provided their culturally heterogeneous populations over which they rule a single identity. Nation and state came to overlap in the forms of nationalism, self-determination, and sovereignty.

Although nationalists often minimize cultural diversity within their countries, all nation-states have ethnic and linguistic minorities. Even in the most classic example of the nation-state, France, had several minorities, including the French Basques, Bretons, and Corsicans. Similarly, Great Britain had Celtic minorities such as the Scots, Irish, and Welsh.

As national identities displaced older feudal ones, the political geography of Europe was steadily redrawn. A key moment in this process was the famous Treaty of Westphalia in 1648, which legitimated the nation-state as the primary unit of international law and relations. The treaty ended the Thirty Years' War (1618–1648), guaranteed the independence of Switzerland, the Netherlands, and Portugal, and allowed the individual states comprising the Holy Roman Empire to choose their own religion, which was critical in the midst of Catholic-Protestant strife. The treaty ensured a state's self-perceived, inalienable right to authority, particularly the use of force, within its boundaries, and the recognition of such authority and boundaries by the international community. The modern world system of nation-states, with clear, inviolable borders, is Westphalian in nature. However, it is simplistic to claim that the nation-state simply emerged fully born from the Westphalian treaty; the institutional bases were laid long before, and some countries, such as England and France, were well on their way to becoming nation-states even before 1648.

The French Revolution of 1789 was another major turning point in the disintegration of the old feudal social order and the rise of the new modern one, creating new forms of political identity, such as citizens. The Napoleonic Wars that ended in 1815 likewise laid much of the geographic basis for the nation-states of the continent. In Western Europe, the old monarchies that characterized feudalism were gradually replaced by constitutional republics, something that did not occur in Eastern Europe until after the First World War. By the twentieth century, the nation-state had become deeply entrenched as the world's dominant form of political organization.

The nation-state and markets were tied together in complex and important ways. The rise of the nation-state was facilitated by the emergence of markets in several ways. Rising levels of wealth, mass literacy and newspapers, growing numbers of intellectuals, political parties, and the beginnings of representative democracies were all elements of this process. Other institutions included the creation of national banks and currency, a military draft, and national rail systems, which tied together the diverse segments of the emerging nation-states and reinforced residents' sense of being part of a community of similar people. Thus, capitalism does not simply create markets, it makes *both* states and markets. Conversely, nation-states facilitated the expansion of markets, including the construction of a transportation infrastructure, public services (e.g., schools, transportation), national currencies, and the protection of domestic producers from foreign competition through tariffs. The close ties between markets and states have led many social scientists to discard the traditional view that markets were born free of the shackles of state in favor of a perspective that maintains markets *required* the state to survive.

It is important to remember that capitalism preceded the nation-state historically and that there is no necessary equivalence between the two phenomena. Capitalism began in the city-states, not nation-states, of northern Italy in the fifteenth century; in contrast, the nation-state was largely the product of the industrialization of the eighteenth and nineteenth centuries. The growth of the nation-state marked a new scale at which capitalist social relations could be contained and managed, including trade and production networks. Indeed, the political geography of capitalism is the interstate system, at both the city-state and nation-state scales. Today, there are hints that globalization is replacing the nation-state with a worldwide necklace of world cities.

2.3 The Industrial Revolution

Capitalism has been a dynamic, ever-changing society since it was born. This rapid pace of change became normalized during the formation of the modern world, particularly in the eighteenth and nineteenth centuries during the Industrial Revolution, which marked a major turning point in global economic history. Notably, the Industrial Revolution occurred long *after* capitalism began; indeed, for most of capitalism's history, it involved preindustrial forms of production, including artisanal types of production in the period known as mercantilism. However, starting in the mid-1800s, an explosive increase in speed and productivity occurred that transformed the worlds of work, everyday life, and the global economy. Later, in the twentieth century, industrialization spread to Eastern Europe and the Soviet Union, and since the 1970s has spread to selected parts of the developing world (Chapter 6).

Industrialization is a process that involves multiple transformations in inputs, outputs, and technologies. Three dimensions are particularly important here: inanimate energy, technological change, and rising productivity.

2.3a Inanimate Energy

Preindustrial societies relied upon animate sources of energy (human and animal muscle power) to get things done; in contrast, industrialization can be defined as the use of inanimate sources of energy. There are several types of inanimate energy. The first to be used historically was running water, which had been used in the late Middle Ages to power mills to grind flour. By the early eighteenth century, some textile producers began to use water-powered mills to run their machines. Running water was widely used in the earliest days of the Industrial Revolution, but it constrained firms to locating near streams and rivers. Many textile factories in southern New England used this strategy. There were severe drawbacks to this approach. Many streams, for example, dry up in the summer if there has been insufficient rainfall in their catchment area. And locating on a stream may put the producer far away from the market, eroding profits.

The great breakthrough in using inanimate energy was the steam engine, which was designed by James Newcomen in 1712; the first operating steam engine was constructed by the Scottish engineer James Watt in 1769. The steam engine was originally designed to pump seawater out of coal mine shafts that penetrated under the ocean. Wood provided the first major source of fuel for this invention, which required heating water into steam in order to drive the engine's pistons. As producers began to cut down forests in Britain in large numbers, deforesting much of the country, wood supplies began to dwindle, and the rising cost eroded profits.

As wood became increasingly scarce, users of steam engines switched to coal, which could be mined in large quantities. Thus, as Britain industrialized, several areas became major coal-producing centers, including Wales and Newcastle. As the Industrial Revolution spread across Europe, the large coal deposits of the Northern European lowlands became increasingly important, including northern France, Belgium, the Ruhr region in Germany, and Silesia in southern Poland. This relation illustrates how nature helps to shape the formation of geographies, one of the themes introduced in Chapter 1. In the United States, Appalachian coal deposits played a key role in the nation's industrialization.

In the nineteenth century, coal was joined by other fossil fuels, particularly petroleum, and to a lesser extent, natural gas (Chapter 5). The abundance of cheap inanimate energy was essential to the process of industrialization, and capitalism became increasingly energy-intensive as a result. This substitution of inanimate for animate energy both freed tens of millions of people from boring, hard manual labor and allowed large numbers to live relatively comfortable lives in the expanding middle class that industrialization produced.

2.3b Technological Innovation

As noted, capitalism is very dynamic: firms and investors, under the lure of profits, create new innovations in order to reduce costs and increase profits. The Industrial Revolution saw an explosive jump in the number, types, and applications of new technologies. A technology is a way of converting inputs into outputs; it can range from very simple to highly sophisticated. The Industrial Revolution gave rise to a whole new series of technologies in agriculture, manufacturing, transportation and communications, and services (Table 2.1).

Many of these new technologies were deployed in a new form of organizing labor, the factory system (Figure 2.4). Before this, work was organized on a small-scale basis, including home-based work. For example, the early textile industry used the "putting out" system of independent contractors who worked at home. By the late eighteenth century, however, firms were grouping large numbers of workers together under one roof. This process effectively created the industrial working class. Never before were so many workers concentrated together, which radically changed how they lived and viewed the world. Inside factories, workers used vast amounts of capital (i.e., machines); put differently, work became much more capital-intensive. The introduction of interchangeable parts, invented by American gun maker Eli Whitney, made machines more reliable and easier to fix. By the early twentieth century, Henry Ford introduced the moving conveyor belt, which further accelerated the tempo of work and the ability of workers to produce.

2.3c Productivity Increases

As a result of the technological changes during the Industrial Revolution, productivity levels surged. Productivity refers to the volume of output generated by a given volume of inputs; productivity increases refer to higher levels of efficiency (i.e., greater levels of output per unit of input [e.g., labor hour or unit of land]) or, conversely, fewer inputs per unit of output.

Productivity levels rose exponentially throughout the Industrial Revolution, making goods significantly cheaper. There were several important repercussions of this phenomenon. As the cost of producing goods declined, standards of living rose. Most workers labored long hours under horrific conditions and endured standards of living that were very low compared to those we enjoy today. Nonetheless, industrialization gradually saw many kinds of goods become increasingly affordable. Because wage rates have been linked historically to the productivity of labor, the

Table 2.1 **Some Major Innovations of the Industrial Revolution**

1708	mechanical seed sower
1712	steam engine
1758	threshing machine
1765	spinning jenny
1787	power loom
1793	cotton gin
1807	steamboat
1828	railroad
1831	electric generator
1834	reaper
1839	photography, vulcanized rubber
1844	telegraph
1846	pneumatic tire
1849	reinforced concrete
1850	refined gasoline
1851	refrigeration; sewing machine
1857	pasteurization
1859	gasoline engine
1866	open hearth furnace
1867	dynamite
1873	typewriter
1876	telephone
1877	phonograph
1878	microphone
1879	electric light bulb
1884	rayon
1886	hydroelectric power plant
1888	camera; radio waves
1892	diesel engine
1895	X-rays
1896	wireless telegraphy
1899	aspirin
1900	zeppelin
1903	airplane
1906	vacuum tube
1925	television

Figure 2.4 A nineteenth-century factory.

working class became better off. Clothing, for example, which was scarce before the Industrial Revolution, became relatively cheap and ceased to be as accurate an indicator of class status as it had been previously.

An important aspect of this process was the industrialization of agriculture. As machines were introduced into farming, including mechanical reapers, harvesters, and ultimately tractors, food became cheaper, and diets improved as more people ate more and better food than ever before. With the notorious exception of the Irish Potato Famine of the 1840s, hunger and malnutrition gradually declined throughout Europe, although they appeared in the aftermath of wars. As diets improved, so did people's resistance to disease, and life expectancies rose as well.

2.3d The Geography of the Industrial Revolution

The Industrial Revolution unfolded unevenly over time and space. While capitalism began in northern Italy, industrialization was very much a product of northwestern Europe. For several reasons, Britain became the world's first industrialized nation. Britain already enjoyed long-distance trade relations with its colonies in North America, India, and elsewhere. Agriculture in Britain was well advanced compared to the European continent. It had a network of canals that reduced transportation costs. Finally, Britain enjoyed large deposits of coal and was the locale where the steam engine was invented.

By the end of the eighteenth century, Britain was essentially the world's only industrial economy, which gave it an enormous advantage over its rivals. Britain's industrial economy, for example, allowed it to defeat France in their eighteenth-century contest for global hegemony. Britain began to flood European markets with cheap textiles. Cities in the Midlands of Britain, such as Leeds, Birmingham, and Manchester, which arose around the textile and steel industries, became known as the workhouses of the world. Others, such as London, Glasgow, and Liverpool, became centers for shipbuilding; in a maritime-based world economy, shipbuilding was an important industry. In many cities, networks of producers in guns, watches, shoes, metals, and light industry formed dense industrial districts in which firms engaged in intricate linkages of inputs and outputs.

A half-century after it began in Britain, the Industrial Revolution spread to the European continent. In France, this process saw the rise of industrial complexes in the lower Seine River and Paris. In northern Italy, the Po River valley became a major producer of textiles and shoes, with cities such as Milan leading the way. In Scandinavia, cities such as Stockholm became major ship producers. In Germany, which was late to industrialize (following unification in 1871), the Ruhr region became a global center of steel, automobile, and petrochemicals firms. By the early nineteenth century, the revolution began to diffuse worldwide, including crossing the Atlantic Ocean, where it industrialized southern New England. Starting in the 1870s, Japan became the first non-Western country to become industrialized as the old feudal order there collapsed following the Meiji Restoration of 1868. Russia, which attempted industrialization under Peter the Great, did not become fully industrialized until the 1920s, when, under Josef Stalin, the Soviet Union leapt to become the world's second-largest economy in the span of a decade.

In the twentieth century, industrialization diffused to some developing countries, particularly in East Asia, where it has had profound consequences for millions of people (Chapter 6). Industrialization of the developing world, which is partial and incomplete, is thus a continuation of a long-standing process.

2.3e Cycles of Industrialization

Because capitalism is constantly changing, the nature of industrialization varied over time. Capitalism is prone to long-term cyclical shifts in its industries, products, labor markets, and geographies, a concept often referred to as Kondratief waves of roughly fifty to seventy-five years duration (Chapter 3). These waves were centered on the rise and decline of different industries at various times. Industrialization was thus not one process but a series of them that varied over time and space.

The first wave of the Industrial Revolution (1770s–1820s) saw the rise of the textile industry. In Britain, as in other parts of Europe, North America, Japan, and the developing world today, textiles have *always* led industrialization. The industry is easy to enter, with low start-up costs and few requirements of skills. Textiles led the creation of industrial landscapes in most of the world. Because the first wave was centered in Britain, it saw it become the leading economic power in the world. The period of British economic rule corresponded with the Pax Britannica, when Britain was the foremost military power on the planet.

The second Kondratief wave, from the 1820s to the 1880s, was centered on heavy industry. In the nineteenth century, sectors such as shipbuilding and iron plants were critical. Large in scale and capital-intensive, these types of firms differed markedly from the light industry of textiles. They required massive investments, were expensive to enter, and quickly moved toward an oligopoly of a few, large producers rather than a competitive market. This was the period in which the US Manufacturing Belt began to form, although most of its growth was after the Civil War of the 1860s.

The third wave, from the 1880s to the 1930s, saw numerous heavy industries appear, including steel, rubber, glass, and automobiles. This period witnessed massive technological change, including capital intensification and the automation of work. As local markets gave way to national ones, most sectors experienced a steady concentration of output and ownership in the hands of a few large firms. Many companies became multi-establishment corporations, with operations in different cities. Simultaneously, this wave saw the rise of robber barons such as Carnegie (steel), Rockefeller (oil), Duke (tobacco), Dupont (chemicals), J.P. Morgan (banking), and John Deere (agricultural machinery).

The fourth wave of industrialization, which started immediately after the Great Depression of the 1930s and lasted until the 1970s, was led by petrochemicals (including plastics) and automobiles. With a relatively stable global economy, this period saw the United States dominate the world economy. During the 1950s and 1960s, the United States produced the bulk of the world's steel, cars, machinery, and other goods.

The fifth wave, which began after the oil shocks of the 1970s, was led by the electronics industry, which was accelerated by the microelectronics revolution. It also saw the explosive growth of producer services (Chapter 10). This epoch saw rapid growth in the productivity of computers and information-processing technologies.

Critically, during each era, the major industry of that time was commonly featured as the high-tech sector of its day. Thus, just as electronics is often celebrated today for its innovativeness and ability to sustain national competitiveness, so too were the textile industry in the eighteenth century and the steel industry in the nineteenth century associated with dynamism, growth, high productivity, and rising wages. What was a leading industry at one moment would become a lagging one at the next historical epoch, as high-wage, high-value-added sectors replaced low-wage ones in the world's core and as low-wage, low-value-added ones dispersed to the world's periphery.

2.3f Consequences of the Industrial Revolution

The Industrial Revolution reworked the social and spatial fabric of many societies, particularly in the societies that now form the economically developed world. No part of the economy, technology, culture, or everyday life was left untouched. Within a century after it started, industrialization transformed rural, poverty-stricken societies into prosperous and urbanized ones. These changes included the following.

Creation of an Industrial Working Class

A major part of industrialization was the reorganization of work in the form of the factory system, in which many workers labored together using machines and inanimate energy. These conditions were quite different from those facing agricultural workers, who were spread over large spaces and relied on animate sources of energy. With the commodification of labor, workers were paid by the hour, day, or week. As firms created a new form of labor, they created a new form of laborer, an industrial working class. Industrial workers labored under brutally exploitative working conditions, typically for ten, twelve, or even fourteen hours per day, six days per week, for relatively low wages. (The eight-hour day and Saturdays free from work were products of the unionization movement in the 1930s.) Often, working conditions were unsanitary and dangerous, even lethal, as workers faced accidents, poor lighting, and poor air quality. Child labor was also common, subjecting those as young as four or five to horrendous conditions.

As industrialization produced a working class, it also gave rise to labor unions. The first of these was the British Luddites in the late eighteenth century. Luddites blamed their working conditions on the textile machines they used, and often destroyed them in attempts to halt their exploitation. By the late nineteenth century, several labor unions arose; in the United States these included the Knights of Labor, the American Federation of Labor (AFL), and in the twentieth century, the Industrial

Workers of the World and the Congress of Industrial Organizations (CIO). Unions played a key role in advancing wages and better working conditions, often leading to strikes. Industrialization was thus often a period of intense class conflict.

Urbanization

Geographically, the Industrial Revolution was located primarily in cities, which grew rapidly, making industrialization and *urbanization* simultaneous and mutually reinforcing processes. Manufacturing firms concentrated in large cities such as the British Midlands (e.g., Leeds, Birmingham, Manchester) or in the US Manufacturing Belt (e.g., Pittsburgh, Cleveland, Detroit, Chicago, and Milwaukee).

The reasons why firms concentrated in cities are important. Often, there is a tendency to attribute this phenomenon to the presence of workers in urban areas. But cities were concentrations of capital as much as they were concentrations of labor. Most cities were very small when the Industrial Revolution began. As Chapter 3 discusses, there are powerful reasons for firms to concentrate, or agglomerate, in cities. Firms benefit by being in close proximity to other firms, which are suppliers of parts and services (input linkages) as well as clients (output linkages). Spatial concentration allows firms to share a specialized infrastructure, information, and labor force. Industrial cities saw the emergence of dense webs of industrial firms tied together with intricate input and output relations.

As they industrialized, societies were transformed from predominantly rural to predominantly urban ones. By the early twentieth century, in Europe, North America, and Japan, the majority of people lived in cities. The growth of cities in industrial societies is frequently illustrated using an urbanization curve, which illustrates the percentage of people living in urban areas over time. In the United States, for example, the first national census of 1790 showed that 95 percent of Americans lived in rural areas. This proportion increased steadily throughout the nineteenth century, and by 1920, 50 percent of the nation's population lived in cities; today, it is roughly 85 percent. This pattern is similar in every country that has industrialized.

Population Effects

Industrialization not only changed the spatial distribution of people (i.e., grouping them in cities), it also altered the growth rates and demographic composition (Chapter 7).

On the eve of the Industrial Revolution, Thomas Malthus famously predicted that rapid population growth would create widespread famine. Yet he was soon proved to be wrong: the industrialization of agriculture generated productivity increases that exceeded the rate of population growth. The formation of a stable and better food supply greatly improved people's diets; as a result, life expectancy rose. Industrialization also lowered death rates, particularly as malnutrition and infant mortality rates dropped. As death rates dropped, the populations of industrializing countries increased dramatically. This change was also accompanied by a shift from the extended to the nuclear family. Eventually, industrialization also led to declines in the birth rate, families had fewer children, and growth rates declined.

Global Markets and International Trade

Yet another impact of the Industrial Revolution was a change in the global economy. Well before industrialization began, capitalism had formed a network of international trade across the Atlantic Ocean. The use of inanimate energy for transportation, such as the steam engine, dramatically accelerated the speed at which ships could cross large bodies of water, forming a significant round of time-space compression. While sailing ships traveled at roughly 10 mph, steamships could reach speeds of 40 mph. On land, railroads could reach 65 mph.

These changes dramatically increased the volume of imports and exports internationally. Europe imported raw materials such as cotton, sugar, and metal ores, and exported high-value-added finished goods, a process that created large numbers of jobs in Europe and raised standards of living. In economic theory, classical theorists such as Adam Smith and David Ricardo advocated for free trade at precisely this historical moment.

In addition, the industrial world economy saw the growth of international finance. British banks, for example, extended their networks to other continents, lending to clients and investing in markets overseas. The American railroad network was largely financed by loans that originated in Britain.

There were important differences between early and late industrializers as they entered the world economy. Early industrializers (e.g., Britain, the United States) faced little competition internationally. Thus, their light industries (i.e., textiles) associated with the first Kondratief wave developed without government assistance. Politically, early industrializers were characterized by laissez-faire politics and minimal government intervention. In contrast, late industrializers faced a significantly more competitive international economy dominated by early industrializers. Countries such as Germany, Italy, and Japan, which did not industrialize until the late nineteenth century, faced significant competition in textiles. Consequently, they tended to experience relatively short periods in which their economies were dominated by light industry and moved rapidly into more capital-intensive sectors such as steel, railroads, armaments, and automobiles. Such heavy industries placed significantly higher demands on the state for their infrastructure, labor training, and trade protection. As a result, their national political cultures tended to look favorably upon state intervention. The same is true today in Asia's Newly Industrializing Countries. Thus, the political culture within countries was strongly affected by the timing of their entry into the international division of labor.

2.4 Colonialism: Capitalism on a World Scale

The rise of capitalism in Europe was closely associated with Europe's conquest of the rest of the planet. Sometimes called the "Age of Exploration," colonialism was the expansion of capitalism on a global scale. Because the geographies of capitalism always involve uneven spatial development, colonialism saw the construction of uneven development on a worldwide scale, with Europe at the center and its colonies on the world's periphery.

Colonialism was simultaneously an economic, political, and cultural process. It was an act of political conquest, in which a small group of Western European empires came to dominate a large group of non-European societies. Culturally, this process involved the emergence of the distinction between the "West" and the "Rest." In conquering the Orient, Europeans discovered themselves as Westerners.

Colonialism was also resisted. Everywhere, colonized people fought back, such as the Inca rebellions against the Spanish, the Zulu attacks on the Dutch Boers, the Indian Sepoy uprising of 1857, and the Boxer Rebellion in China in 1899–1902. Yet Western powers, armed with guns, ships, and cannon, proved to be militarily superior to those they conquered. The only nation to escape colonialism altogether was Japan, which, under the Tokugawa Shogunate, closed itself off from the world for 250 years until 1868.

Colonialism had profound implications for the world economy, which is why it is important to contemporary economic geography. Globally, it created the division between the world's developed and less developed countries, a theme explored in Chapter 6. Colonialism also made Europeans wealthy and powerful. Prior to the rise of capitalism, Europe was relatively poor and powerless compared to the Arab Empire, India, or China; afterward, Europe became the most powerful group of societies on the planet.

2.4a The Temporal and Spatial Unevenness of Colonialism

Colonialism unfolded in differing ways depending on the historical period and the places involved. Temporally, there were two major waves of colonialism, one that occurred during the preindustrial mercantile era and the second during the Industrial Revolution. During the first wave, from the sixteenth until the early nineteenth century, Western economic thought was characterized by mercantilism, in which state protection of private interests was justified as necessary for the national well-being. During this time, the primary colonial powers were Spain and Portugal, and their colonies were primarily located in the New World and parts of Africa.

After the Napoleonic Wars, which ended in the Treaty of Vienna in 1815, European powers were relatively weak. This condition created an opportunity for nationalist movements in Latin America, led by Simon Bolivar, to throw off the Spanish yoke and become independent. Thus, the number of colonies declined sharply in the early nineteenth century in history's first wave of **decolonization**. However, during the second, industrial phase of industrialization in the nineteenth century, the number of colonies grew again. Britain, the world's superpower, and France colonized large parts of Africa and Asia. Finally, the number of colonies declined after the Second World War amid the era of decolonization.

Spatially, colonialism was also uneven. Different European empires controlled widely varying parts of the world (Figure 2.5). The British Empire encompassed one-quarter of the world's land surface, including parts of every continent, such as large parts of eastern Africa, parts of the Middle East, the Indian subcontinent, Australia and New Zealand, and Malaysia. The French ruled over Quebec, western Africa, and Indochina. The Portuguese had Brazil, Angola, and Mozambique, Goa in India, Timor in Indonesia, and Chinese Macau. The Belgians possessed Congo, the private holdings of Leopold II. Italy occupied Libya and Ethiopia for a brief time. Even Germany, late to unify and to industrialize, controlled parts of Africa (Togo, Namibia, Tanganyika) and New Guinea.

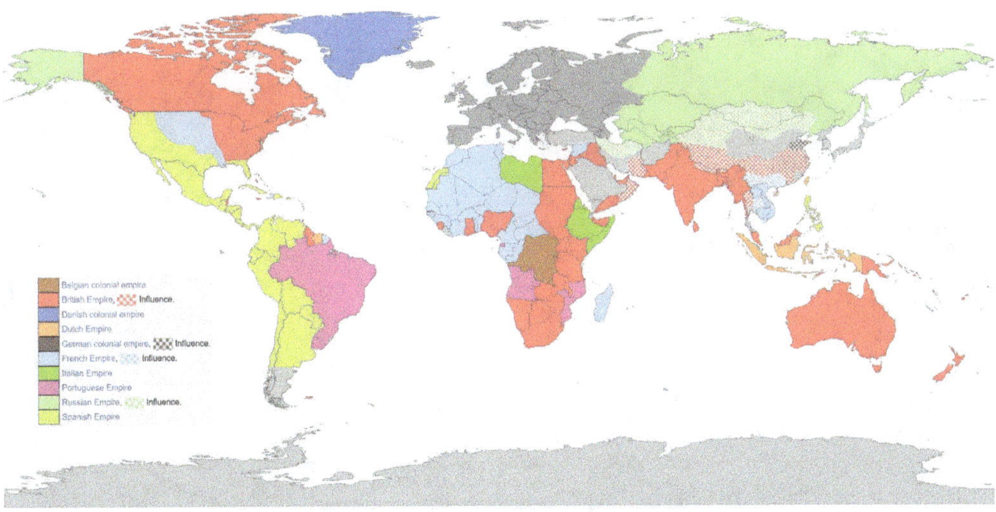

Figure 2.5 European colonies at various dates of colonization.

2.4b How Did the West do it?

How did Europe conquer the planet? There are no simple answers to this question. Jared Diamond (1999), in his Pulitzer Prize-winning book *Guns, Germs, and Steel*, argues that Western societies enjoyed a long series of technological advantages. Agriculture in the West, centered on wheat, was productive and could sustain large, dense populations. The long Western history of metal-working produced weapons such as guns and cannons. By the sixteenth century, Europeans had become highly skilled at building ships and navigating the world's oceans. And with the Industrial Revolution, the West discovered how to use inanimate energy, which offered numerous economic and military advantages. Additionally, the Europeans unleashed diseases (if unintentionally), particularly smallpox, on the New World, which provided them with an unintended advantage. Others argue that the West's advantages were not simply technological, but political. The Western legal system stressed secular laws and the importance of property rights. Western governments typically protected their capitalist and merchant classes, building infrastructure and using protective trade policies. Another argument is that, unlike Mughal India or China, Europe was never united politically. When one European power attempted to conquer the others, it was invariably defeated, such as when France under Napoleon attempted to conquer Europe in the early nineteenth century, and Germany twice in the twentieth century. The lack of centralized political control allowed dissent and experimentation that would not be possible under a unified state.

2.4c The Historical Geography of Conquest

To understand colonialism, we must appreciate how it occurred in different times and places, that is, how it was historically and geographically specific.

Latin America

Home to the Inca, Mayan, and Aztec civilizations, Latin America was the first major world region to be taken over by Europeans. Two years after Columbus arrived in 1492, Spain and Portugal struggled over control of the New World. The rivalry was settled by the pope at the Treaty of Tordesillas in 1494 (Figure 2.6). Shortly thereafter, Spanish conquistadors in Mexico and Peru destroyed the

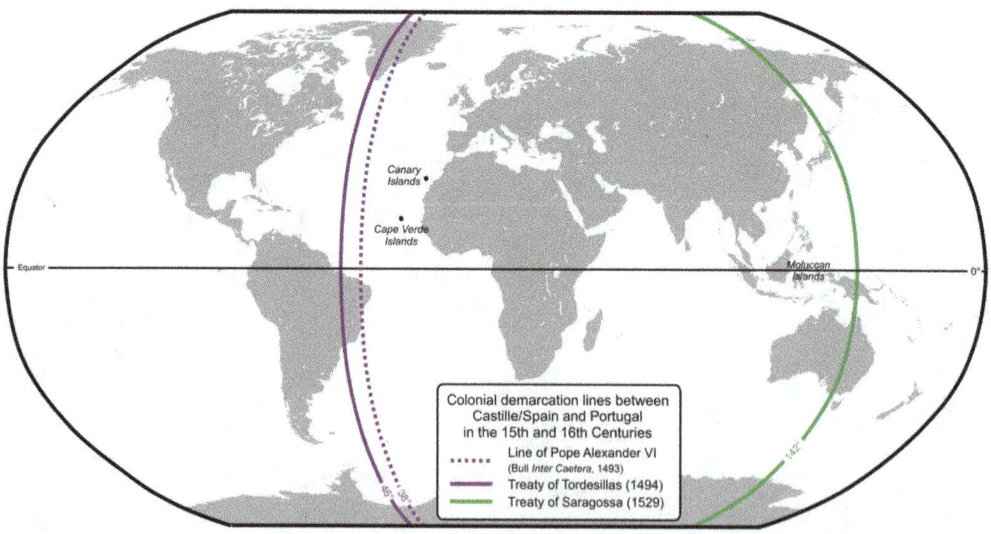

Figure 2.6 The Treaty of Tordesillas in 1494 divided the world between Spain and Portugal.

Aztec and Incan states, respectively. In large part, they managed this through the introduction of smallpox, which killed fifty to eighty million people within a century of Columbus's arrival, depopulating the Western Hemisphere in the largest act of genocide in human history.

The economic philosophy of mercantilism stressed the accumulation of bullion, or precious metals, as the key to national wealth. Accordingly, the Spanish took home enormous quantities of silver from the New World, notably from mines in Mexico and Bolivia, using Indian slave labor. Silver was taken back to Spain in galleons, making Spain the wealthiest country in the world in the sixteenth century. Silver financed Spain's attack on England in 1688, its wars with the Ottoman Turks, and paid for Europe's trade deficits with China.

Spain also introduced the *encomienda* land grant system into the New World, in which a small, landed aristocracy owned enormous holdings of farmland and ranches. This pattern continues today, indicating how colonialism shapes the geographies of the contemporary world.

The Spanish Empire in the New World abruptly came to an end with the wars of independence in the 1820s that followed the Napoleonic Wars (although it was the Spanish-American War of 1898 that finally ended it). The Spanish Empire broke up into a series of independent countries stretching from Mexico to Argentina. The Portuguese empire in Brazil did not fragment in the same way, leaving that country the giant of Latin America.

North America

In North America, numerous colonial powers were present. The Spanish controlled Florida and the Southwest. The French took over Quebec and the St. Lawrence River valley, only to lose that region to Britain in the eighteenth century. France also controlled the Mississippi River valley, with the key port of New Orleans, but sold it to the United States with the Louisiana Purchase of 1803. The Russians seized Alaska but sold it to the United States in 1867. The Dutch established the colony of New Amsterdam, but the British captured it in 1664 and renamed it New York. Britain eventually became the dominant power in North America, controlling New England and the southern states along the eastern seaboard.

British colonialism in North America centered on a string of port cities on the East Coast, typically located at the mouths of rivers (e.g., Boston, New York, Philadelphia). After the independence of these colonies in 1776, settlers moved west, across the Appalachian Mountains in the early nineteenth century and across the Great Plains somewhat later. In the late nineteenth century, railroads connected this region to East Coast cities, allowing it to create a comparative advantage in wheat, beef, minerals, and lumber. This process involved the eradication of Native American peoples, theft of their land, and commodification of their territory.

Canada was a British colony controlled by the famous Hudson's Bay Company, then the world's largest, which controlled much of the fur and fish trade. The company was one of a series of chartered monopolies that used private capital to take over lands in the name of the crown. Furs, wood, cod, and other raw materials long formed the basis of the colonial Canadian economy. In 1867, Canada acquired independence.

Africa

Colonialism in Africa included slavery, the kidnapping of roughly twelve million people, and exporting them to the New World (Figure 2.7) under brutal conditions, where they were used to compensate for the labor shortages brought on by the decimation of Native Americans. The capture of slaves robbed these societies of young adults in their prime working years and often with the assistance of local kings who sought to profit from the trade. Slavers from Portugal, Britain, and the United States gave African kings textiles, guns, and whiskey in return for slaves.

Slaves were generally taken from western Africa, and the largest numbers were brought to work the sugar **plantations** in Brazil and the Caribbean. (Notably, there was also an Arabic tradition of slavery in eastern Africa and the Indian Ocean.) Others were brought to work the cotton and tobacco plantations of the American South. Slavery was one part of the broader Triangular Trade system that connected Britain, Africa, and North America.

Moreover, slavery helped to give rise to ideologies that naturalized European, or more broadly, Western hegemony. Colonialism was a racist process that portrayed Whites as superior due to their military and economic power. In this way, the expanding colonial world system and racial subjugation were intricately tied together.

Africa is exceptionally rich in minerals. In the nineteenth century, when European powers branched out from the coastal areas into the interior of the continent, mines for copper, gold, and diamonds soon opened using slave labor. These resources remain the basis of many African economies today, an illustration of how colonial impacts continue to the present day.

An important colonial impact on Africa was the political geography that Europe created. Following the famous Berlin Conference of 1884, when European powers drew maps demarcating their respective areas of influence, the resulting boundaries collapsed roughly one thousand tribes into fifty colonies (Figure 2.8). Some groups were separated; many others, with widely different religions, cultures, and economies, were grouped together. The results of this mapping continue to this day: since African states gained independence in the 1950s and 1960s, they have been wracked by civil wars and tribal conflicts in which millions have perished.

The Arab World

The Arab world had been colonized long before the Europeans arrived by the Ottoman Empire. Turks, who are not Arabs, dominated the Arab peoples of the Middle East and North Africa. Gradually, starting in the nineteenth century, European powers encroached on this vast domain. In the nineteenth century, the French took over Morocco, Algeria, and Tunisia. In 1798, the French seized Egypt from the Ottomans, only to lose it to Britain four years later. The British used Egypt as a source of cotton, growing large plantations along the Nile River and building the Suez Canal in 1869.

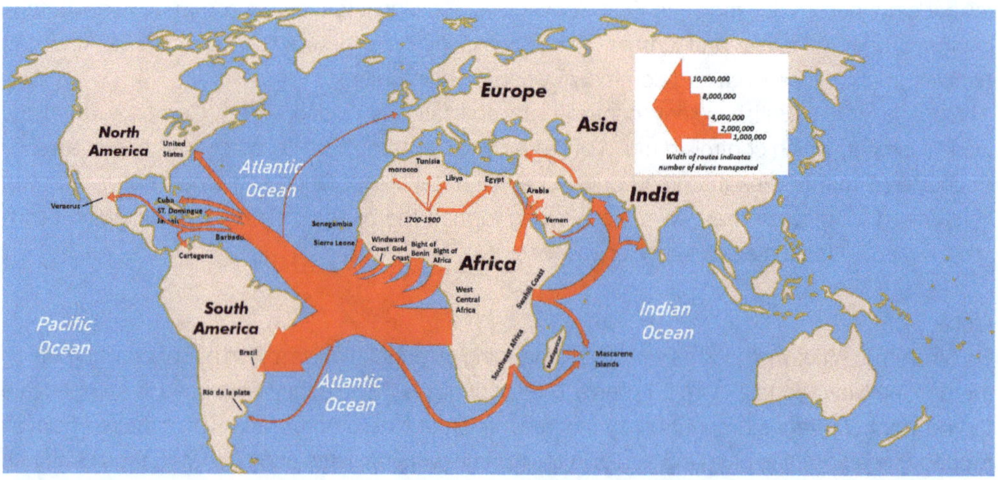

Figure 2.7 The slave trade was an integral part of colonialism in Africa and the conquest of the Western Hemisphere.

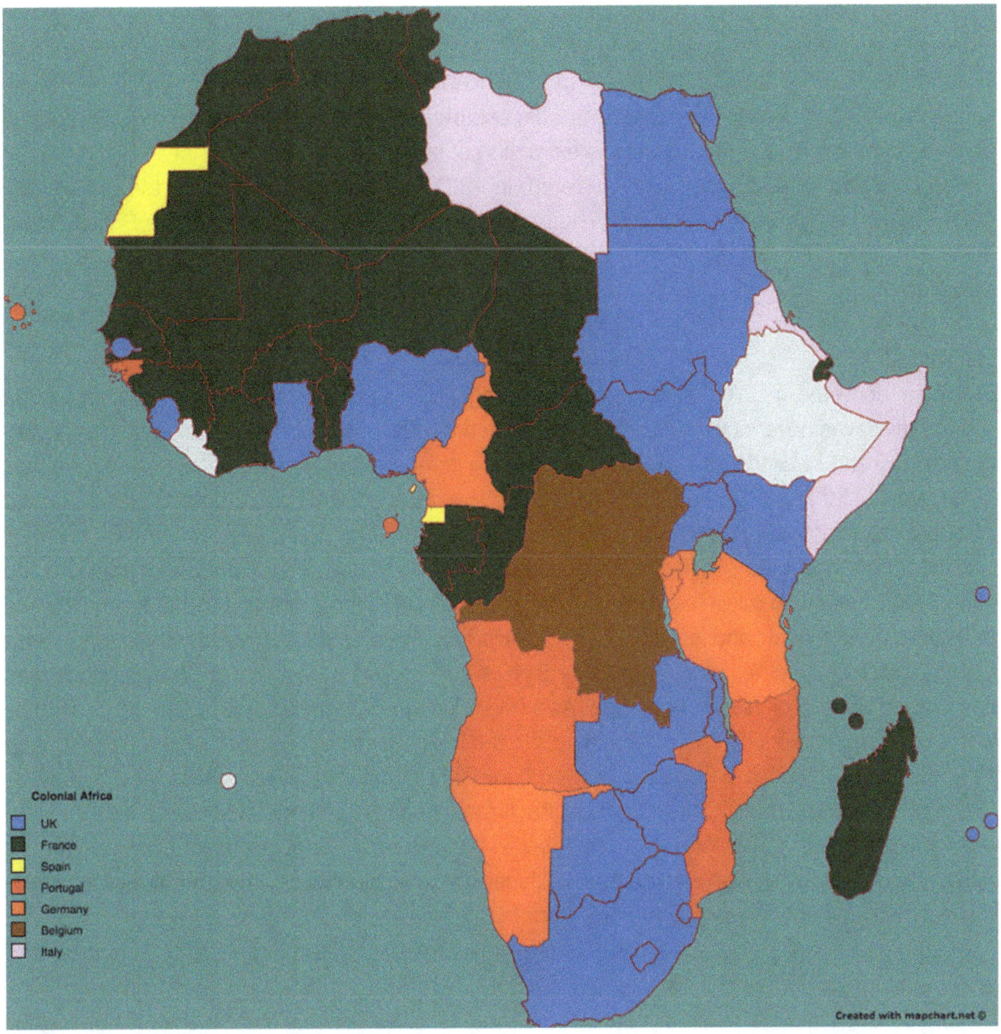

Figure 2.8 European colonies in Africa.

After the First World War, the Ottoman Empire collapsed, and the British and French divided up the Middle East with the signing of the Sykes-Picot Agreement in 1916. The British assumed control over Jordan and Palestine, much of which became Israel in 1948, as well as Iraq and the Persian Gulf. Many Arabs, who initially welcomed the Europeans as liberators, quickly learned that the new colonial powers were very similar to the old Turkish rulers in the ruthless suppression of their liberties.

South Asia

The Indian subcontinent was the jewel in the crown of the British Empire. Starting in the seventeenth century, the British East India Company established footholds in this domain, starting with the city of Kolkata (formerly Calcutta). India, which is predominantly Hindu, had long been controlled by the Muslim Mughals, whose power was gradually undermined by the Europeans. A vast land that stretched from Muslim Afghanistan in the east through Hindu India to Buddhist Burma, the Indian colony was the largest colonial possession on earth.

Britain had enormous economic impacts on South Asia. In the nineteenth century, British textile imports flooded India, destroying the Mughal textile industry. Indian laborers were exported throughout the British Empire, including the Caribbean, eastern and southern Africa, and some Pacific islands such as Fiji. Tea plantations were established in Bengal and Ceylon (now Sri Lanka). Britain also built railroads to facilitate the extraction of Indian wealth.

As noted, colonialism was always resisted. In 1857, a mass uprising against the British took place, known as the Sepoy Rebellion. The revolt eventually failed, but it forced the British crown to assume direct control rather than administer it through the East India Company, which was dissolved.

East Asia

East Asia, comprising China, Korea, and Japan, also had a unique colonial trajectory. Japan, as noted earlier, was never colonized. When it emerged from a long period of isolation, Japan rapidly westernized and industrialized, gradually expanding its power through northeast Asia. Korea was taken over by Japan in 1895 and annexed in 1905, and Japan held it until the end of the Second World War. A similar situation occurred in Taiwan.

China, however, was a different story. Under the rule of the Manchus, foreigners from the north (Manchuria) who ruled the Qing dynasty from 1644 until the revolution of 1911, the Chinese government was weak and corrupt. Except for a few cities along the coast, China was never formally colonized; rather, Europeans relied on a subservient government. Chinese labor was exported to British colonies in Southeast Asia (the "Overseas Chinese") and to the United States, where Chinese workers constructed railroads in the western states. British, French, German, and American trade interests purchased vast amounts of Chinese tea, silks, spices, and porcelain. Britain purchased so many Chinese goods that it ran a negative balance of trade in the eighteenth and early nineteenth centuries, which it corrected by introducing large amounts of opium. By the 1830s, opium addiction was widespread in China. Opium is highly addictive, and the introduction of mass quantities generated severe problems in Chinese society.

In a rare moment of defiance, the Manchu government resisted British opium imports, and Britain and China fought the two Opium Wars of the 1840s, which the British won easily. As compensation, the British seized "treaty ports," coastal cities such as Hong Kong and Shanghai, where Western, not Chinese, laws applied. Britain held Hong Kong until 1997.

Britain was not the only European power in China. The Portuguese had long been in Macau. The French, who controlled Indochina, encroached on southern China. Russia did so from the north. Japan threatened from the east. Even the United States was involved, although it did not claim formal colonies there.

Chinese resentment against the Manchus culminated in the Taiping Rebellion, a huge uprising in the southern part of the country. Lasting from 1851 to 1864, this rebellion led to the deaths of more than twenty million people but was ultimately crushed by the Chinese government with Western backing. The shorter Boxer Rebellion of 1899–1901 was more explicitly anti-Western. These revolts set the stage for the successful nationalist revolution of 1911, which ended Manchu rule and initiated the Republic of China.

Southeast Asia

The peninsula of Indochina and the islands of Southeast Asia were conquered by a variety of different European powers (Figure 2.9). The Philippines was Spain's only Asian colony and served as the western terminus of the trans-Pacific galleon trade. Administered as part of Mexico, the Philippines was heavily affected by Spanish rule, including land use patterns (such as sugar

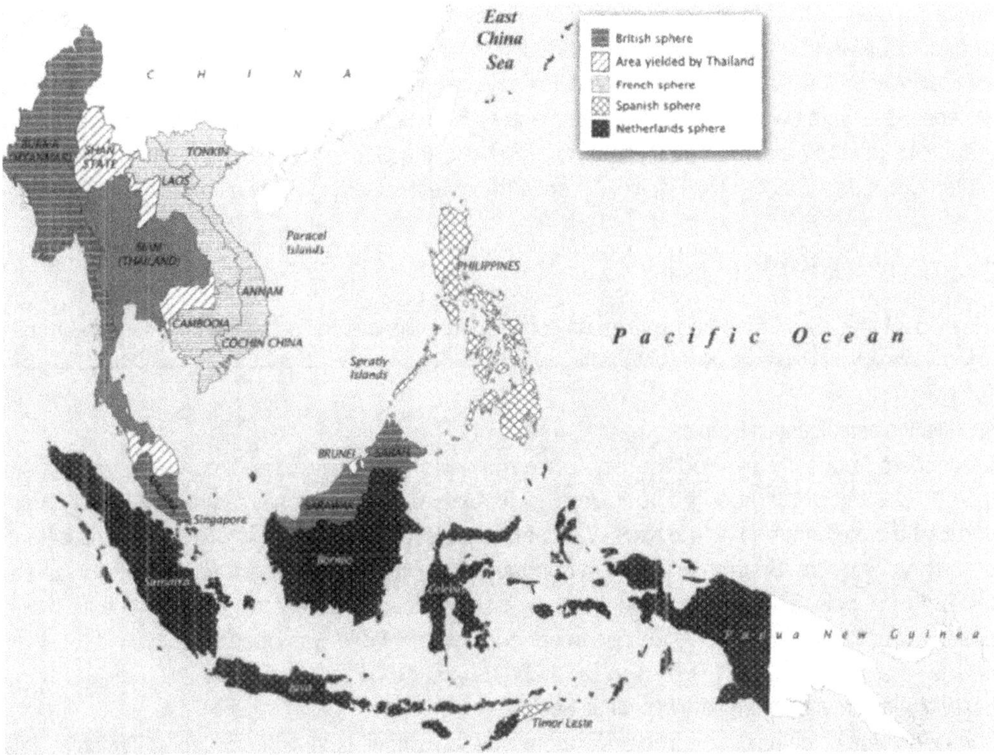

Figure 2.9 Colonial territories in Southeast Asia.

plantations) and its language and religion (it became the only predominantly Catholic country in Asia). Spanish rule ended in 1898, when the United States took over, and it became independent after the Second World War.

The French controlled much of Indochina, including Vietnam, Laos, and Cambodia. French rule shaped the design and architecture of cities such as Saigon, and large numbers of Vietnamese converted to Catholicism. French domination did not end until 1954, with the defeat at Dienbienphu, an event that led to American military involvement there. Britain was also a major power in Southeast Asia, controlling Burma (now Myanmar) and, only informally, the economy of Thailand. These were made into rice exporters for other parts of the British Empire. They founded the city of Singapore as a naval and commercial center to exercise control over this region. Malaya, like other colonies in the area, became a major producer of rubber products as well as timber and tin. Indonesia was dominated by the Dutch for several hundred years. Dutch rule, starting with the founding of the Batavia colony on Java in the eighteenth century, gradually expanded to include the other islands. The primary institution involved was the Dutch East India Company, a chartered crown monopoly similar to the Hudson's Bay Company and the British East India Company. Indonesia became a significant source of spices, tropical hardwoods, rubber, cotton, and palm oils.

Oceania

Australia, New Zealand, and the Pacific Ocean islands, which are called Oceania, formed yet another domain of colonialism. Australian Aborigines were largely eradicated by the British. The continent originally served as a penal colony for criminals. In the nineteenth century, it became a significant exporter of wheat and beef. New Zealand, also a British colony, had a much larger native population, the Maori. The introduction of refrigerated shipping in the late nineteenth century turned this island nation into a major producer of lamb and dairy products. Finally, the countless islands of the Pacific Ocean, home to varying groups of Polynesians, Micronesians, and Melanesians, were conquered by the British and French. Fishing and whaling interests used these islands to refuel in the nineteenth century. Following the Second World War, when the United States drove Japan out of the Pacific, America became the leading political force in the area.

2.4d The Effects of Colonialism

It is evident that colonialism had enormous consequences on colonized places. It is worth summarizing these more explicitly and systematically.

Annihilation of Indigenous Peoples

Colonialism frequently involved the wholesale murder of conquered peoples. In Australia, more than 90 percent of the Aboriginal population was exterminated. Colonial famines in Egypt, India, and China were common. In the New World, diseases such as smallpox led to the deaths of tens of millions. Both the Spanish and British conquerors used guns, dogs, and cannon to obliterate millions of people. The African slave trade devastated tribal societies on that continent. These observations remind us that the European conquest of the world was often violent.

Restructuring around the Primary Economic Sector

The conquest of colonies incorporated them into a worldwide division of labor. Typically, this process involved turning them into centers of primary economic activity concerned with the extraction of raw materials from the earth, including logging, fishing, mining, and agriculture. The plantation system was the leading edge of the global commodification of agriculture. In the United States, this took the form of tobacco and cotton plantations. In Latin America, Africa, and Southeast Asia, cash crops such as sugar, cotton, fruits, rubber, and tobacco were grown for sale abroad with the plantation system. Silver, tin, copper, gold, diamonds, and other ores were mined using slave labor or peasants working in slave-like conditions. Mercantilist trade policies worked to suppress industrial growth in the colonies by reducing imports through stiff tariffs, such as the ones the British imposed on Indian textiles. This process is largely responsible for the fact that many developing countries today export low-valued goods and must import high-valued ones.

Creation of Dual Societies

Colonialism produced great inequality in places (many of which already had great contrasts between the ruling elites and the masses). Often, colonial administrations used a small native elite to assist them in governing the colonies, typically drawn from a religious or ethnic minority. For example, the French utilized the Alawites in Syria, an offshoot of Shiism. The Germans and Belgians favored lighter-skinned Tutsi over darker-skinned Hutu in Rwanda and Burundi. The British relied on the Muslim Mughal rulers to govern Hindu India.

For the vast majority of people, colonialism entailed declining economic opportunities. Traditional patterns of agriculture were disrupted, often with disastrous effects. Land use patterns favored colonialists, who appropriated the best farm land. The commodification of food made many dependent on the market, and vulnerable. People living in dry climates in western Africa, for example, coped well with periodic drought until the British forced them into a system of cash cropping. Huge famines shook India, Egypt, and China in the nineteenth century.

Polarized Geographies

As colonial societies became polarized, so too did the spaces they inhabited. Ports were central to European maritime trade and control, and became important centers of commerce, often to the detriment of traditional capital cities further inland. In Peru, Lima displaced the Incan city of Cuzco; in western Africa, the famous trading city of Timbuktu declined as new maritime routes flourished; in India, coastal cities such as Calcutta, Bombay (now Mumbai), and Madras displaced the Mughal capital of Delhi; in Burma (now Myanmar), Mandalay gave way to coastal Rangoon; in Vietnam, the imperial capital Hue declined as Saigon flourished; and in Indonesia, the traditional capital of Jogjakarta was marginalized by the Dutch port city of Batavia, later Jakarta. As cities flourished and offered more opportunities, millions of people left the poorer rural areas in waves of rural-to-urban migration.

From the coasts, colonial railroads extended into the interior, often reaching into mineral-rich regions or plantations. A long-term consequence of this pattern is that the road and rail networks of developing countries often bear little resemblance to the distribution of the population that lives there; rather, they were constructed to facilitate the export of raw materials to the colonizing country, and today, to the global economy.

Transplantation of the Nation-State

The nation-state was fundamentally a European creation. However, it was widely dispersed around the world as colonies were transformed into states. In Africa, this process led to the formation of unstable states with highly artificial borders drawn by the Treaty of Berlin in 1884. Similarly, Burma and Afghanistan were both creations of the British. Even India, which is more culturally diverse than Europe, was welded into one country; surprisingly, it eventually divided into only three (including Pakistan and Bangladesh). Unlike Europe, where states contained some degree of ethnic similarity, the states of much of the developing world were too diverse to be seen in the same terms. In such countries, where local religious, ethnic, and tribal loyalties supersede nationalism, political conflicts often impair economic growth and development. Of course, the United States is also very diverse ethnically and culturally, but it emerged under very different historical circumstances than did the former European colonies in the developing world.

Cultural Westernization

Because colonialism is not simply an economic or political process, Western economic and political control was accompanied by the imposition of Western culture.

Racism emerged on the heels of colonialism as a way to dehumanize non-White populations, notably with the slave trade. Because Europeans were generally White, their conquests of non-White peoples generated ideologies that viewed White superiority as natural and inevitable. Thus, race relations and racial inequalities are central to the dynamics of the capitalist world system.

Other dimensions of Westernization concerned European religions and languages. Missionaries spread Christianity throughout the colonial world, sometimes successfully (e.g., Latin America, parts of Africa, the Philippines) and sometimes not (e.g., China). European languages such as Spanish, English, French, and Portuguese became widespread in different regions of the colonized world. School systems in colonies, generally set up to benefit the ruling elite, offered students extensive instruction in the history and culture of the colonial country, but little about the society in which the students lived. More broadly, colonialism may be seen as one chapter in the broader process by which global capitalism homogenizes lifestyles, values, and role models around the world, turning disparate sorts of peoples into ready consumers.

2.4e The End of Colonialism

The European empires lasted almost half a millennium, but ultimately they collapsed. In Latin America, this process began after the Napoleonic Wars in the early nineteenth century, which weakened the world's core sufficiently to allow the periphery to break away. In Africa and Asia, the end of colonialism came later, after Europe self-destructed in the First and Second World Wars, which was similar to the Napoleonic conflicts of the early nineteenth century in their impacts.

The international environment following the Second World War provided an ideal opportunity for nationalist and independence movements in the Arab world, Africa, India, and Southeast Asia. Sometimes communists were involved. The Japanese occupation of Indochina, the Philippines, and Indonesia destroyed the myth of European invincibility. Independence movements were occasionally led by intellectuals educated in the West, such as Kwame Nkrumah in Ghana, Ho Chi Minh in Vietnam, and Mohandas Gandhi in India. Meanwhile, the Cold War rivalry between the United States and the Soviet Union allowed political leaders in the developing world to play the superpowers off against one another.

Independence movements succeeded, sometimes peacefully, as in India, and often violently, as exemplified by the Vietnamese and Algerian defeats of the French. As a result of this process of decolonization, the number of independent states multiplied rapidly in the 1950s, 1960s, and 1970s. Today, there are very few official colonies remaining. Whether colonialism is truly dead, however, is another matter; it may well persist in the form of neocolonialism, in which former colonies acquire the trappings of independence but face control by foreign firms and capital.

2.5 Summary

This chapter has laid out the historical foundations of the modern world economy. To understand the world's economic geography today, one must appreciate how it originated and came to be. This means that we must know something about capitalism, the type of society that emerged in the sixteenth and seventeenth centuries and now dominates the entire globe.

First, the chapter sketched a portrait of feudal society, both to show what a noncapitalist society looks like and to reveal the historical context in which capitalism emerged. Then it described the fundamental characteristics of capitalism, which, uniquely, is dominated by markets as the primary (but not only) way in which resources are organized. Markets are not unique to capitalism, but their importance definitely is. However, even in market-based societies, the state plays a role in the distribution of resources, such as through public goods and services. Capitalism also has its own set of class relations, including commodified labor and a working class. Capitalism and finance were bound together from the beginning. Spatially, capitalism also creates landscapes of uneven territorial development as capital flows from one region to another in search of the highest rate of profit.

Next, the chapter explored the Industrial Revolution, which, during the eighteenth and nineteenth centuries, saw an enormous transformation in capitalist social relations. The ability to harness inanimate sources of energy led to innumerable technological innovations and productivity increases. Industrialization created the factory system, and with it, the working class. The Industrial Revolution produced radically new geographies that were largely urban in nature. Industrialization began in Britain and catapulted that country to become the most powerful in the world. It also generated a new world economy characterized by significantly higher levels of international trade and investment, which were facilitated by the steam engine and its use in both land and ocean transportation. The higher speeds of transportation and communication unleashed great waves of time-space compression. Industrialization also saw long-term cycles of investment, innovation, and technological change that periodically restructured markets, goods, and places.

The third part summarized colonialism, the expansion of capitalism on a global scale. It traced the European advantages that allowed them to conquer many societies across the planet. The chapter examined various world regions conquered by different European powers, noting that colonialism entailed different impacts across the globe. It summed up the impacts of colonialism, which produced the division between the world's economically developed and underdeveloped nations. Finally, the chapter noted the breakup of European empires, particularly after the Second World War.

Glossary

Bubonic plague: a type of plague that ravaged late medieval Europe and Asia in the fourteenth century
Burghers: the wealth merchants of early capitalism who dominated their societies economically and politically
Capitalism: the type of society that centers on markets and the private ownership of production
Colonialism: the process by which capitalism went global and European powers dominated the world
Commodification: the process of transforming goods and services into commodities
Commodities: goods and services bought and sold on a market for profit
Decolonization: the process that ended European colonialism
Feudalism: the type of society that dominated Europe for 1500 years before capitalism; also found in Japan, China, and elsewhere
Guilds: the associations of craftsmen in medieval Europe that organized skilled trades
Hanseatic League: a network of trading cities in early capitalism that extended across the Baltic and North Seas
Market: a mechanism for linking the buyers and sellers of commodities
Plantations: large farms erected to produce a single commodity using peasant or slave labor
Serfs: rural inhabitants of feudal Europe who were bound to the land by law and custom
Uneven spatial development: the differential landscapes of wealth and poverty that are an integral part of capitalist geographies

Study Questions

1. Why is historical context important to the analysis of contemporary economic geography?
2. What is feudalism, and how did it differ from capitalism?
3. What was the fourteenth-century plague, and what impacts did it have?
4. How is capitalism a unique type of economy and society?
5. Are there classes under capitalism? Which ones?
6. Is capitalism the only type of society to have markets? Why or why not? How does the role of markets in capitalism differ from that in noncapitalist societies?
7. What territorial changes accompanied the emergence of capitalism?
8. How did capitalism unleash new ideologies and ways of looking at the world?
9. How is the nation-state related to the growth of capitalism?
10. When and where did capitalism begin?
11. What is industrialization?

12 When and where did the Industrial Revolution begin?

13 What are some major economic, social, and geographic impacts of the Industrial Revolution?

14 How did Europe manage to colonize the rest of the world?

15 How did colonialism differ among Latin America, Africa, and Asia?

16 What are five ways in which colonialism affected the societies and geographies of the colonies?

17 When did colonialism come to an end, and why?

Suggested Readings

Abu-Lughod, J. 1989. *Before European Hegemony: The World System A.D. 1250–1350*. Oxford University Press.

Banaji, J. 2020. *A Brief History of Commercial Capitalism*. Haymarket Books.

Beckert, S. 2025. *Capitalism: A Global History*. Penguin.

Cassidy, J. 2025. *Capitalism and Its Critics: A History: From the Industrial Revolution to AI*. Farrar, Straus and Giroux.

Chirot, D. 1985. The rise of the West. *American Sociological Review* 50:181–95.

Diamond, J. 1999. *Guns, Germs, and Steel*. W.W. Norton.

Hart, J. 2024. The rise and fall of national capitalism. *Economic Anthropology* 11(1):134–44.

Ingram, R., and J. Vaughn (eds.) 2025. *Capitalism: Histories*. Boydell Press.

Landes, D. 1969. *The Unbound Prometheus: Technological Change and Industrial Development in Western Europe from 1750 to the Present*. Cambridge University Press.

Mielants, E. 2007. *The Origins of Capitalism and the "Rise of the West"*. Temple University Press.

Perelman, M. 2000. *The Invention of Capitalism: Classical Political Economy and the Secret History of Primitive Accumulation*. Duke University Press.

Sheppard, E. 2015. Thinking geographically: Globalizing capitalism and beyond. *Annals of the Association of American Geographers* 105(6):1113–34.

Thompson, E. P. 1966. *The Making of the English Working Class*. Vintage Press.

Wallerstein, I. 1979. *The Capitalist World-Economy*. Cambridge University Press.

Yazdani, K., and D. Menon (eds.) 2020. *Capitalisms: Towards a Global History*. Oxford.

3 Theoretical Foundations of Economic Geography

Theory is a lens for looking at the world, an explanation of a problem, a means of providing answers to questions, and a way of making sense of the relationships among variables. It is what separates description from explanation. A theory allows us to establish causal relations, test hypotheses, justify arguments, and make claims to truth. Theory is not a choice, because it is inescapable. Economic geographers use several types of theory to understand how economic landscapes are structured and how they change over time.

This chapter offers an overview of some of the major conceptual issues in economic geography. It begins with a discussion of the factors of location that influence the locations of corporations, including labor, land, capital, and transportation costs. Second, it delves into economic base analysis to understand how exports drive regional economic growth. Third, it explores the interrelations among production techniques, scale of output, and geographic location. Fourth, it discusses how and why firms grow over time and space. Fifth, it turns to the geographical organization of firms, moving from simple single-establishment ones to large, multi-establishment, multiproduct corporations. Sixth, because capitalism is inherently unstable and changes constantly over time, it turns to boom and bust cycles and how they shape uneven spatial development. Finally, the chapter concludes by reciting some of the many ways the state, or government, plays a critical role in building, changing, and reproducing the economic geographies of capitalism.

3.1 Factors of Location

There are many variables that influence the location of firms and industries, which are aggregations of firms. The locational decisions of corporations are thus quite complex, and many companies spend significant time and effort in choosing the optimal location. Good locations help to generate profits; bad locations can be disastrous. For this reason, corporate decision making is a rational process, if an imperfect one, and is subject to the laws of **market** competition. Firms cannot choose locations arbitrarily or they will be forced out of business by their more rational competitors. The major factors that shape firms' locations include labor, land, capital, management, and transportation costs.

3.1a Labor

For most industries, labor cost and quality are the most important factors in choosing a location, especially at the regional, national, and global scales (Figure 3.1a). Firms often begin the process of making location decisions by examining the geography of labor availability, cost, productivity, and skills. The degree to which firms rely on labor, however, varies considerably among different sectors of the economy, may adopt different production techniques.

Figure 3.1abc: Whether in the fields, factories, or offices, labor is the most important factor of production, and frequently the greatest cost to firms.

Labor is essential for all forms of economic production, but the relative contribution of labor to the cost structure varies considerably among industries. For example, the contribution of labor costs is high in the textile industry and services but extremely low in the petroleum industry. The demand for labor depends on how labor-intensive or capital-intensive a given production process is. In highly capital-intensive industries such as petroleum or mining, labor costs are almost irrelevant. Thus, it is a mistake, but a common one, to assume that all industries seek

Figure 3.1a Farm workers: Labor in the primary economic sector.

Figure 3.1b Factory workers: Labor in the secondary economic sector.

Figure 3.1c Office workers: Labor in the tertiary economic sector.

out cheap labor. Over time, most industries have become increasingly capital-intensive; that is, they have substituted capital for labor, particularly when production in large quantities justifies the investments involved. The use of tractors in agriculture, machinery and robots in manufacturing, or computers in office work are examples of capital intensification, which historically has helped to generate high rates of productivity growth.

The supply and demand of labor in a given region greatly affect its cost. In countries with high birth rates, the supply tends to be relatively high, and labor costs are therefore low. The supply of labor also reflects factors such as cultural attitudes toward women working outside of the home; in very patriarchal societies, the supply of female labor is low. In economically advanced countries, late in the demographic transition (Chapter 7), birth rates are low, and labor is relatively expensive. These trends shape the willingness of firms to become more capital-intensive: when labor is cheap, there is little incentive to mechanize.

There are also great variations among industries in terms of their preferences for different groups of workers. Some firms demand particular types of workers in terms of their age or sex; the fast-food industry, for example, prefers young workers who will work for minimum wage. Finally, since labor is mobile over space, migration (or internationally, immigration) also shapes local supplies of labor. In regions that can attract labor easily, wage rates tend to be low. When the supply of labor is limited by immigration restrictions, wage rates tend to go up. For example, the US Congress passed quotas on immigration after the First World War, which contributed to the rise in wages in the 1920s. At the local level, high housing costs can also constrain the supply of labor if workers cannot find affordable places to live, an increasingly common problem in many large cities.

In theory, because labor is relatively mobile, regional differences in the supply and demand for labor, and therefore wages in a given industry, should move toward equilibrium over time. A high demand for labor in one place, and thus high wages, and an excess supply in another, and thus low wages, should be equalized by labor migration. Examples are the nineteenth-century migration of people from rural areas to the city in Britain and the United States, and the twentieth-century movement of labor from the Third World to the wealthy countries of northern Europe: from the Caribbean and South Asia to Britain, from North Africa to France, and from Turkey to Germany (Chapter 7).

The response of labor to changing and uneven wage rates is not instantaneous. In the short run, labor has a relatively high degree of inertia: most people are reluctant to leave familiar places, even if jobs are plentiful in other areas. They may sit out periods of unemployment or accept a lower income than could be earned elsewhere. The lack of instantaneous adjustment in labor demand and supply results in variations in the cost of labor within and among countries. In the United States, wages are often higher in the more industrial states and in highly unionized environments. In addition, the cost of labor is shaped by the state, including the minimum wage, which varies among states, as well as unemployment compensation.

It is a common but erroneous notion that companies always seek the cheapest labor possible. Labor cost is only one of many factors; equally important is productivity. Productivity is largely a function of the skills possessed by the local labor force, or ***human capital***, which in turn are provided by educational systems, on-the-job training, and experience working. Firms will pay high wages for skilled, productive labor if the cost is justified by higher output, lower error rates, and higher profits. If labor costs alone were critical to the locations of firms, then very low-wage countries, say Nicaragua, should attract vast quantities of capital investments, which they don't, and high-wage countries, such as the United States or Canada, should see a rapid exodus of jobs. The geography of labor is much more complex than wages alone, and involves national and local labor markets in which jobs are being simultaneously created and destroyed; skills are produced, reproduced, and changed; new technologies are introduced; and other cultural, political, and economic forces interact. What firms seek as much as cheap labor is productive labor.

The skill level of occupations greatly affects the size of their labor markets. Generally, skilled labor markets tend to be geographically larger than unskilled ones. Workers may migrate long distances for well-paying positions, and the market for many skilled jobs today is global in scope, including corporate management, professional services, and university professors. Unskilled positions, in contrast, typically draw from a relatively small area; few people would travel across the country, for example, to take a job as a janitor or cashier.

Labor is not simply economic in nature, but political as well: the labor process is saturated with politics. Labor is the only factor input that is conscious of itself, and it alone is able to organize unions, go on strike, or engage in slowdowns. Unionization rates vary widely across the United States, reflecting in part "right-to-work" laws that allow (or do not allow) unions to compel workers at a given establishment to join. Differences in the ability to unionize contribute to the geography of labor costs. In the United States, the South is generally much less unionized than the North. Besides the cost of labor, firms also take into account working conditions, health and safety standards, pensions and health benefits, vacations and holidays, demands for worker training, and the role of labor unions, all of which shape local wage rates and productivity levels.

At the global scale, developed countries have much higher wage levels than developing countries (Figure 3.2). This pattern reflects the legacy of colonialism (Chapter 2), differences in birth rates and thus the supply of labor, and the quality of educational systems and thus the supply of skilled labor. Government policies in some developing countries keep wages low in order to entice foreign investors. Unionization is generally more prevalent in developed countries than in the newly industrializing ones.

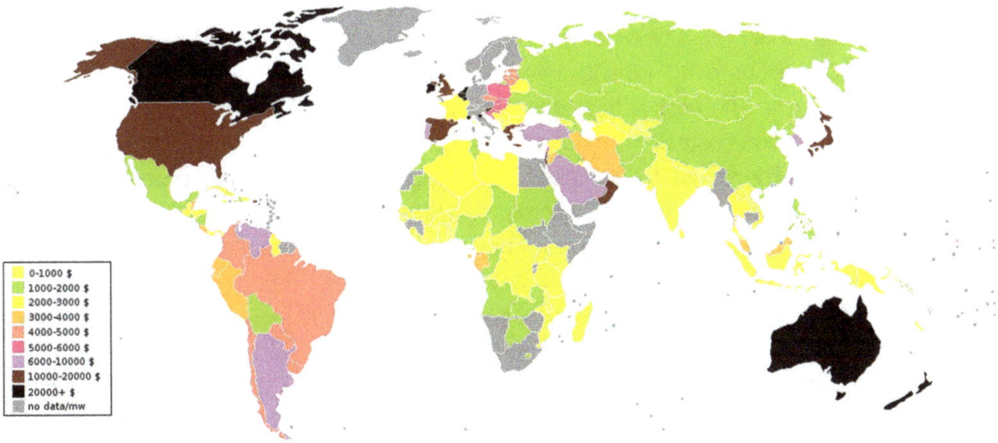

Figure 3.2 Average wages vary widely around the world, reflecting the national supply and demand for labor. Wealthy countries obviously have much higher wages than do poor ones, but also more skilled and productive labor forces.

3.1b Land

At the local level (within a particular metropolitan commuting area, in which labor costs are relatively constant), land availability and cost are the single most important locational factors affecting corporate location decisions. Different types of firms require different quantities of land. Generally, larger firms, particularly in manufacturing, require more land and are thus more sensitive to its costs. In some sectors, such as **producer services** that require centralized locations to operate, firms pay very high land costs (in rent or by purchasing a site). Firms often engage in intensive examinations of several possible sites before settling on an optimum location.

The primary determinant of the cost of land is its accessibility. Transport costs (the measure of accessibility) determine the location rent of parcels at different distances from the city. Thus, because land downtown is the most accessible, it is by far the most expensive; in most cities, land costs decline rapidly away from the city center. Space-extensive land uses thus tend to be located on the urban periphery.

However, not all firms necessarily seek out cheap land. The need to do so depends on the trade-off between land and transportation costs that all firms must make. Some firms must have very accessible locations: generally labor-intensive firms must maximize their accessibility to labor, to each other, to specialized pools of information, and to urban services such as legal firms. Such companies will pay very high rents in order to locate near the city center. In contrast, firms that do not require access to clients, suppliers, and services, on the other hand, such as large manufacturing firms in suburban industrial parks, locate on the urban periphery where land costs are low but transport costs are higher. The demand for land and the need to agglomerate are thus inversely related. Within metropolitan areas, there has been a steady drift of manufacturing to suburban properties since the 1970s.

3.1c Capital

Under capitalism, capital plays a dominant role in structuring the production process. Capital takes one of two major forms: **fixed capital** and **financial capital**. Fixed capital is tangible, and includes machinery, equipment, and buildings; it ranges from simple tools to sophisticated computerized machines. Most types of manufacturing require large amounts of fixed capital to establish their operations or to expand, retool, or replace outdated equipment, or to branch out into new products. The cost of this capital must be paid from future revenues. Besides the purchase and installation costs, firms must plan for maintenance and repair, and the depreciation of capital as it ages. The age of the capital stock of a region shapes its overall productivity levels: places with newer fixed investments are more productive. For example, after the Second World War, Japan and Germany, whose capital stock had been destroyed, started over with new, highly productive machinery.

In contrast to fixed capital, financial capital is intangible, including corporate profits, savings, loans, stocks, bonds, and other financial instruments. The rate of capital formation reflects variables such as corporate profitability (including market prices, production costs), savings rates, interest rates, and taxation levels. Financial capital is by far the most mobile production factor. It can be transmitted almost instantaneously over fiber optic lines. Fixed capital is much less mobile than liquid capital; for example, capital invested in buildings and equipment is immobile. Any functional firm has an assured supply of financial capital to draw upon. Interest rates—the cost of capital—hardly vary within individual countries, but do vary among them.

Investment capital has a variety of sources: personal funds; family and friends; lending institutions, such as banks and savings and loan associations; and the sale of stocks and bonds. Most capital in advanced industrial countries is raised from the sale of stocks and bonds, although American firms rely on this approach more than firms in Europe, where banks play a larger role in industrial financing. The total supply of investment capital is a function of total national wealth and the proportion of total income that is saved. Savings become the investment capital for future expansion.

Whether a particular type of industry, or a given firm, can secure an adequate amount of capital depends on several factors. One factor is the demand for capital, which varies from place to place and from time to time. Of course, capital can always be obtained if users are willing to pay high enough interest rates. Beyond supply-and-demand considerations, investor confidence is the prime determinant of whether capital can be obtained at an acceptable rate. Investor confidence in a particular industry may exist in one area but be lacking in another. It is often expressed in the form of a firm's credit rating.

Capital is important as well because firms can substitute capital for labor in the process of capital intensification. The history of capitalism is largely one of steady capital intensification, particularly in agriculture, in which only a tiny segment of the labor force in industrialized countries now works. Capital intensification can increase productivity, but it may also displace workers. Only if the cost of goods drops sufficiently to increase real incomes can it generate job growth in the long run. Historically, this has generally been the case.

One of capital's crucial advantages over labor is geographic mobility; it can generally make use of distance and differentiation in a way that labor cannot. Corporations often take advantage of such flexibility by shifting production to low-wage regions, setting up plants in areas with low levels of worker organization, or establishing plants in areas that offer incentive policies. Even the threat of mobility may be used in bargaining with workers.

3.1d Management Skills

Managerial skills are also required for any type of production. Management involves the nuts and bolts of corporate decision making, including how to allocate its scarce resources, raise capital, keep abreast of the competition and government regulations, make investment decisions, hire and fire workers, marketing, and similar functions. Corporate management structures reflect the organizational structure of the firm, including its ownership and therefore how decisions are made. Corporate management forms may range from sole proprietorships to partnerships, and firms may be either publicly owned (i.e., by stockholders) or privately owned (typically family owned).

Within firms, management forms an important part of the corporate **division of labor**. Corporate headquarters decide a firm's competitive strategy, what markets and products to focus on, labor policies, where to invest and how much, and deal with legal issues, marketing, advertising, public relations, and financing (Figure 3.3). These tend to be very skilled, well-paying, white-collar jobs. Most headquarters are located in large, urbanized areas (Table 3.1).

Technical skills are necessary for the continuous development of new products and processes. These skills are often called research and development (R&D). The R&D required for new products is typically a long and expensive process, involving long lead times between invention and mass production, a process that is often beyond the scope of small firms. R&D is concentrated in a few major research-university clusters and established areas of innovation such as Silicon Valley, Boston, and Route 128, and the Research Triangle of North Carolina.

For most of the twentieth century, the United States was the world's leader in technological innovation. Starting in the 1970s, however, the number of new inventions and granted patents declined. Factors that slowed US innovation included a decline in federal government support for basic research, a widespread lack of interest in science education among the public, an increased corporate interest in quick returns rather than long-run growth, and difficulties in obtaining venture capital. Outlays for R&D rose in the late twentieth century as foreign competition and high labor costs forced firms to automate in order to reduce costs and increase productivity. In 2020, US expenditures for R&D represented 3.0 percent of the GDP, compared to Japan (3.1 percent), China (2.1 percent), and the European Union (2.0 percent), which, as a group, dominates the world in the number of R&D scientists and engineers and in the amount of R&D expenditures.

3.1e Transportation Costs

For firms to make profits, they must purchase inputs, combine them with labor and capital, and ship their outputs to market. Since all of these steps involve moving goods over space, they entail transport costs, the most expensive part of corporate logistics. The degree to which transport costs affect the costs of the final product varies considerably (Table 3.2). For heavy, bulky, and low-priced goods such as soft drinks or stone products, they can be considerable.

There are also different types of transport costs, such as terminal costs (e.g., loading and unloading goods at a port or railyard) and line-haul costs, the costs that increase as a function of

Table 3.1 **Location of Fortune 500 US Corporate Headquarters, 2024**

San Francisco	18
Minneapolis	18
Houston	21
Dallas	22
Chicago	33
New York	65

Table 3.2 **Transport Costs as a Percentage of Product Prices**

Stone, clay, and glass	27
Petroleum	24
Lumber and wood	18
Food	13
Furniture	12
Paper products	11
Primary metals	9
Textiles	8
Fabricated metals	8
Transport equipment	8
Rubber and plastics	7
Tobacco	5
Machinery	5
Instruments	4
Apparel	4
Printing and publishing	4
Electronics	4
Leather products	3

Source: Compiled from U.S. Department of Commerce national input-output tables.

distance. Transport costs vary by mode of transport (shipping, trucking, air). In terms of volume, shipping is by far the most important: most of the world's goods move by ship because it is cheap. Thus, the geography of the world's ports reflects the changing locations of production; as East Asia has become the world's largest center of manufacturing, the largest ports are located in Hong Kong and Singapore. Some high-valued, expensive goods move by air, which is expensive; this holds for products such as pharmaceuticals, medical equipment, office equipment, and some consumer electronics. However, as a share of the volume of total trade, air is a modest fraction.

The first economic geographer to theorize the role of transportation costs was Alfred Weber, younger brother of the famed sociologist Max Weber, who developed a transport cost minimization model in 1909. Weber noted that firms incur costs for their **backward linkages**, or purchases of inputs, as well as costs for their **forward linkages**, or outputs. To maximize profits, firms will seek to minimize the aggregate sum of their transport costs. Many Weberian models of transport cost minimization have been developed to help firms find optimum locations. The Weberian location model became the basis for much industrial **location theory** and was enormously influential. It is still used today, often in conjunction with Geographical Information Systems (GIS).

Because modern economies require large, expensive **infrastructures** to move goods and people, such as ports, airports, and rail networks, the transportation infrastructure is typically built by the state. Building the infrastructure costs enormous sums, which no private firm can afford, and can take decades. This process allows the profits of production to be retained privately while the costs are socialized.

Finally, it should be noted that transport costs have steadily declined over time, which has rendered them increasingly less important. Containerization, or putting goods in standardized metal containers, was introduced in the 1950s and greatly reduced shipping costs. Containers can be easily loaded and unloaded mechanically, reducing the need for labor at ports; they can also be put onto trucks and trains quickly. Rail lines began to introduce double stacking of containers in the 1980s, which further reduced transport costs. However, on occasion, such as when the global price of petroleum rises, transport costs can surge in importance.

3.2 Economic Base Analysis

Because they are always part of a broader capitalist division of labor, countries, regions, and cities are always integrated into a wider environment that both reflects and shapes their structure, location, and behavior in time and space. This integration often takes the form of linkages among firms, which buy and sell parts, goods, and services to one another. Put differently, corporate backward and forward linkages are, in part, inter-regional. Firms thus simultaneously shape and are shaped by their local, regional, and global surroundings. Economic base analysis (sometimes also called export base theory) offers a means of understanding how cities and regions are integrated with one another.

An important part of the geography of capitalism is the tendency for cities and regions to specialize in the production of some outputs and not others (i.e., develop a comparative advantage, which is discussed in Chapter 4). The history of American cities offers numerous examples of urban specialization. New York was long known as a garment-producing center and today is the capital of finance; Detroit arose around automobile production; Akron was once the world's capitol of rubber manufacturing; Pittsburgh, Buffalo, and Youngstown were major centers of the steel industry; Minneapolis originated as a flour-milling center; Corning, New York, has long been an important glass-producing city; Cincinnati was known as "porkopolis" for its numerous pork-processing firms; Memphis was a cottonseed oil center; Los Angeles is the leading film-producing center of the planet; and Seattle was long closely tied to aerospace through the Boeing Corporation. Similarly, the Great Plains became the breadbasket of North America, Appalachia was a coal-producing area, and the Pacific Northwest exported timber and lumber. Often, smaller towns arise as agricultural processing, timber and lumber, or mining centers.

The economic base (or export base) of a country, region, or city is the part of its economy that exports to other places, linking it to markets in other regions and countries. The economic base sells products to a wider market and derives revenues in return. The term "export" in this context does not necessarily mean foreign exports: regions and cities that sell their output to clients in other locations in the same country are effectively exporting their output. For example, Dalton, Georgia, exports carpets to every state in the nation, and New York exports financial and advertising services to clients across the United States. The economic base is the engine that drives a place's economic activity, and is thus vital to the health of a city's or region's economy.

A common notion is that cities and regions only export agricultural and manufactured goods. Often, planners view services as only produced and consumed locally. Yet this idea is mistaken; services are routinely bought and sold interregionally and internationally. Given that the vast majority of the labor force is involved in services, it would be surprising if these services were *not* traded across places. The United States derives one-third of its foreign revenues from selling services to other countries. New York sells financial, advertising, and legal services around the world; Pittsburgh, Birmingham, Alabama, and Gainesville, Florida, export health care services to clients in other cities and states; and Los Angeles exports television and film services. University towns export educational services when students move there from other places to attend school. Whenever a radio station in one city sells advertising time to a client located in another, it exports a service. When tourists go to Walt Disney World, New Orleans, Miami, Las Vegas, or San Francisco, those places export services. Even Washington, DC (as well as state capitols) exports government services in the sense that the costs of producing these are paid for by other locations. Thus, countries, regions, and cities certainly can and do have services as an export base.

Economic base theory divides the local economy into two broad sectors: the economic base, or basic sector (B), and the nonbasic sector (NB). The basic sector is export-oriented (selling agricultural or manufactured goods or services); in contrast, the nonbasic sector recycles corporate and personal incomes to meet local demand. Nonbasic economic sectors include retail trade, eating and drinking establishments, real estate, and personal services (e.g., hair salons, lawyers, repair shops). Essentially, these functions cater to local demand. From the perspective of economic base analysis, therefore, the nonbasic sector occupies a less important role in shaping the local economy than does the economic base. Of course, the division between these two sectors is not always so simple. Some cities may export retail services when people go there to shop, and some sectors, such as legal services, may be both basic and nonbasic (e.g., a law firm that handles both corporate mergers in another city and local divorces).

We can understand these relations using a simple economic base model, in which total employment (T) equals the sum of the basic (B) and nonbasic (NB) sectors, or

$$T = B + NB.$$

We now introduce the concept of a multiplier, m. The multiplier reflects the degree to which the basic sector drives the total economy. There are several types of multipliers and different ways of estimating them, but here we define the multiplier (m) as the ratio of total to basic sector employment, or

$$m = T/B.$$

Substituting ($B + NB$) for T in the preceding equation yields

$$m = (B + NB)/B = 1 + NB/B.$$

This approach allows us to estimate the degree to which changes in the basic sector (ΔB) cause changes in the total economy (ΔT), that is,

$$\Delta T = m\Delta B.$$

Consider a simple example. Say that a small town has 1,000 people in the labor force. (Total population will be higher, depending on the labor force participation rate.) Of that 1,000, 400 work for a local textile mill. If the mill hires an additional 100 people, what will be the impact on the community? We can answer this question simply enough by defining B = 400; therefore, NB = 600. Thus, the paper mill's multiplier is

$$m = 1 + 600/400 = 2.5.$$

The change in total employment that an additional 100 in the basic sector generates is thus

$$\Delta T = m\,\Delta B = 2.5 \cdot 100 = 250.$$

The additional 100 jobs in the textile mill created 250 jobs in the community (including the 100 new jobs in the basic sector). Because the multiplier is always greater than zero, changes in the basic sector always create larger changes in the rest of the local economy; that is, since $m > 0$, $\Delta T > \Delta B$.

How did this process operate? The economic base model holds that when the basic sector expands (or contracts), those changes reverberate through the local economy in a series of interfirm linkages as well as changes in consumer spending. The total changes can be decomposed into three constituent parts: direct, indirect, and induced effects. When the economic base changes (increasing or decreasing employment), it induces direct effects; second, there are changes in employment in firms that sell goods and services to the export base firm(s) through subcontracts, or indirect effects. For example, when the textile mill in the preceding example expands, it may increase its purchases of equipment, repair and maintenance services, office machinery, advertising services, legal services, and other inputs. Thus, the size of a multiplier—and thus the impacts of the basic sector on the rest of the urban economy—is closely tied to the linkages among firms. Every firm purchases inputs from other firms; that is, every purchase is a sale, depending on whether one takes the perspective of the buyer or the seller. Firms that supply the textile mill with inputs of goods and services will enjoy increases in sales as their client buys more inputs from them. These are the indirect effects.

The size of the multiplier in many ways reflects the spatial distribution of backward linkages from the basic sector, that is, the geography of subcontracts. To the degree that the basic sector firm subcontracts with suppliers located far away, many of the benefits of its growth will be exported as well as "leakages." Conversely, the more the firms in the basic sector subcontract locally, the higher the local multiplier effect will be. Sometimes governments require that public sector subcontracts (e.g., for construction) only use local firms for this reason.

Finally, the third part of the total changes resulting from multiplier effects are changes in consumer spending resulting from changes in the basic sector. When the exporting firm and its subcontractors hire more workers, there is more disposable income circulating in the local economy. These expenditures, or induced effects, increase jobs in the nonbasic part of the economy. Thus, higher wages will lead to higher expenditures on real estate, food, entertainment, transportation, and other consumer goods and services. The size of induced effects will depend on the number of workers affected, their salaries, and their spending habits and preferences.

It is important to note that multipliers are double-edged razors; that is, they may work against as well as for a region. Changes in the basic sector can be negative as well as positive, such as when a steel plant or auto assembly plant shuts down. Thus, if

$$\Delta B < 0, m\ \Delta B \text{ will be} < 0.$$

For example, if a textile mill closes down, or if an automobile factory reduces its labor force (perhaps as part of a strategy of industrial reorganization and relocation overseas), then the multiplier effects will take hold, amplifying the initial losses. Subcontracts to supplying firms dry up, reducing the indirect effects. Lower workers' incomes in turn lead to negative induced effects, causing local businesses to suffer and real estate prices to decline. For this reason, the downtowns of many deindustrialized cities often have numerous storefronts that have been closed and housing that is relatively cheap.

Finally, the diversity of the local economy is important. Typically, the basic sector declines and the nonbasic sector rises as cities increase in population. This pattern reflects the fact that large cities usually have a diversified economic base. The Los Angeles region, for example, has an economic base in the film industry as well as finance, software, the port, and aerospace. Conversely, small towns often have very narrow economic bases centered on one industry, and often only one firm, such as agricultural processing centers, mining towns, and lumber-dependent cities.

3.3 Capital Substitution, Economies of Scale, and Agglomeration Economies

Any corporate operation involves the interdependent decisions concerning scale—the size of the total output—and technique—the combination of inputs that are used to produce an output.

As we saw earlier, firms combine inputs such as land, labor, and capital to produce their final good or service. Within economic and technological limits, firms can substitute among inputs to minimize their total costs: capital may be substituted for labor. The greater the cost and productivity differences among inputs, the more the incentive that firms have to substitute among them.

The most evident trend has been the substitution of fixed capital (machinery) for labor. This trend is most evident in agriculture, which has changed over time from very labor-intensive to very capital-intensive. Many manufacturing firms, which use sophisticated computerized technology to improve the efficiency of production, are reducing their use of labor. In services, capital intensification is also evident in the computerization of the office. Whether or not firms substitute between production factors depends on the relative costs and productivity of the inputs. If, for example, labor costs rise at a given location, the firm may choose to substitute capital for labor at that location, or it may opt to change locations to take advantage of lower labor costs and thus maintain the same labor-to-capital ratio. Typically, as an economy becomes more capital-intensive over time, the productivity of labor rises as well.

3.3a Economies of Scale

Economic scale, or the level of output, is also very important. A critical part of the growth of mass production was the increasing division of labor, or specialization of work. Workers who perform one task over and over are much more efficient than those who are responsible for many different jobs. The division of labor not only increases the efficiency of production but also facilitates the use of relatively unskilled labor. This process was important in the early growth of the US automobile

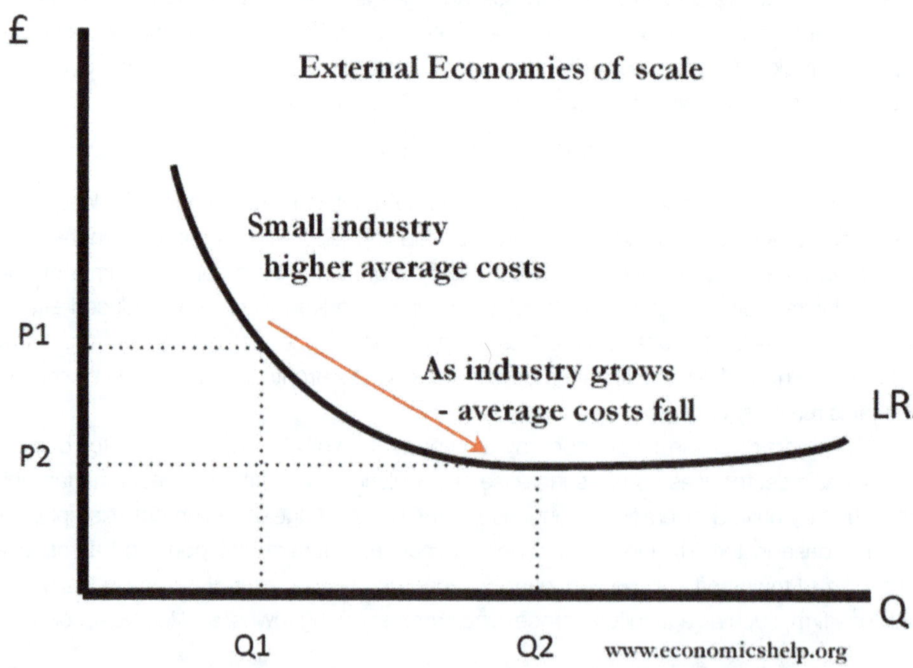

Figure 3.3 A long-run average cost (LRAC) curve illustrating economies of scale.

industry in the system pioneered by Henry Ford. Division of labor, however, requires a relatively large scale of production because a large pool of workers is necessary.

Large firms generally pay much less per unit for their inputs than small firms do. For example, Ford Motor Company can obtain tires for a much lower unit price than a small company can because Ford buys millions of tires a year. Increasing scale, in other words, generally lowers the unit cost of inputs.

A common way of portraying scale economies is the long-run average cost (LRAC) curve, which graphs the unit costs as a function of scale of output (Figure 3.3). Notice that unit costs decrease, reach an optimum point, and then begin to increase. The rise in the curve is termed **diseconomies of scale** (diminishing marginal returns to scale) and occurs when a firm becomes too large to manage and operate efficiently.

Scale economies have important implications for economic geography. They imply that larger firms are more efficient than smaller ones, and may out-compete them, leading to few firms in an industry (oligopolization). In industries with just a few, large firms, production tends to be more centralized in a few locations than when it takes place among many small ones.

3.3b Vertical and Horizontal Integration and Disintegration

In addition to scale economies, firms may change their patterns of purchasing inputs. Some firms purchase the sources of their raw materials: Ford Motor Company at one point founded a city in Brazil called Fordlandia to provide rubber for its tires. Other firms may purchase the companies that provide them with parts or services. This process is called ***vertical integration*** and occurs when the firm controls more stages "up and down" in the production process. Some early large automobile firms, for example, owned their own iron and coal mines and produced their own steel: they were highly vertically integrated. Over time, however, auto producers have become more vertically disintegrated, i.e., the production of different parts has been taken over by specialized firms. Today, automobile firms are mostly just assembly operations for parts made by other companies. Large oil companies are often vertically integrated; they control exploration, drilling, refining, and retailing. In contrast, **horizontal integration** occurs when a firm gains an increasing market share of a particular industry, often through mergers and acquisitions. This process also makes markets become oligopolistic, i.e., controlled by a handful of large firms.

The degree of vertical integration affects where firms choose to locate. Firms that are vertically disintegrated have many ties to suppliers and clients, i.e., numerous backward and forward linkages. Such companies face high transport costs and therefore tend to choose centralized urban locations. As firms become more vertically integrated, the costs of shipping their inputs decline as they produce more inputs "in-house." As their transport costs decline, they tend to shift away from urban cores to peripheral areas where land is cheaper.

3.3c Agglomeration Economies

So far, we have focused on scale economies within one company. However, scale economies also apply to clusters of firms in the same or closely related industries—for example, the computer firms localized in California's Silicon Valley. By clustering, firms can lower unit costs in ways that could not occur if they were spatially separate. These economies are called **agglomeration economies** and were first described by the famed British economist Alfred Marshall in the late nineteenth century. By clustering, distribution and assembly costs are reduced. This process is central to the economic functions of cities and is responsible for the dense downtown regions of metropolitan areas, where business services conglomerate, as well as industrial districts.

The advertising industry in New York offers a great example of agglomeration economies. Advertising agencies must cluster within a short distance of Madison Avenue in order to avail themselves of the dense networks of interfirm linkages to be found there, including information and gossip on the latest trends, markets, clients, hires, and products. For a firm to be successful, it must be near this complex; otherwise, it might as well be on the other side of the moon.

Cities provide markets, specialized labor forces and services, transportation connections, and access to specialized information. Clusters of production thus arise from the numerous inter-connections of people, goods, services, and information that tie together firms, their supplier networks, ancillary services, and related institutions. In addition to firms, other entities in clusters may include government agencies and offices, public/private partnerships, trade associations, universities, legal services and patent attorneys, and related services. Clusters arise because the creation of new products is a constant process that typically takes place within relatively small geographic areas. Agglomeration is essential to the creation of knowledge, which takes place through the interactions of skilled individuals, and to the production of synergies, the benefits of interaction such as new ideas, processes, and products that would not arise from firms or individuals acting in isolation of one another. Organizational learning and knowledge transfer among and within firms occur most effectively when firms are in close physical proximity to one another. The agglomeration of firms that both compete and cooperate with one another is essential to the formation of an entrepreneurial and creative local economy. The processes of discovery and innovation are made possible by collaborative relationships, networking, and tight spatial linkages among firms and individuals. In essence, the formation of knowledge spillovers relies on frequent, repeated, and sustained interactions among individuals and firms in a given location. An important outcome of this process is the creation of synergies, i.e., interactions that generate benefits (i.e., ideas) that would not be possible if actors (firms, individuals) operate in isolation of one another (i.e., the combined effect is greater than the sum of the parts). The synergistic benefits of agglomeration are often called "external **economies of scale**" (i.e., external to an individual firm) in that they lower production costs in ways that would not be possible if firms and workers were geographically dispersed. Thus, in most economic sectors, research and intellectual activity are concentrated in or near large metropolitan areas, whereas unskilled, low-wage occupations that involve little creative activity (e.g., branch plants, back offices) tend to disperse to smaller towns.

3.4 The Product Cycle

The **product cycle** is a long-standing notion that highlights technological considerations in corporate spatial organization. The product life cycle begins with a product's development and ends when it is replaced with something better. It is important geographically because products at different stages of production tend to be produced at different places. These places might be within one country or in different countries.

The famous economist Simon Kuznets developed the concept of the international product cycle (Figure 3.4). In Stage 1, innovators discover, develop, and launch a new product. Capitalism is dynamic and generates new products all the time: consider electric cars, solar panels, Amazon's Alexa, and smartwatches. Firms at this stage of production tend to be vertically disintegrated and cluster in urban areas where access to specialized labor and ideas is easy. Producers of new products benefit from a temporary monopoly and all the high profits that result. Innovation begins in an advanced industrial country. These countries have the science, the technology, and the internal market to justify R&D. Developed countries are the innovators of the world economy and less

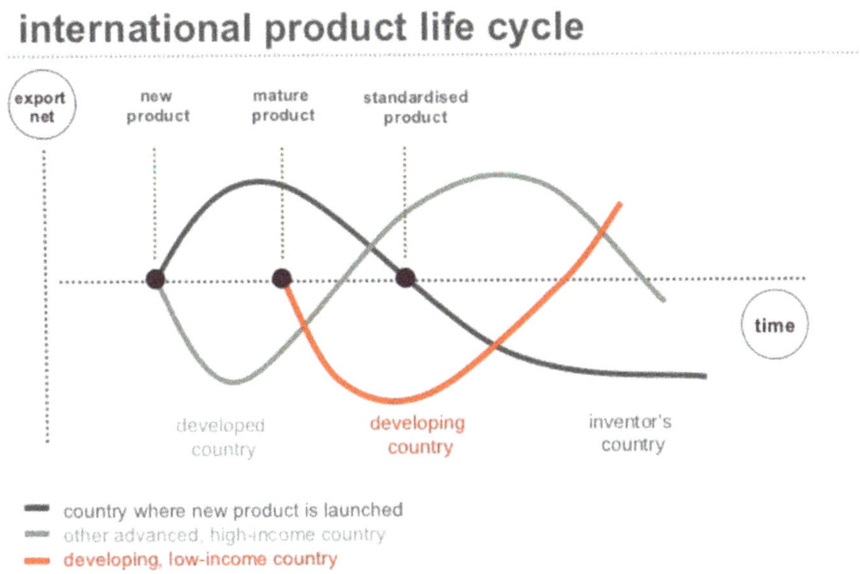

Figure 3.4 The international product cycle describes how and why production of new products arises in developed countries and disperses to less developed ones.

developed countries specialize in transforming raw materials into commodities. As a result, they also have an international advantage, and they export their product around the world.

In Stage 2, particularly after the patent expires, competitors emerge, which forces producers to adopt low-cost, standardized, mass-production technologies. Sales of the product increase for a while, but the initial high returns diminish. But as the technology becomes routinized, production tends to shift to the developing world, where labor is cheaper. By Stage 3, the product begins to be superseded as other products emerge. Markets are lost to newer products, and manufacturing capacity is reduced.

3.5 Business Cycles and Regional Landscapes

Capitalism is notorious for its instability and constant change. The history of capitalist economies is filled with boom and bust periods, periods of rapid growth, high profits, and low unemployment, and subsequent periods of crisis, economic depression, bankruptcies, and high unemployment. Why does this instability exist? How does this instability affect national, regional, and local economies?

The most famous depiction of capitalist cyclicality is **Kondratief cycles**, named after the Soviet economist Nikolai Kondratief, who first identified them in the 1920s. Examining historical data on output, wages, prices, and profits, Kondratief hypothesized that industrial countries experienced successive waves of growth and decline since the start of the Industrial Revolution. Each cycle, he argued, is based on the emergence of key industries and new technologies. These cycles last roughly fifty to seventy-five years (Figure 3.5). The reasons that underlie this periodicity of these waves center on the rate of capital formation and depreciation; as fixed capital investments reach the end of their useful economic life, the impediment on productivity they create generates incentives to innovate and find new technologies.

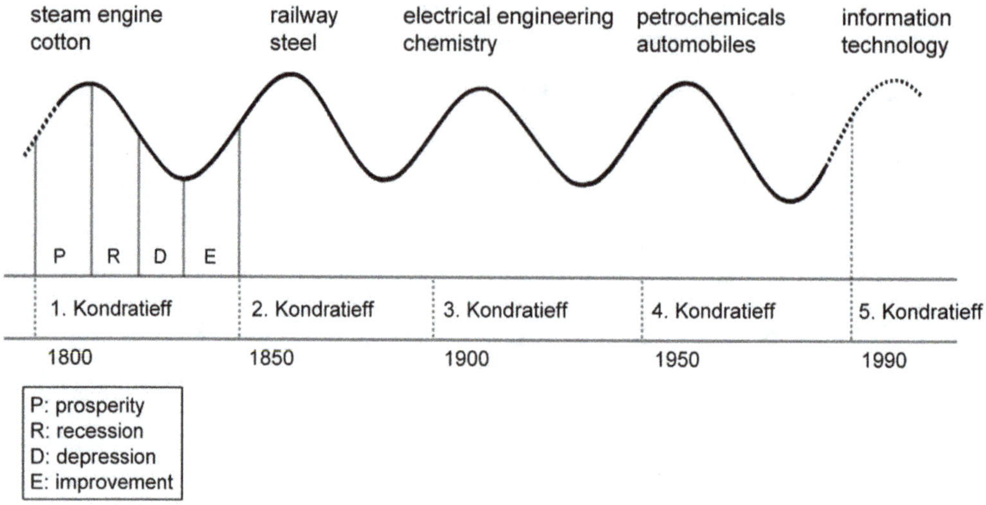

Figure 3.5 Kondratief waves depict long-term cycles of capital accumulation, technological changes, and fluctuations in output, employment, wages, and profit.

The first Kondratief wave began with the Industrial Revolution and was centered on the textile industry and the use of coal. It lasted from the 1770s to the 1820s, when European economies underwent a series of recessions, bankruptcies, and bank failures. The second wave was focused on the steam engine, railroads, and the iron industry; it originated in the 1820s and ended in the enormous round of changes in the 1890s, particularly the depression of 1893, the second worst in world history. The third wave was associated with industries such as automobiles but also included electricity and chemicals; it started at the end of the nineteenth century, peaked around the First World War, and collapsed suddenly during the Great Depression of the 1930s. The fourth wave peaked in the 1960s and witnessed the postwar wave of growth; it centered on propulsive industries such as petrochemicals, including plastics, and aerospace, and ended with the oil shocks of the 1970s. The fifth wave, which began in the 1980s and centered around microelectronics, information technology (e.g., the internet), and producer services. In the twenty-first century, more Kondratief waves are likely to emerge, centered on industries such as biomedical science (e.g., genetically modified foods, the human genome, and molecular biology).

3.5a Business Cycles and the Spatial Division of Labor

Kondratief waves have important implications for economic geography. They emphasize that capitalist landscapes are constantly undergoing change, and that uneven development in time and space is intimately related. Uneven development in space is produced by uneven flows of capital investment and disinvestment. These change the conditions of profitability in different places at different times, generating localized comparative and competitive advantages in regions at different historical moments (Chapter 4). Because of capitalism's dynamism, no region or country enjoys its comparative advantage indefinitely. Uneven development in time unfolds when a region's or nation's comparative advantage is created and eventually lost as capital creates and destroys regions over successive Kondratief waves.

The loss of comparative advantage, e.g., through deindustrialization (Chapter 9), is devastating for firms and workers there, leading to widespread plant closures and layoffs. Ultimately, however, the region may become attractive to firms once again: such places offer unemployed, and therefore low-cost labor, and an infrastructure. In short, regions abandoned by capital in one Kondratief wave may be ready to be reused by a new industry in the next wave. From the perspective of individual regions, therefore, these cycles resemble successive waves of investment and disinvestment. Over time, different waves of investment may locate in the same region, each reshaping the local landscape, creating an imprint that cannot be erased overnight. In each wave, firms create a labor force, invest in equipment and buildings, and press the state to create an infrastructure to suit their needs.

An excellent example of this process is New England. In the early nineteenth century, the textile industry settled there, creating small mills located on streams and, later, large factories in urban areas. This industrialized landscape persisted long after the textile industry abandoned New England for the South in the 1930s during the depression. For the next half-century, New England was a relatively poor region with high unemployment rates. In the 1980s, however, as the fifth Kondratief wave began, new wave of investment centered on the region based on the electronics industry. Computer manufacturing and software firms found the Boston metropolitan region attractive, including the Route 128 corridor. The geography of the new industry was shaped by the residues of the older, textile-based one: often electronics firms would refurbish old textile plants for their use.

In short, each round of capital investments in a given region is influenced by earlier rounds. As they gain and lose their comparative advantage over time, regions accumulate a series of different imprints from successive Kondratief waves: each period of change is shaped by and transforms the remnants of previous waves. Thus, every region is a unique combination of layers of investment and disinvestment over time: landscapes can only be explained through their economic histories (see theme 1 in the Introduction). This type of approach is a hallmark of geographic inquiry, combining broad processes with the unique set of attributes in individual areas.

3.6 The State and Economic Geography

Despite much received opinion, capitalist landscapes are not the products of "free markets," but also reflect the role of the state (government in all its forms and functions). In precapitalist societies, particularly feudalism, in which markets were minuscule, the state was essentially the only major means by which resources were allocated; essentially, there was no division between the public power of government and the private power of the aristocracy. Under capitalism, however, markets are the *primary* means through which allocations of labor, land, and capital are made, but not the *only* means. The state does what no single firm can do, addressing predicaments too large for private firms and providing socially necessary but unprofitable public services. Because there has been a long history of state intervention in market-based economies, it is likely that capitalism could not survive without the state. Even the simplest, most unfettered of markets, such as a garage sale, depends on the existence of public goods such as money, property rights, and an infrastructure such as a road system.

The degree of government intervention varies historically and spatially, but it is never zero. In the nineteenth century, particularly in laissez-faire economies such as the United States, state intervention was considerably lower than today: there were few public services, although the federal government subsidized railroads and erected tariffs against imports. During the Great Depression of the 1930s, when the collapse of the market generated suffering for millions of people, the role of the state changed radically. The rise of the so-called welfare state offered numerous social protections (e.g., Social Security, the minimum wage). Since the late twentieth century, under the pressure of globalization, the welfare state has withered around much of the world, and has been replaced with a deregulated, "neoliberal" state based on the view that the market is the optimal way of allocating public as well as private resources. Any understanding of economic geography, therefore, must include the role of the state.

One way that states shape economies is through the legal system. Laws act economically to protect property rights: without the right to buy and sell, to have assets secured against theft, markets could not exist. The rise of capitalism thus entailed a secular legal system (Chapter 2). Laws and regulations both constrain behavior and protect some parties from the actions of others, including, for example, a handful of health and safety regulations (e.g., environmental protection, workplace safety rules, restaurant inspections), antidiscrimination laws, and antitrust prohibitions. The state enforces laws through the police, judiciary, and the military, and maintains a monopoly over the legal use of violence.

The state also determines fiscal and monetary policy. Fiscal policy concerns how the state raises funds and how governments spend money, and has significant effects on regions. In 2021, the US federal government collected $4 trillion in taxes and spent $6.8 trillion. Governments collect revenues through several types of taxes and fees, including, most important, through individual income taxes, but also the social insurance payroll tax and corporate income taxes.

Conversely, government expenditures, which are geographically uneven, have large impacts on local areas, generating jobs, wages, subcontracts, and corporate revenues. These outlays primarily consist spending on social services (e.g., Medicare, Medicaid, Social Security, veterans' benefits), defense, and interest on the federal debt. In some places, such as Washington, DC, entire metropolitan regions exist due to these expenditures. Public outlays include transfer payments such as Social Security, veterans' benefits, and other entitlement programs. The government also subsidizes producers of many goods, particularly agricultural and dairy products, spending far more on corporate welfare than aid to poor people.

Similarly, the government controls the money supply, which in turn affects inflation, interest, and exchange rates. Inflation rates affect the cost of living; interest rates influence corporate borrowing; and exchange rates are important to international trade. Because the state influences financial markets, these actions have spatially uneven consequences. Almost all countries have a national bank that works to manage their money supply; in the United States, this role is played by the Federal Reserve.

The state also constructs and maintains the infrastructure, including the transportation and communication network (roads, highways, bridges, airports, ports), water and electrical supply systems, hydroelectric dams, sewers, and the like. Without the infrastructure, the circulation of people and goods that reflect how markets are stretched out geographically would be impossible. For example, the Federal Interstate Highway System (Figure 3.6), the largest project the federal government ever built, is vital to connecting the nation's cities.

Similarly, government services include public education, hospitals, fire and police departments, libraries, trash and snow removal, public transportation, public housing, and parks. In the United States, most public services are provided at the municipal or county level; federal government services (as opposed to transfer payments) primarily consist of defense, which consumes the majority of congressional discretionary spending. Net federal government expenditures per capita—the difference between total taxes collected from a state and spending in it—varies widely across the country. The federal government tends to collect the most per capita from wealthy, urbanized states such as California, Illinois, and New York and spend more heavily per capita in rural, lightly populated ones such as North Dakota, Kentucky, Alabama, and South Carolina.

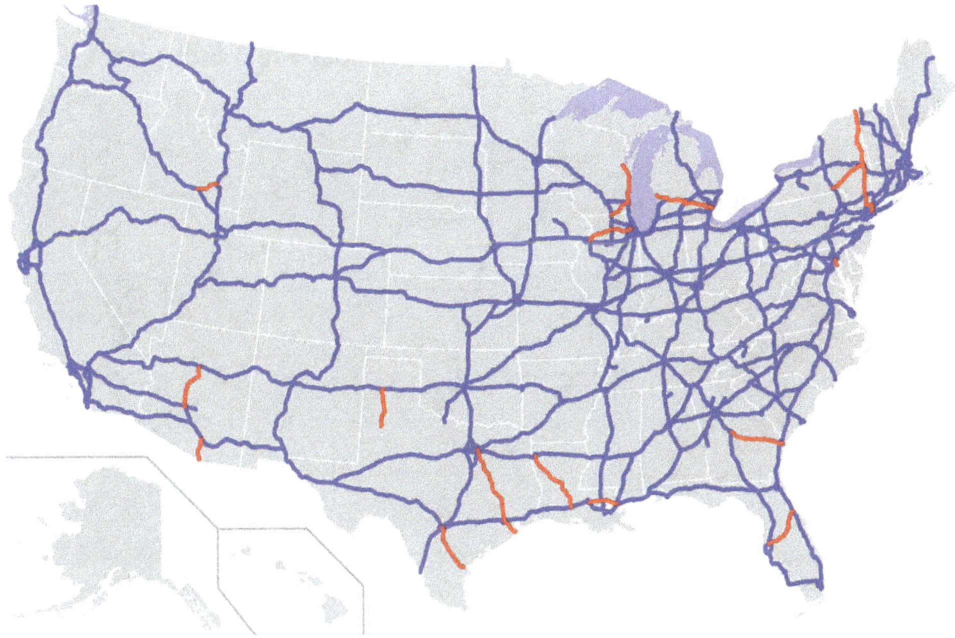

Figure 3.6 The US Interstate Highway System.

The state also shapes the labor markets of capitalist societies, both directly and indirectly. In the United States, the federal government is the nation's largest single employer, with 2.1 million people in 2021 out of a national labor force of roughly 165 million. State and local government employment is even higher, amounting to approximately 20 million in 2021. Additionally, the state affects the supply and demand for labor through the minimum wage, health and safety codes, antidiscrimination laws, and regulations concerning benefits, overtime, and vacations. The state offers training grants and generates human capital through the public education system, which in turn affects private employers.

Housing markets are another area that the state affects. The demand for housing is significantly affected by interest and inflation rates, which in turn are influenced by the money supply. Public housing, which in the United States amounts to only 2 percent of the total housing stock, is an important source for low-income people. At the local level, zoning codes that determine permissible land uses affect urban population densities, minimum lot sizes, and separate incompatible land uses. In some cities, the municipal government imposes rent control, which provides housing at below-market prices. In addition, the federal government owns 40 percent of the area of the United States (Figure 3.7), primarily in the western part of the country, for national parks and military facilities.

Finally, the state in the international arena. Governments shape trade in many ways (Chapter 4) through tariffs and nontariff barriers. Many governments manipulate exchange rates to make their exports competitive and to improve their trade balance. The state also controls international movements of labor, putting restrictions on immigration, which affects the supply of workers domestically. These policies have uneven consequences across the national landscape because some regions are more connected to the international economy than others. All of these examples undermine the myth of the "free market," revealing that in capitalism, *both* markets and the state are in operation, although not to an equal extent.

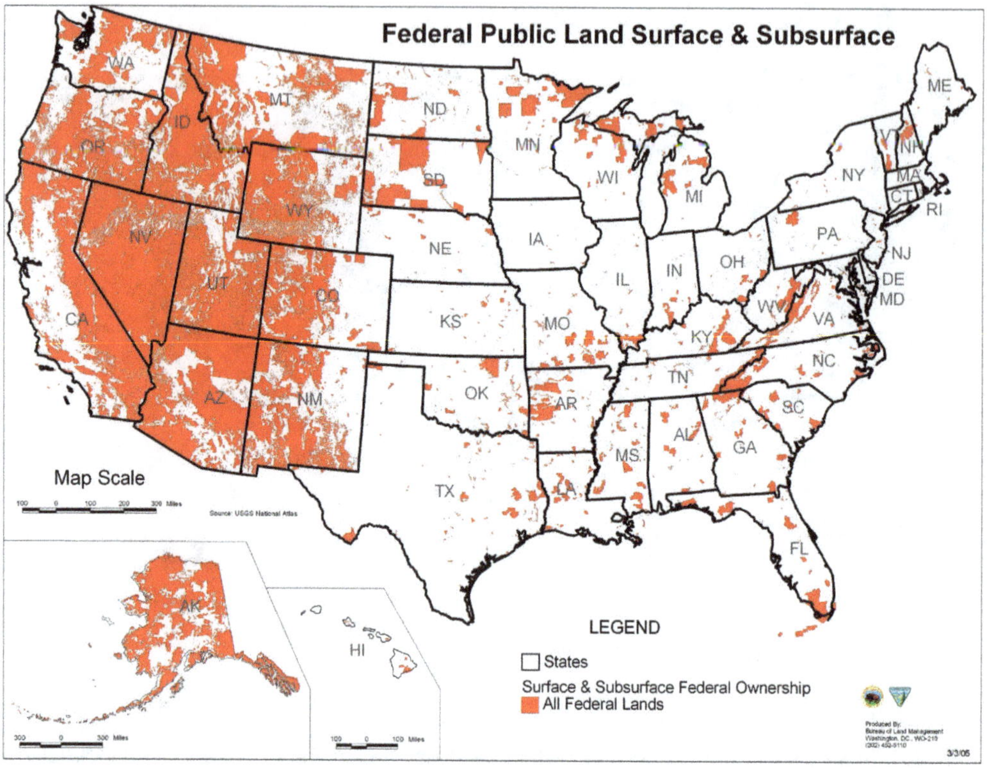

Figure 3.7 Federally owned lands in the United States.

3.7 Summary

This chapter examined several conceptual approaches to comprehend how economic landscapes are constructed. It began by discussing the major factors of location that influence the locations of firms, including labor, land, capital, management, and transportation costs. Labor productivity is as important as its cost. Labor, being conscious, can change working conditions through slowdowns, strikes, and unions. The state actively shapes labor markets, including the minimum wage, right-to-work laws, and occupational and safety regulations. Capital includes both fixed and financial capital. Management decisions made by skilled personnel in headquarters are vital to a firm's long-run health. Transport costs may constrain the locations of firms, especially for those producing heavy goods; firms typically minimize the combined transport costs of their inputs and outputs. The uneven geographic distribution of these factors produces uneven patterns of location as firms minimize costs and maximize profits.

Second, the chapter summarized economic base analysis (or export base theory). Because regions, cities, and countries specialize in the production of some products and services and not others, their sales to clients elsewhere form an economic base. Hence, local exports (which may be within the same country) are the driving force behind the economic vitality of places and regions. In contrast, the nonbasic sector, such as retail trade, recycles these revenues locally.

Third, the chapter noted that firms change their mix of inputs over time. Usually, this entails substituting capital for labor. As firms spread out their fixed costs (e.g., for land, equipment, and buildings) over a larger volume of output, they reduce the per-unit costs, generating economies of scale. In addition, many firms gravitate to be near other firms: by locating in close proximity to one another, they reduce transport costs, create a pool of specialized labor, share information, and use a common infrastructure. The process of clustering results in agglomeration economies that make it cheaper and more profitable to group together than to produce in isolation of one another. Agglomeration economies are the main advantage of cities in capitalist economies.

Next, the chapter summarized the product cycle, a famous model of how production changes over time. Innovative firms developing new products tend to cluster in developed countries with abundant supplies of skilled labor. As products mature, or move into later phases of the cycle, they are produced on a standardized basis. During this phase, firms substitute capital for labor, realizing economies of scale. The demand for skilled labor begins to drop, and the firm can move its most capital-intensive operations to the developing world, where labor costs are low.

Because capitalism is highly dynamic and changes continually, the economic advantages of regions are always ephemeral. To understand this, the chapter stressed the cyclical nature of capitalism, particularly long-term periods of growth and decline, or Kondratief waves. Historically, there have been five such waves tied to different propulsive industries such as textiles, coal, iron and steel, automobiles, electronics, and producer services. Each of these was the high-tech sector of its day. The present world economy is in the midst of a fifth wave based on a cluster of microelectronics and information technologies. Business cycles periodically reshape local landscapes through waves of investment and disinvestment; in individual regions, this leads to the successive imposition of different layers of investment over time.

The chapter concludes by noting that capitalism has never been synonymous with the "free market" because markets are inevitably shaped by the state, or government. The state makes and enforces laws that protect property rights; collects and spends large sums of money; affects interest, inflation, and exchange rates; hires people and creates jobs through subcontracts; regulates working conditions; zones land uses; builds the infrastructure and provides public services; affects housing markets; and intervenes internationally in trade, exchange rates, and immigration. In these ways, the state is an extremely important in the formation of economic geographies.

Glossary

Agglomeration economies: the benefits that firms accrue by clustering together, including reduced transport costs and access to specialized labor and information

Backward linkages: firms' purchases of inputs from other firms and labor

Capital substitution: the replacement of labor by machinery

Diseconomies of scale: the disadvantages and higher costs that set in when production exceeds the optimal limits of a given system

Division of labor: the economic and spatial specialization of production that raises efficiency and productivity

Economic base theory: the view of local economic development that emphasizes the sales of goods and services to clients located elsewhere, generating revenues

Economies of scale: the benefits firms accrue from producing in large quantities, notably by spreading fixed costs over a larger volume of output

Financial capital: funds held in financial institutions that are used for investments, including savings, stocks, and bonds

Fixed capital: investment capital that is material in nature, including buildings, machinery, and equipment

Footloose industries: firms that have low capital investments in a region and are relatively free to move to other locations

Forces of production: the set of factors that determines where firms locate, including labor, land, transport costs, the infrastructure, and state regulations

Forward linkages: firms sales to clients, which are frequently other firms

Horizontal integration: the enlargement of market share in a given industry through the acquisition of other firms

Human capital: the skills and education of the labor force

Infrastructure: the network of transportation, communications, and water control structures in a society, including roads, bridges, tunnels, airports, telecommunications facilities, dams, and aqueducts

Kondratief cycles: long-term economic boom and bust swings that are accompanied by associated changes in prices, profits, technologies, and products

Location theory: the branch of economic geography concerned with understanding and modeling the spatial decisions of firms

Market: the institutions that link buyers and sellers of commodities

Producer services: knowledge-intensive services and expertise that are sold to firms

Product cycle: a view of products and the industries that produce them as moving through a designated cycle of production and consumption

Vertical integration: the growth of firms forward and backward along supply chains

Study Questions

1. How is labor different from other inputs firms use?
2. What determines an industry's demand for labor?
3. Do firms always pursue the cheapest labor? Why not?
4. How does land figure into the locational calculus of firms?
5. What is capital? Differentiate between fixed and liquid capital.
6. What is the Weberian model of firm location?
7. What are scale and agglomeration economies?
8. What are forward and backward integration?
9. What are multiplant and multiproduct enterprises?
10. What is the product life cycle model?
11. What are Kondratief cycles?
12. How do business cycles shape local landscapes over time?
13. What are six ways the state affects economic landscapes?

Suggested Readings

Dicken, P. 2015. *Global Shift: Industrial Change in a Turbulent World*, 7th ed. Guilford.

Dicken, P., P. Kelly, K. Olds, and H. Yeung. 2001. Chains and networks, territories and scales: Towards a relational framework for analyzing the global economy. *Global Networks* 1(2):89–112.

Francis, N., M. Owyang, and D. Soques. 2022. Business cycles across space and time. *Journal of Money, Credit and Banking* 54(4):921–52.

Herod, A. 2017. *Labor*. John Wiley & Sons.

Herod, A. 2022. Adventures with Marx. *Human Geography* 15(1):137–46.

Hyötyläinen, M. and R. Beauregard (eds.) 2022. *The Political Economy of Land*. Routledge.

Knox, P., Agnew, J., and L. McCarthy. 2014. *The Geography of the World Economy*, 6th ed. Edward Arnold.
Marianov, V., and H. Eiselt. 2024. Fifty years of location theory: A selective review. *European Journal of Operational Research* 318(3):701–18.
Porter, M. 1990. *The Competitive Advantage of Nations*. Free Press.
Rickard, S. 2020. Economic geography, politics, and policy. *Annual Review of Political Science* 23(1): 187–202.
Rodrigue, J.P. 2020. *The Geography of Transport Systems*. Routledge.
Taylor, M., and B. Asheim. 2001. The concept of the firm in economic geography. *Economic Geography* 77(4):315–28.
Vorley, T. 2008. The geographic cluster: A historical overview. *Geography Compass* 2(3):790–813.

4 International Trade and Investment

Over the last century or so, almost all national economies around the world have become more specialized and integrated within the global division of labor. This process reflects several factors: the decline in **protectionism**, lower transportation and communications costs, and the steady growth of **transnational corporations**. The global flows of people, goods, services, information, and capital have reached unprecedented heights. Ever larger numbers of companies invest in different countries to acquire resource materials, open up new markets, and use cheap labor. International trade has grown, and its composition and patterns are changing. Enabled by the digitization of money and a global fiber optics network, gargantuan quantities of capital circulate across the globe.

This chapter examines the concepts and patterns that underlie international business. It begins with an overview of international trade. Next, it reviews the major theories that explain the existence and benefits of trade, **comparative advantage**, and competitive advantage. Third, it summarizes international capital markets, including **exchange rates** and **foreign direct investment** (FDI) by transnational corporations. The chapter then considers protectionism, or detriments to trade, particularly **tariffs** and **nontariff barriers** such as **quotas**. Finally, it turns to the mechanisms that have made protectionism decline, such as trade organizations like the **World Trade Organization** and regional trade agreements such as the **European Union** and the North American Free Trade Agreement.

4.1 International Trade

Trade among countries has long been a central part of capitalism (Chapter 2). World trade jumped from $2 trillion annually in 1980 to over $28 trillion today. Most international trade occurs among the world's wealthiest regions, Europe, North America, and East Asia, which have the highest incomes and largest production capacities.

Several reasons explain the growth of world trade over the last three centuries. Average incomes have risen greatly, allowing more people to afford imports. Transportation costs and times have declined significantly, which reduces the costs of imported goods. Government attempts to limit imports, or protectionism, have declined since the Second World War. As Figure 4.1 illustrates, trade has grown most rapidly for countries that have seen the highest gains in their GDP, meaning that trade both reflects and reproduces patterns of wealth and inequality.

Why are almost all of the world's countries so involved in international trade? The answer lies in the benefits that trade generates: jobs in export-related industries, cheaper consumer goods, and the ability to acquire what a given country cannot produce for itself. Trade offers a partial remedy to

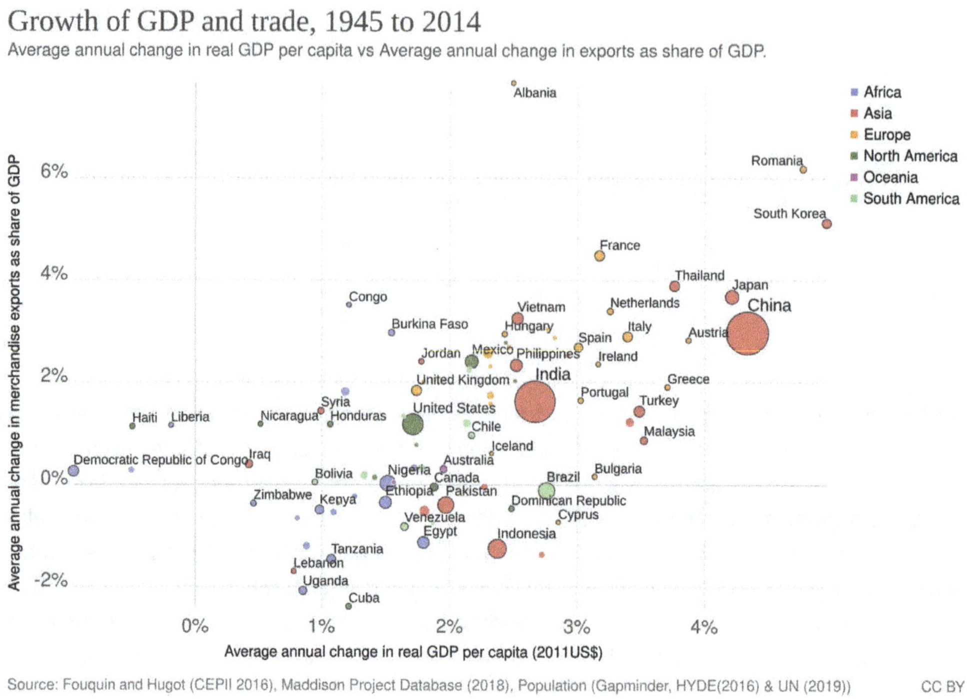

Figure 4.1 Growth of GDP and merchandise exports, 1945–2014.

the persistent problem of uneven spatial development. There are also costs to trade, such as job displacement in sectors that compete with imports.

Production factors—labor, capital, technologies, skills, entrepreneurship, land, and natural resources—vary enormously among countries. These reflect the uneven spatial development of capitalism. Some countries have populations large enough to support significant domestic markets; others do not. Some have pools of skilled labor, others oceans of unskilled labor. Some have vast areas of farmland and ideal climates for growing crops, while others do not. The imbalance in natural and human-made resources explains much of why international trade takes place.

4.2 Comparative and Competitive Advantage

Since the emergence of capitalism, different regions have specialized in the production of different types of goods. In Europe, Britain became a major producer of textiles, ships, and iron; France exported silks and wine; Spain, Portugal, and Greece produced citrus fruits, wine, and olive oil; Germany, by the early twentieth century, was a major exporter of iron and steel products and chemicals; the Czechs exported glass and linens; Scandinavia had furs and timber; and Iceland and Canada exported dried cod to the growing middle classes. Within the United States different places acquired advantages in some goods and not others: New England and New York City dominated light industry, particularly textiles; the Manufacturing Belt became the center of heavy industry; Appalachia developed a large coal industry to feed the factories of the industrial core; the South grew cotton and tobacco; the Midwest became the largest agricultural producer in the world; the Rocky Mountains produced coal and copper; and the Pacific Northwest relied on the timber and lumber industry.

Why regions and countries specialize in some goods and not others has long been a topic of scholarly thought. The two major approaches to explaining this topic were the classical theory of comparative advantage and the more contemporary theory of competitive advantage.

4.2a Comparative Advantage

When a region or country specializes in the production and export of certain goods or services, it enjoys a comparative advantage. This notion was first introduced by the first great economist, Adam Smith, and then adapted to a geographic context by David Ricardo, one of the most famous figures in the history of economics. Ricardo, like all classical political economists, believed in the labor theory of value, which holds that all value is derived from labor and that the prices of goods reflect the amount of socially necessary labor time that goes into their production. Ricardo's approach ignored demand and foregrounded labor productivity as the main driver of trade. He concluded that countries specialize in the production of goods that they can produce using the least labor compared to other nations.

Ricardo's model of comparative advantage is illustrated in Table 4.1, which depicts the allocation of labor, two long-time trading partners, England and Portugal, before and after they specialized and traded. Both indicate the labor hours per unit of wine or cloth that each country uses in the production of a unit of each good. In the first table, Portugal has an ***absolute* advantage** in both goods (that is, it can produce both of them with fewer labor hours than England). If Portugal is more efficient in both goods, using fewer hours to produce them, is it in Portugal's best interest to trade? The answer is yes, because efficient producers benefit from trade. In the second table, each country allocates its resources to the good it can produce most efficiently (i.e., the good in which it holds a comparative advantage). Thus, England produces two units of cloth (at a hundred hours each), and Portugal produces two units of wine (at eighty hours each). In the process of trading and specializing, i.e., of producing for a market that consists of both economies together, each country frees up some resources that would otherwise have been dedicated to making a good in which it did not have a comparative advantage. England saves twenty labor

Table 4.1 The Ricardian Model of Comparative Advantage

England and Portugal (labor hours/unit)				
Before Specialization:				
		Wine Cloth	Total	
England	120	100	220	
Portugal	80	90	170	
units	2	2	390	
After Specialization:				
		Wine Cloth	Total	Savings
England	0	200	200	20
Portugal	160	0	160	10
units	2	2	390	30

hours, Portugal saves ten, and the two combined save thirty hours, which can be reallocated toward other purposes (the model is static and says nothing about change over time).

The Ricardian model—the simplest of many more sophisticated notions of comparative advantage—has important implications. First, it demonstrates that trade affects as well as reflects regional economies: it allocates resources to the most efficient (i.e., profitable) ends at minimum cost. In this model, the only costs of free trade are borne by inefficient producers: English wine growers and Portuguese textile producers. Second, the Ricardian model reveals that specialization reduces the costs of production: trade improves efficiency even without reallocating resources. Even ignoring technological change (which is absent in this model), the increasingly specialized spatial division of labor has long been critical to capitalism's growth and success. For these reasons, almost all economists strongly favor free trade as beneficial to all parties concerned. Third, the Ricardian model points out that large markets allow more specialization than do small ones; thus, the combined market of England and Portugal allows more specialization than either could achieve alone. Adam Smith noted that "the division of labor is governed by size of the market." In this case, when the market expands from one country to two, it allows firms to specialize and become more efficient in the process. Large markets allow firms to develop economies of scale (Chapter 3) and become more efficient. Small economies, with limited domestic markets, thus tend to be inefficient ones.

4.2b Transport Costs and Comparative Advantage

Just as trade requires regional specialization, trade invariably incurs transportation costs. Goods must be moved from producer to consumer, and the cost of transportation must ultimately be borne by consumers. To the extent that transport costs increase the delivered price of products, they reduce consumers' willingness to buy them and therefore the competitiveness of the regions that export them. If transport costs are low, their impacts will be low. However, particularly for heavy and bulky goods, transport costs may increase the market prices of exports/imports prohibitively.

Throughout the history of capitalism, steady long-run declines in transport costs have made it easier for regions to realize their comparative advantages; in this way, lower transport costs have contributed to declining production costs and higher standards of living. For example, in the nineteenth century, New Zealand became a major producer of lamb following the introduction of refrigerated shipping. After the completion of the transcontinental railroads in the 1870s, the Pacific Northwest began to export vast quantities of wood and paper to the cities of the Midwest and East Coast, and the Great Plains became a major world producer of grains with the opening up of the Mississippi River.

4.2c Terms of Trade

The phrase "**terms of trade**" refers to the relative prices that a country receives for exports relative to the prices it pays for imports. Poor terms of trade are characteristic of most less developed countries, which typically export raw materials such as agricultural commodities, lumber, fish, and minerals. Primary economic sector commodities account for about 70 percent and 47 percent of the total exports of the world's low- and middle-income countries (excluding China and India), respectively. The proceeds from these exports pay for imports of manufactured goods, which are vital for industrialization and consumer well-being.

Declines in the relative prices of goods can change the relative worth of the exports of less developed countries dramatically. This means they earn lower foreign reviews in exchange for the same volume of imports. The situation is compounded because many low-income countries are dependent on a single commodity for much of their revenues, such as petroleum or mineral ores, which leaves them vulnerable to fluctuations in the price of that good. Over the last forty years, less developed countries have experienced a worsening in their terms of trade due to declines in the prices of primary commodities and increases in the prices of manufactured goods. There are numerous producers of primary sector goods globally that compete with one another, and productivity gains in mining and agriculture have increased the supply much more than demand has risen, leading to falling prices, which is good for consumers but bad for exporters.

Another factor that may create worsening terms of trade is technological advances in developed economies. Advanced technology enables them to reduce the amount of raw materials in final products and produce substitutes for existing primary sector products (e.g., synthetic rubber for naturally grown rubber). These changes are irreversible. Increases in the price of a raw material create an incentive for industrial research aimed at reducing the use of the commodity, or substituting something else for it, or producing it in the importing country.

4.2d Competitive Advantage

The traditional Ricardian model of comparative advantage is simplistic and unrealistic. It does not explain why labor productivity varies, ignores capital inputs, is static with respect to time, ignores consumption, economies of scale, and agglomeration, and is silent about the nature of competition as well as public policy. Some of these issues have been addressed by neoclassical models in more sophisticated and complex models of comparative advantage.

A different, much more contemporary model, famously developed by management guru Michael Porter, is called the theory of competitive advantage. Unlike the Ricardian model, Porter's approach focuses on the social creation of innovation in a knowledge-based economy.

Porter begins by dismissing two common myths about national competitiveness: cheap labor and abundant natural resources. Is cheap labor essential to economic success? Cheap labor is, on a global scale, ubiquitous. Countries that have economically succeeded, such as Germany or Japan, have done so with labor costs well above those of their competitors. Their labor is productive and well educated, and their economies endow workers with sufficient capital, which generates high rates of productivity. In contrast, countries with the cheapest labor, say most of Africa, have done poorly in the world economy. Neither are abundant natural resources necessary for economic success. For example, Japan has done well despite having very few resources; conversely, many developing countries with resources are trapped in low-wage economies that export raw materials (Chapter 6). In a global economy, supplies of oil, minerals, and agricultural products are available anywhere.

What, then, determines economic success? The key is productivity growth: over the long run, rising productivity lowers prices and generates wealth for everyone, if not equally. Productivity growth in turn reflects many factors, including the education and skills of the labor force, supplies of capital and technology (including the age of equipment), government policies, infrastructure, and the presence of scale economies. Porter emphasizes that competitive advantage, unlike the Ricardian view, is dynamic and changes over time. The goal of national development strategies is to move from low-wage, low-value-added goods into high-value-added, high-profit, high-wage industries as quickly as possible. To accomplish this goal, firms and countries strive to sell high-quality goods at premium prices in differentiated markets. Quality is a key variable here; countries often acquire reputations for producing high- or low-quality goods, earning (or not earning) brand loyalty as a result. Finland is well known for its production of cell phones, for example, just as

Japan is highly regarded for its automobiles. When moving into high-value-added goods, nations should seek to automate low-wage, low-skill functions and retain highly skilled, knowledge-intensive ones, such as computer software and engineering.

Although the global economy is increasingly integrated, competitive advantage is created in highly localized metropolitan areas. Globalization does not eliminate the importance of a home base. Countries that successfully compete internationally do so because a few select regions within them move into advanced products and processes. For example, the propulsive regions of the US economy include Silicon Valley, North Carolina's Research Triangle, and New York City's financial and producer services firms; in Europe, they are Italy's Emilia-Romagna, Germany's Baden-Wurttemberg, Denmark's Jutland peninsula, and the Cambridge region of the UK; in Japan, the government has actively constructed a series of technopolizes that play a similar role.

The determinants of competitive advantage include the following factors:

1. Skilled labor, large supplies of human capital, good educational systems, and adequate technical training.
2. Agglomeration economies, which are typically generated in clusters of similar firms and generate pools of expertise, webs of formal and informal interactions, strategic alliances, trade associations, and networks of suppliers and ancillary services.
3. A culture of innovation that prioritizes experimentation, risk-taking, and entrepreneurship. This is linked to heavy levels of corporate and public scientific research and development, and the continual upgrading of capital and skills. In successful countries, corporations engage in organizational learning and anticipate changes in markets; in contrast, rigid corporate bureaucracies, like public ones, lead to complacency and short planning horizons.
4. Competitive markets at home generate innovation and keep prices low; in contrast, uncompetitive markets dominated by oligopolies exhibit little innovation. To serve increasingly sophisticated buyers requires a constant upgrading in the quality of output.
5. Adequate financing and venture capital, which require deregulated capital markets.
6. Public policies that encourage productivity growth, including subsidized research (e.g., at universities), export promotion, high-quality educational systems, and well-developed infrastructure (i.e., airports, telecommunications networks).

The theory of competitive advantage argues that four attributes of countries combine to increase or decrease their global competitive advantage and world trade: (1) factor conditions concerning the quality and price of inputs; (2) demand conditions, or the structure of markets; (3) supporting industries; and (4) firm strategy, structure, and competition.

Factor conditions include land, labor, capital, technology, and entrepreneurial skills:

1. Human resources: The quantity, skills, educational level, productivity, and cost of labor.
2. Physical resources: Raw materials and their costs, location, access, and transport costs.
3. Capital resources: The availability of funds to finance the industry and trade, including the amount of investment capital available; the savings rate; capital markets and banking; government policies that affect the money supply and thus interest, inflation, and exchange rates; **trade deficits**; and public and international debt.

4. Knowledge-based resources: Research and development, capacities, the scientific and technical community within the country, rates of innovation (e.g., patents), and the likelihood for future technological support.
5. Infrastructure: The public goods and services that provide a country with a competitive advantage, such as transportation and communications systems, water, and electricity.

Demand conditions refer to the structure of markets in a country, including levels of competitiveness, scale economies, and standards of quality, service, and durability. Demand conditions can cause firms to become innovative and therefore to produce products that will sell not only domestically but also in the world market.

To be competitive internationally, firms require access to networks of other firms specialized in different tasks. For example, large financial institutions require law firms, software engineers, marketers, and advertisers. Often, large companies use management consultants or similar business services, subcontracting tasks that require heavy investments in human capital. Access to these services, which generally provide expertise, is typically done through face-to-face contacts. Firms' competitive strategies reflect the conditions under which they originate and expand, which vary among countries. For example, because stockholders demand that US companies show short-term profits, American corporate performance may be less successful than if it were judged over a longer time period, as is Japanese and German firms. State support of corporate strategy and performance is important. For example, a country can regulate taxes and incentives that affect the levels of investment by firms. In addition, competition within a country can impose demands on company performance; new business formations often pressure existing firms to improve products and lower prices, and thus increase competitiveness. Finally, the competitiveness of exporters is affected by labor laws, tax laws, and patent laws.

4.2e Inadequacies of Trade Theories

Trade theories, such as the Ricardian model, are often based on assumptions that limit their applicability. They generally ignore scale economies and transportation costs. Scale economies improve the ability of a country to compete even in the face of higher factor costs: thus, large economies tend to be more productive and competitive than small ones. Trade theories assume perfect knowledge of international trading opportunities, when in fact firms face great levels of uncertainty. Other inadequacies of trade theories include the assumptions of homogeneous products, perfect competition, the immobility of production factors, and freedom from governmental interference. Products are not homogeneous: Japan and the United States sell cars to one another. Oligopolies exist in many industries, which raises prices and reduces competition. Production factors such as capital, technology, management, and labor are mobile to varying extents. Governments manage trade through protectionism and subsidies; they can raise barriers to the movement of goods and services, as well as labor, and affect interest, inflation, and exchange rates, which in turn affect the prices of imports and exports.

The most important shortcoming of trade theories, however, is their failure to incorporate the role of multinational corporations, which operate from a multinational perspective rather than from a national one. When international trade occurs between different affiliates of the same company, it is referred to as ***intra-corporate trade***, which comprises a large share of global trade but about which little is known.

4.3 International Money and Capital Markets

Capital markets, or financial markets, form another dimension of the international financial system. Given the financialization of capitalism over the last forty years, finance has become the "tail that wags the dog." As a result, stock exchanges, currency speculation, futures exchanges, and tax havens have proliferated. American, European, and Asian multinational corporations take advantage of **tax-haven countries** where taxes on income or capital gains are low or nonexistent.

Capital movements take two major forms. The first type involves lending and borrowing money. Lenders and borrowers may be in either the private or the public sector. Private lenders include banks and investment firms, while public ones include governments, the **World Bank**, and the **International Monetary Fund**. The second type of capital movement involves investments in the equity of companies, including foreign direct investment. Multinational corporations are the epitome of direct investors.

Optimally, financial markets would produce an efficient distribution of money and capital throughout the world, helping lenders make a profit and borrowers get access to investment funds. In practice, this is not often the case. There are many barriers to the optimal distribution of funds, including the preferences of investors, practices of investment bankers, and government controls that direct money and capital movements to some areas and not to others, even though the need in neglected areas may be greater.

International banks have existed for centuries; for example, banking houses such as the Medici and the Rothschilds helped to finance companies, governments, voyages of discovery, and colonial operations. British banks were important in financing the American railroad network in the nineteenth century. The banks of the colonial powers—Britain and France—have long been established overseas. American, Japanese, and other European banks went international much later, after the Second World War. Major American banks moved into international banking in the 1960s, and the Japanese banks and their European counterparts in the 1970s. Today, these are being followed by international banks headquartered in China, Korea, Singapore, and elsewhere.

Banks were enticed into international banking because of the explosion of foreign investment by industrial corporations after the Second World War. The banks of different countries "followed the flag" of the corporations from their respective nations. Once established overseas, many found international banking highly profitable. From their original focus on serving their domestic customers' international activities, banks evolved to service foreign customers as well, including foreign firms and governments.

Eurocurrencies are bank deposits that are not subject to domestic banking legislation. With relatively few exceptions, they are held in outside countries, "offshore" from the country in which they serve as legal tender. They have generated a large part of the growth of world trade since the late 1960s. The Eurocurrency market is attractive because it provides funds to borrowers with few conditions; it also offers investors higher interest rates than can be found in domestic markets. At first, Euromarkets involved US dollars deposited in Europe; hence, they were called Eurodollar markets. Although the dollar still represents about 80 percent of all Eurocurrencies, other currencies, such as the deutsche mark and yen, are also vehicles of international transactions. Therefore, Eurocurrencies is preferred to the less accurate term Eurodollar. However, even the term Eurocurrencies is misleading: only 50 percent of the market is in Europe, the major center of which is London. Other Eurocenters have developed in the Bahamas, Panama, Singapore, Bahrain, and Abu Dhabi.

4.4 Exchange Rates and International Trade

In international trade, buyers in the importing country must purchase goods in the currency of the exporting country. For example, if a retailer in the United States wants to buy televisions or cell phones from Japanese firms, it must convert US dollars to Japanese yen in order to satisfy the terms of the purchase. The seller receives payment in the currency of its own country, not in the currency of the purchasing country.

The value of a currency compared with that of another country is called the exchange rate, the amount of one currency required to purchase one unit of another. Exchange rates are never stable but fluctuate constantly over time, with important impacts on the prices of imports and exports. For example, the US dollar strengthened steadily against the Mexican peso in the 1990s and early twenty-first century, making American exports to Mexico more expensive and imports from Mexico cheaper, thus weakening the US trade balance with that country.

When the dollar appreciates in foreign exchange markets relative to other currencies, it can buy more foreign currency and, therefore, more imports from abroad. As a result, American retailers import more goods to the United States. However, the appreciated US dollar means more expensive American goods overseas and therefore less demand for them. Exports from America decline when this happens. Thus, a strong dollar means that the dollar can buy more units of foreign currency, but this is not always optimal for US trade. Conversely, when the dollar depreciates, it can buy fewer units of foreign currency and therefore fewer goods and services; imports usually decline under these circumstances.

Exchange rates fluctuate for several reasons. First, as countries become wealthier, they typically import more goods from abroad. The result is that the increased demand for foreign currency raises the exchange rate of their currencies and decreases the value of the home country's currency internationally.

A second factor is the inflation rate. If the inflation rate of one nation increases more quickly than that of its trading partners, its currency will depreciate compared with the currencies of its trading partners. Consequently, the products of the trading partners will be more attractive to consumers in the country with high inflation. For example, when the US inflation rate increases, the demand for Canadian dollars increases as investors seek a currency that will not lose its value so quickly. This demand raises the US dollar relative to the Canadian dollar and makes Canada's exports to the United States cheaper. But as US exports to Canada increase in price, Canadians demand fewer of the comparatively higher-priced US goods.

Third, the level of domestic demand also affects exchange rates. Real income growth and the relative price levels between countries determine domestic demand. However, domestic demand also depends on consumer tastes and preferences. For example, Americans willingly pay higher prices for specialty items and technologically advanced foreign goods than for comparable products made at home, such as electronic consumer products from Asia, French wines, German automobiles, and Italian shoes.

Fourth, a country's currency may appreciate on foreign exchange markets if its domestic interest rates increase and provide a higher yield to foreign investors. When the US dollar appreciates, foreign investors increase their demand for dollars and thus supply more of their foreign currency in exchange for dollars. When domestic interest rates are low, foreign investments yield lower rates of return than in other nations and do not attract foreign investors. Conversely, as interest rates rise, a country's currency becomes more attractive to investors elsewhere.

Fifth, currency speculation affects exchange rates worldwide. Speculation is a prime cause of the financial "bubbles" that have affected finance-dominated capitalism for the last several decades. Speculation is driven by expectations of future returns, such as hopes that prices in a market will rise or fears of political disasters, which encourage individuals to sell the currency of the affected country to buy foreign ones.

4.5 US Trade Deficits

As the world's largest economy, the US trade situation deserves a closer look. The United States enjoyed a trade surplus throughout most of its history. Starting in the 1970s, however, the volume of its imports began to exceed the volume of exports. The merchandise trade deficit was $25 billion in 1980, but by 2020, it jumped to $800 billion. There are several causes of the US trade deficit: an overvalued dollar, which makes imports cheap and exports expensive; and the American consumer's voracious demand for imported goods, often fueled by consumer debt (Chapter 12). While the US demand for imported goods remains strong, its exports have not kept pace, leading to severe trade deficits of more than $800 billion annually, or 6 percent of GDP. In short, Americans import significantly more products from other countries than they export to them. Many households and corporations finance these purchases through debt, as do many corporations and the federal government.

The US deficit had several consequences: an increase in the volume of imports pushes prices of domestically produced goods down as they compete with cheap imports. This is good for consumers but not good for producers. For example, US automobile manufacturers, steel manufacturers, and textiles producers have been seriously affected by the trade deficit.

In addition to its trade deficits, the United States is a large net debtor, owing foreign governments much more than they owe it. Foreign debt includes federal government debt held by foreigners

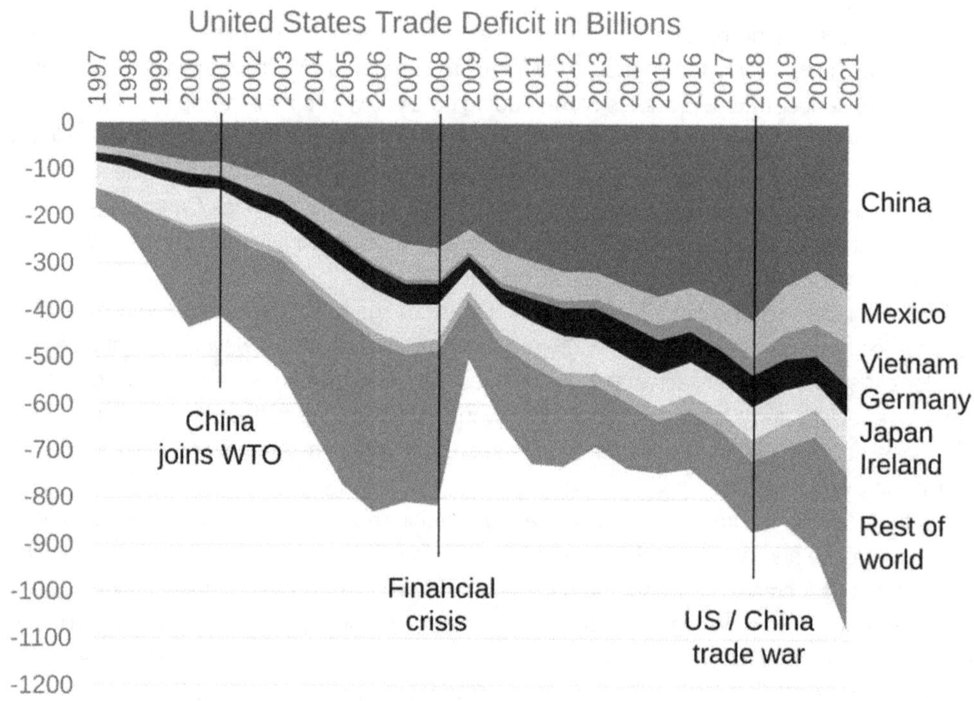

Figure 4.2 United States Trade Deficit over Time.

(largely Asian and West European investors), primarily in the form of Treasury Department securities, as well as private debt issued by corporations (bonds). The total US foreign debt was roughly $20 trillion in 2022 (about 90 percent of GDP), making the United States by far the largest debtor nation in the world. The reverse is true for its lenders, including particularly Japan and China. This situation, however, is not likely to be tenable in the long run. Large federal budget deficits and huge balance of trade deficits have led to the so-called selling of America to foreign investors. Foreign investors now own 26 percent of America's total domestic assets.

One bright spot in the US trade balance is services. Services comprise roughly one-quarter of international trade, and the United States has a distinct comparative advantage in this regard. In contrast to manufacturing, in which the country runs large deficits, the United States enjoys a modest surplus in services (Figure 4.2), which accounts for about one-third of its foreign revenues.

4.6 Foreign Direct Investment

Closely related to trade flows is the flow of capital, including foreign direct investment (FDI). Given the massive scale of FDI today, an understanding of such investments is important.

Three strategic profit motives drive a firm's decision to operate abroad. One motive for many direct investments is to obtain natural or human resources, including low-cost labor that is also sufficiently productive. A second motive is to penetrate foreign markets. The third goal is to increase operating efficiency by locating different operations in different countries. These three motives are not mutually exclusive. Some segments of a transnational corporation's operation may be aimed at obtaining raw materials, whereas other segments may be aimed at penetrating markets for the products made from the raw materials or semi-finished parts. Automobile producers such as Ford, for example, rely on a global network of parts suppliers.

There may be strong motivations for a firm to internationalize, but there are also important constraints. Prominent among these are the uncertainties of investing or operating in a foreign environment. Consumers' incomes, tastes, and preferences vary from country to country. Japanese consumers, for example, are wary of foreign products, at least those that are not name brands. Cultural differences in business ethics and protocol complicate the task of conducting business in two or more languages. Added to these barriers are differences in laws, taxation, and governmental restrictions. Foreign exchange fluctuations may also increase the uncertainty of investing abroad.

4.6a World Investment by Transnational Corporations

Transnational corporations (TNCs) are the leading sources of foreign investment in the world and one of the most prominent dimensions of globalization. TNCs have a long history that stretches to the dawn of capitalism in the sixteenth century, when their roots were to be found in the chartered monopolies such as the British East India Company that played a crucial role in European colonialism (Chapter 2). In the twentieth century, the number of TNCs grew exponentially, as did their economic and political power. Today, they generate trillions of dollars in flows of cross-border financial assets and FDI (Figure 4.3). Despite common views that TNCs always invest in developing countries where labor costs are lower, the reality is that most FDI is concentrated in the developed world. Most of the world's FDI is found in China, the United States, and Europe, the leading centers of the world economy. So rapid has their growth been in the late twentieth and early twenty-first century that the total values of their assets, sales, and exports are an order of magnitude larger than they were in the early 1980s.

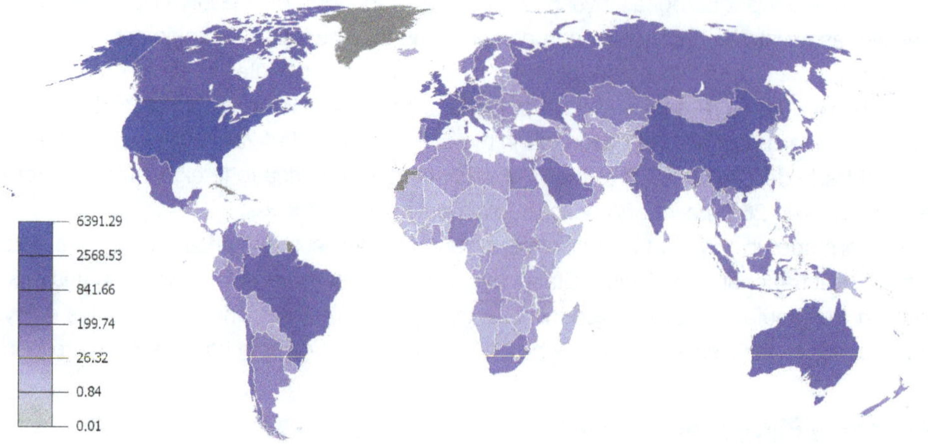

Figure 4.3 Inward foreign direct investment.

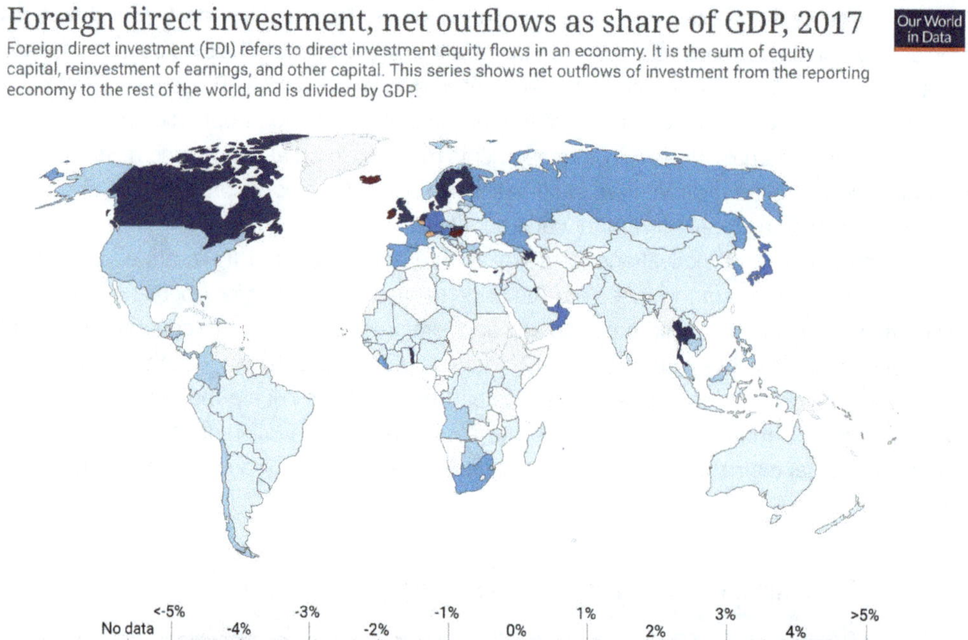

Figure 4.4 Foreign direct investment outflows as percentage of GDP.

Most TNCs and FDI originate in economically developed countries, which have the surplus capital to invest abroad (Figure 4.4). American firms lead the world in FDI, but their share of the total is declining. The rate of increase has been most rapid for companies from Western Europe; however, some developing countries have also increased their outflow of FDI, such as Brazil, Singapore, South Korea, and Taiwan.

4.6b Investment by Foreign Multinationals in the United States

Foreign firms invest in the United States and in other developed economies to gain access to their domestic markets and to avoid protectionist measures. For example, cars produced by Japanese firms operating in the United States are exempt from quotas. FDI in the United States grew rapidly from 1970, when it was a skimpy $13 billion, to 2020, when it amounted to $19.7 trillion. In fact, it increased more rapidly than did US foreign investment overseas. The popularity of investing in the United States was a result of the stability of the country economically and politically, and the relatively inexpensive dollar in world markets.

Trade deficits and FDI are closely linked. The US trade deficit led to the outflow of American dollars into foreign hands. This pool of funds allowed foreign governments and corporations to buy American real estate and factories. As a result, foreign investment in the United States increased sharply. Because of cultural affinities and the lack of language barriers, the UK maintained its lead as America's chief FDI investor. FDI in the United States is centered on manufacturing, chemicals, electrical machinery, electronics, pharmaceuticals, and services.

4.6c Effects of Foreign Direct Investment

Is FDI always desirable? Should the operations of multinationals be controlled? There is no unanimity of opinion, particularly when FDI occurs in less developed countries. Some argue that multinational firms offer great potential to aid the economic development process. In this view, the TNC is an efficient social, economic, and political institution that accomplishes, raises, and reallocates capital, transfers technology, creates jobs and upgrades the local labor force, and provides a market for local goods and services.

This view stands in marked contrast to those who argue that multinational corporations are counterproductive to development (Chapter 6). Foreign firms can bankrupt local producers, who may be small or undercapitalized, and establish local monopolies. Transnationals that use extensive networks of foreign suppliers have few local linkages and low employment multipliers (Chapter 3). They may not improve the host country's balance of payments because their profits may be repatriated to the home country. For this reason, some countries restrict the repatriation of profits to keep the benefits of investment at home. Although the balance-of-payments problem could be avoided in part if multinational firms reinvested more of their profits in the host country, it is uncertain that the national interest would be served. Reinvestment leads to growing foreign control of the economy and the denationalization of local industry. Some argue that the multinational firm is an assault on political sovereignty, including TNCs' demands for government subsidies, training programs, and tax breaks. This latter issue is particularly important given the long history of MNC involvement in the politics of developing countries, ranging from bribes to helping military officers overthrow democratically elected governments.

To be sure, multinationals are imperfect organs of development in developing countries, and their ability to exploit poor countries is tremendous. There is, therefore, an inherent tension between the TNC's desire to integrate its activities on a global basis and the host country's goals of integrating foreign corporate affiliates with its national economy. Maximizing corporate profits is not necessarily the same as maximizing national economic objectives.

The impacts of TNCs depend on both the characteristics and bargaining power of the corporation and the host country. Both larger and wealthier countries and TNCs have more bargaining leverage. Similarly, the degree of host-country control varies across industries and states. Manufacturing industries with advanced and dynamic technologies are more difficult to control than firms in the raw-materials area. There is a wide variety of host governments' policies, ranging from open hostility to open encouragement. Corporations prefer to invest in countries that follow an outward-oriented, export-led development strategy, impose few regulatory controls, offer incentives and subsidies, and appreciate the employment, skills, exports, and import substitutes that foreign investment can bring, an outlook exemplified by the newly industrializing countries.

4.7 Barriers to International Trade and Investment

Much as trade can reduce geographic inequalities in the distribution of resources, so too can international flows of the factors of production, such as labor and capital. Whereas trade *offsets* differences in factor endowments, factor movements *reduce* these differences. International trade and factor flows would occur more frequently if barriers did not exist. The primary barriers are management, distance, and government.

4.7a Management Barriers

Several managerial characteristics of firms reduce trade and the expansion of investment. These include lack of awareness of opportunities, lack of skills, and inertia.

Firms may have the potential to expand but fail to do so because they are satisficers rather than optimizers—they settle for less than the best possible outcome. For example, many US firms paid little attention to foreign markets until the massive round of globalization that unfolded in the late twentieth century. They were satisfied with the large domestic US market. Moreover, firms may have the will to go international but may lack knowledge of potential markets and investment opportunities. While the Internet has reduced uncertainty, often the possibilities of foreign markets are precluded by a lack of awareness. Most national and local governments actively promote increasing international awareness and exports.

Firms may have the will and ability to invest and sell abroad, but they may be thwarted by the complexity of international business, which is a complex and dynamic arena. Knowledge of intermodal rate structures, freight forwarders, shipping rates, and customs-house brokers (firms that contract to bring other companies' imported goods through local customs) is vital for the conduct of international business. Just as necessary is a knowledge of foreign cultures. For example, US firms had great difficulty penetrating the Japanese retail market, partly because of a failure to understand Japanese culture.

4.7b Tariffs and Nontariff Barriers

No country permits a completely free flow of trade across its borders; in this sense, complete free trade is a myth. Governments have erected barriers to achieve objectives regarding trade relationships. Trade barriers include *tariffs*—taxes or duties levied on products when they cross national borders—and *nontariff barriers*—quotas, subsidies, licenses, and other restrictions on imports and exports. These kinds of obstacles are the most pervasive barriers to trade.

Tariffs are by far the most common of all trade barriers, and they can be levied on a product when it is exported, imported, or in transit through a country. The tariff structure established by the developed countries after the Second World War works to the detriment of underdeveloped countries. The underdeveloped countries encounter low tariffs on traditional primary commodities, higher tariffs on semi-manufactured products, and still higher tariffs on manufactured ones. These higher rates are intended to encourage firms in industrial countries to import raw materials

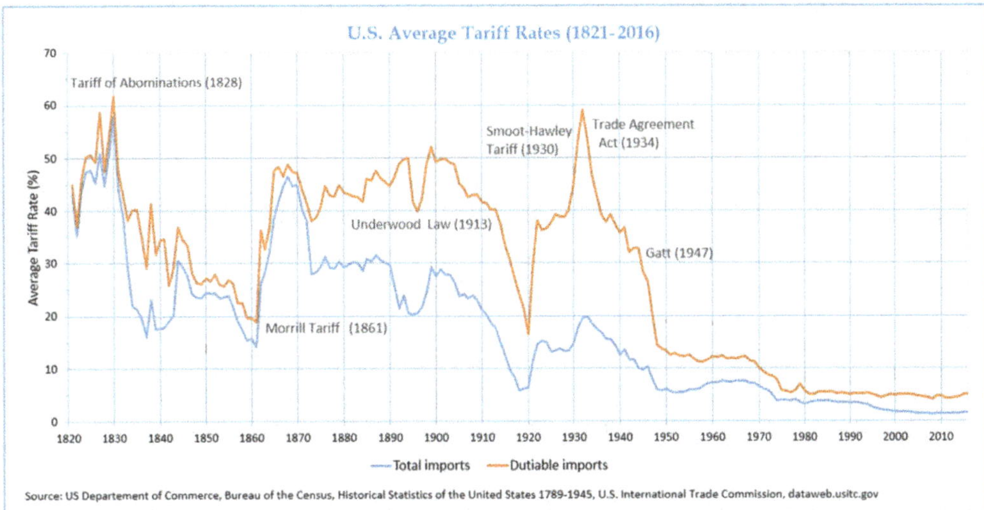

Figure 4.5 US tariff rates over time.

and process them at home. They also discourage the development of industry in the developing world.

Throughout the history of the United States, tariff rates have been high (Figure 4.5). In the nineteenth century, tariffs were the federal government's primary source of income, a situation that changed with the adoption of the federal income tax in 1913. Tariffs soared again during the Great Depression of the 1930s with the Smoot-Hawley Tariff of 1930, which ignited a trade war with Europe.

The reasons behind tariffs are clear. Special-interest groups that stand to gain economically from tariffs and quotas press the government for protection through the use of high-powered lobbyists. Typically, it is industries under duress from foreign competition that advocate most ardently for tariffs. The public, which must then absorb these tariffs and quotas as a surcharge on all imported products, is typically not politically informed and not well represented in Congress.

In 1947, however, the United States and twenty-five other countries signed the **General Agreement on Tariffs and Trade** (GATT), which led to widespread reductions of tariffs and import quotas. In 1995, GATT became the World Trade Organization (WTO) (discussed in more detail later).

In recent years, the importance of tariff barriers has declined, whereas nontariff barriers have gained significance. The simplest form of nontariff barrier is the *quota*—a limit on the volume of imports permitted. A prominent example in developed countries is textiles and clothing: since the early 1970s, textiles and clothing have been subject to quotas under Multifibre Arrangements (MFAs). These created a worldwide system of managed trade in textiles and clothing in which quotas severely restrict developing countries' exports, which has hampered the growth of the industry in China, Indonesia, and Bangladesh. Another common nontariff barrier is the **export restraint agreement**, in which governments coerce other governments to accept "voluntary" limitations.

Other nontariff barriers include: discretionary licensing standards; labeling and certificate-of-origin regulations; overly detailed health and safety regulations, which are especially common on foodstuffs; and packaging requirements. Increasingly, loose, or break-bulk, cargo is unacceptable to developed-country mechanized transportation handlers. These examples represent only a few of the many nontariff barriers devised by governments. Generally, these barriers placed by developed countries are higher for exports from developing countries than they are for exports from other developed countries.

4.7c Effects of Tariffs and Quotas

There are several economic effects of tariffs and quotas on the host country, all of them negative. Protectionism leads to the development and expansion of inefficient industries that do not have comparative advantages. They penalize industries that are relatively efficient and that do have comparative advantages. The result is less international trade and consumers who pay higher prices than necessary.

Figure 4.6 shows the economic effects of a protective tariff. If the imported price is P_{world}, the tariff raises it to P_{tariff}. As a result of the higher price, consumers reduce their demand from Q_{c1} to Q_{c2}. Consumers are hurt by the tariff because they buy fewer goods at a higher price.

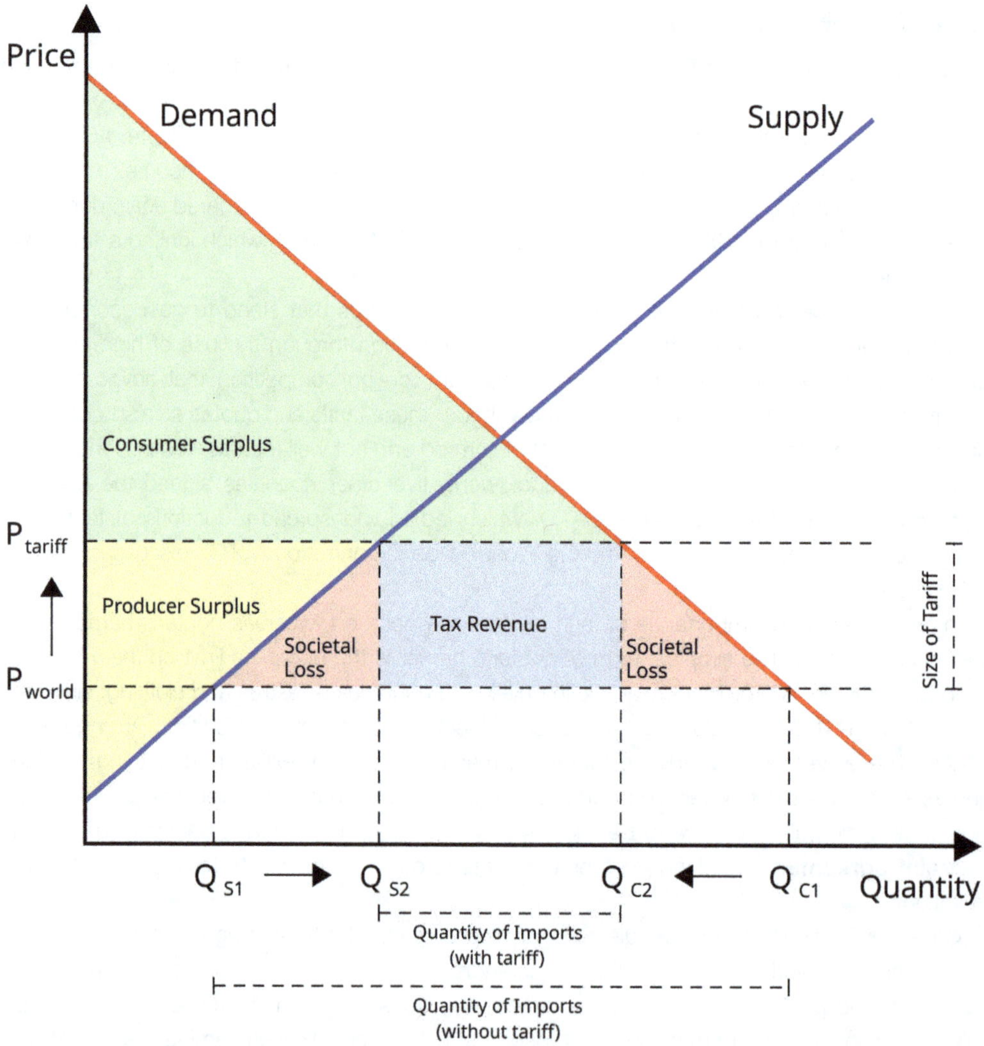

Figure 4.6 Effects of a tariff.

Meanwhile, domestic producers, shielded from foreign competition, increase their supply from Q_{s1} to Q_{s2}. This explains why domestic producers send lobbyists to Washington to pass tariffs that give the producers a relative advantage in the market. Protectionism drives up the costs of imports, and with them import-competing products, thus adding to inflation, and invites retaliation on the part of trading partners. The result of levying a tariff is reduced world trade and reduced efficiency of the international economic system, which hurts foreign suppliers, aids domestic producers, and costs.

4.7d Arguments in Favor of Protectionism

If free trade benefits almost everyone, why would anyone advocate in favor of protectionism? The most common argument is that protectionism saves jobs by keeping cheap foreign imports out. This view is usually put forward by industries having difficulty competing internationally. This argument contradicts the principle of comparative advantage: by purchasing cheap imports, consumers have more money to spend on other things. Conversely, protectionism takes money away from consumers, which generates jobs in consumer-oriented sectors such as retail trade and entertainment, and transfers it to inefficient industries that cannot compete internationally. Each job "saved" in a declining, protected sector comes at enormous expense to consumers in the form of more expensive goods. Thus, protectionism does not generate jobs, it *redistributes* them, imposing on cost on large numbers of people for the benefit of a few. Typically, firms in trouble are well organized politically, while consumer groups are not.

A second argument is that protectionism is necessary to reduce dependence on foreign supplies of critical goods necessary for national defense. Usually, this applies to strategically important minerals and petroleum. However, deeming what goods are, and are not, essential for national defense is often arbitrary. This is a political, not an economic, decision that reflects the entrenched power of lobbyists. Moreover, it is more efficient to store critically important goods than to use tariffs to keep foreign suppliers at bay. This line of thought led, for example, to the formation of the Strategic Petroleum Reserve, large salt domes in Louisiana and Texas, where the federal government stores large amounts of oil in case of a national emergency.

Still another argument is that tariffs can be used to protect an industry that is just getting off the ground and is less efficient than a well-established counterpart in another country. The *infant-industry argument* was invoked to justify protectionist policies in nineteenth- and twentieth-century America and nineteenth-century Germany, as well as that found in less developed countries after the Second World War. There are problems with this strategy, however. If grounds exist to protect an **infant industry** until it has grown large enough to take advantage of scale economies, protection could be given through a subsidy rather than through a protective tariff. Moreover, if an infant industry is protected indefinitely, why should it ever "grow up?" In each case, protectionism makes for appealing political stands that cater to troubled industries and regions, but poor economic arguments.

4.8 Trade Agreements

Since the Second World War, multiple efforts have been made to reduce some of the trade barriers that were erected in the past. These include GATT, the World Trade Organization, and multilateral agreements involving several parties.

4.8a The General Agreement on Tariffs and Trade

The most notable effort to reduce trade barriers was a multilateral effort known as GATT, which started in 1947. GATT grew over time to include more than a hundred countries, a process that resulted in a substantial reduction of tariffs. Services, which now account for about 30 percent of world trade, were not covered by GATT at all, meaning it had no effect on issues such as **intellectual property rights**. Moreover, as tariffs declined, many countries resorted to nontariff barriers, with which GATT was ill-equipped to cope.

Another major problem with GATT was European and US opposition to reducing and phasing out agricultural subsidies (Chapter 8). The EU wanted to maintain **export subsidies**, which are government payments that reduce the prices of goods to buyers abroad. In the United States, agricultural subsidies mostly take the form of direct payments to farmers according to their production levels in order to subsidize their output. The result of these subsidies is increased domestic food output, which provides unfair competition for less developed countries' agricultural products on the world market. Both types of subsidies are artificial barriers that reduce prices on the world market and provide advantages to farmers in rich countries.

4.8b The World Trade Organization

In 1995, the last round of GATT negotiations established the World Trade Organization (WTO), to which most of the world's countries now belong (Figure 4.7). Unlike GATT, the WTO is a permanent standing body. Until the WTO, countries with conflicts over trade had to resolve their own problems. The WTO provides for arbitration to settle conflicts between nations over trade disagreements. Its judgments are enforced through retaliatory trade sanctions by all members of the WTO. Thus, joining the WTO entails some degree of loss of sovereignty for member states.

Under the WTO, most quantitative limits on imports (quotas) are illegal. Intellectual property rights are another important aspect of the WTO, which is important to states with a strong

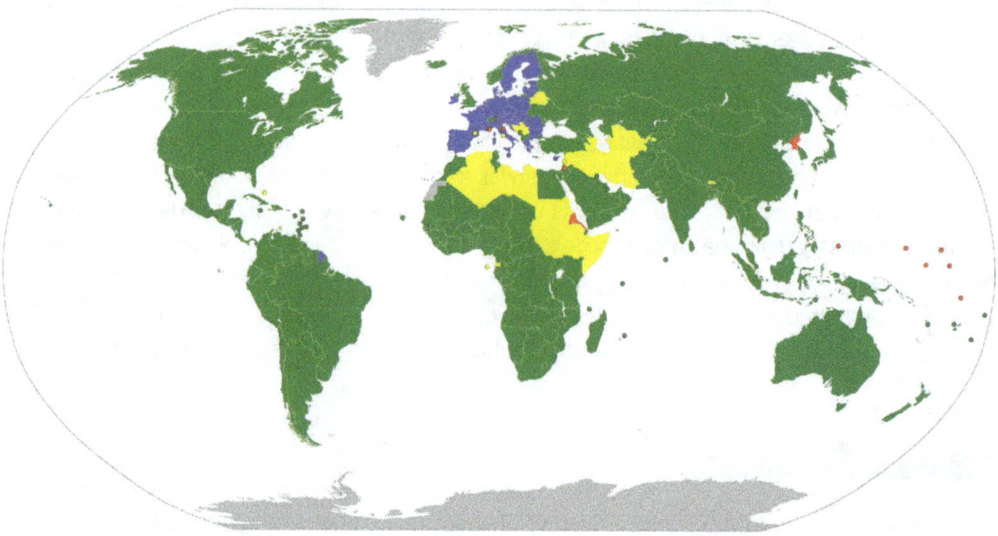

Figure 4.7 Map of World Trade Organization members.

competitive advantage in this arena, such as the United States. All signatories of the WTO are required to protect patents, copyrights, trade secrets, and trademarks. This measure is designed to end the wholesale pirating of computer software, movies, musical recordings, books, and prescription drugs widely practiced in some developing countries (particularly China). The WTO also prohibits members from requiring a certain proportion of content in products to be manufactured within their borders, a practice widely employed as a device to limit the use of imported parts and components and bolster local employment.

4.8c Multinational Economic Organizations

Another mechanism for reducing protectionism is multinational organizations, which are loosely connected associations entailing little or no surrender of sovereignty on the part of member nations. Some of these international organizations are global in scale. The most inclusive is the United Nations (UN), with 193 member states that account for more than 99 percent of the world's population. Much of the UN's work is accomplished through approximately two dozen specialized agencies, such as the World Health Organization (WHO), UNICEF, UNESCO, and the International Labor Organization (ILO). Other international organizations have a regional character; for example, the Association of Southeast Asian Nations (ASEAN) and the Asian Development Bank (ADB). Other international organizations are relatively narrow in focus—mostly military, such as the North Atlantic Treaty Organization (NATO), or economic, such as the **Organization of Petroleum Exporting Countries** (OPEC). The Organization of Economic Cooperation and Development (OECD) is a group of thirty-eight wealthy countries that serves as a forum to study problems, share data, compare experiences, and exchange best practices.

The International Monetary Fund (IMF) and the World Bank have independent, multinational authority, performing functions that individual states cannot perform by themselves. The IMF and the World Bank were originally established by the Bretton Woods conference of 1944 to prevent a recurrence of the depression of the 1930s. The World Bank is the world's largest lender and funds infrastructure projects around the world. Starting in the 1980s, the IMF and the World Bank became advocates of neoliberalism, which advocates unfettered free markets and minimal government intervention. Under pressure from the United States, these institutions adopted adamantly market-oriented positions and imposed them on the governments of less developed countries that required their assistance. The neoliberal "Washington Consensus" that these organizations uphold revolves around requiring less developed countries to engage in tight monetary policy to combat inflation, which serves investors and bondholders well but raises interest rates for consumers, as well as a variety of deregulatory programs in finance, transportation, and communications. The results have included enormous suffering for the poor, as when governments are forced to reduce subsidies for public transportation, kerosene, or cooking oil. Privatization policies encourage or require governments to sell off public assets to private (often foreign) investors, often at reduced prices; all over the world, formerly state-owned or operated power plants, hydroelectric facilities, bus routes, airlines, telecommunications firms, and other assets are being sold to the private sector on the assumption that it is more efficient than the public sector. Yet public services often exist precisely to overcome market failures, particularly the inability of markets to provide adequate services to the poor. Similarly, IMF policies require trade liberalization, including the end of tariffs, nontariff barriers, and subsidies.

Critics of these policies deride them as neoliberal "market fundamentalism." By the IMF's own admission, its policies have often exacerbated the problems of less developed countries. Tight monetary policies can generate recessions and lead to high unemployment. The IMF's macroeconomic models are often oversimplified and underestimate the complexity of local countries' political and social contexts. Many less developed countries lack a proper institutional environment for deregulation and privatization to work, including bankruptcy procedures, protection of property rights, and debt repayment programs. In addition, the Washington Consensus increases inequality in less developed countries by protecting and rewarding investors at the expense of the poor. In pursuing policies that emphasize economic stabilization over job creation, the IMF protects bankers but not the impoverished masses. Finally, market fundamentalism tends to be overly optimistic about the private sector (i.e., markets) and overly pessimistic about the public sector (i.e., governments).

4.9 Economic Integration

Regional integration occurs when a group of countries forms a single economic region to promote investment, trade, and growth. Four levels of **economic integration** are possible. At progressively higher levels, member states make more concessions and surrender more sovereignty. The lowest level of economic integration is the ***free-trade area***, in which members agree to remove trade barriers among themselves but continue to retain their own trade practices with nonmembers. A *customs union* is the next higher level of integration, in which members agree not only to eliminate trade barriers among themselves but also to impose a common set of trade barriers on nonmembers. The third type is the ***common market***, which, like the customs union, eliminates internal trade barriers and imposes common external trade barriers, but in addition permits free production factor mobility of labor and capital. At a still higher level, an ***economic union*** has the common-market characteristics plus a common currency and a common international economic policy.

There are a variety of trade organizations in effect throughout the world (Figure 4.8). Many developing countries turned to regional integration strategies in the late twentieth century. The primary

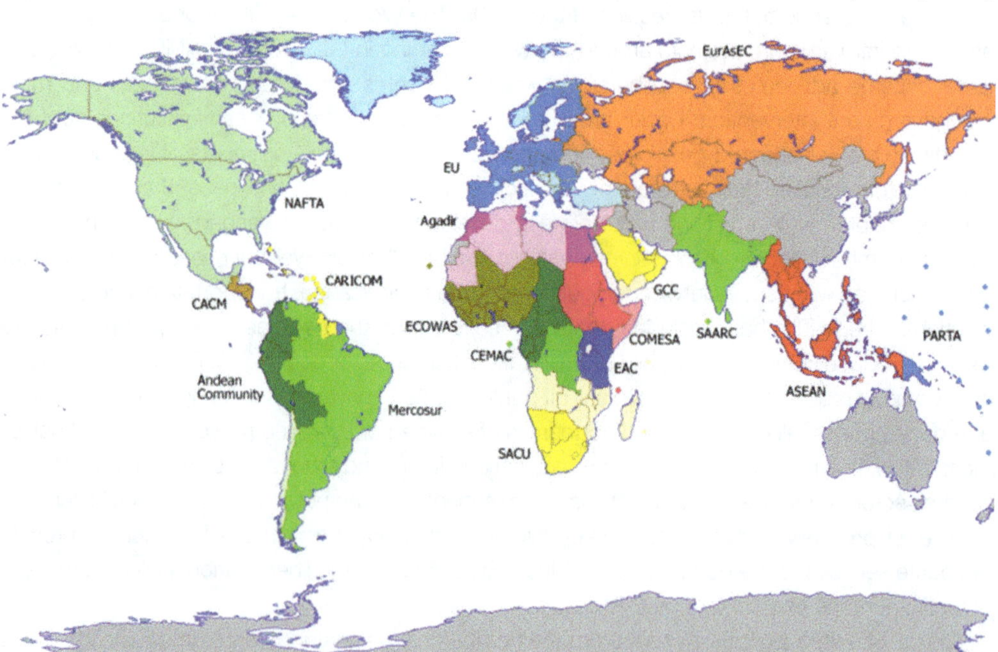

Figure 4.8 The world's major trading blocs.

reasons for doing so included access to larger markets, to obtain more bargaining power with the developed countries, and to strengthen their position in negotiating with transnational corporations. These groups range from loosely integrated free-trade areas such as the Latin American Free Trade Association (LAFTA) to common markets such as the European Union. There are North-South ties between the EU and LAFTA, and South-South ties between LAFTA and ASEAN. Most of these links are bilateral, that is, agreements between nations within different regions.

4.9a The European Union

The most successful example of economic integration is the European Union (EU). The EU started in 1957 with six countries: France, West Germany, Italy, Belgium, the Netherlands, and Luxembourg, and since then has added most Western European countries; for example, the UK, Ireland, Denmark, Greece, Spain, and Portugal. In addition, most Eastern European nations have joined or applied for membership in the EU, and several have been accepted, so that by 2022, it had twenty-seven members and a total population of more than 500 million (Figure 4.9). (It used to have twenty-eight members, but the UK left the union in 2021.) The EU today is the largest single trade bloc in the world and accounts for 40 percent of international trade, which is three times its

Figure 4.9 The twenty-seven members of the European Union, 2023.

world share of population.

The intent of the EU was to give its members freer trade advantages while limiting imports of goods from outside Europe. It aspired to: (1) the establishment of a common system of tariffs applicable to imports from outside countries; (2) the removal of tariffs and import quotas on all products traded among the participating states; (3) the establishment of common policies with regard to agriculture and transportation; (4) free movement and access of capital and labor among EU countries; (5) transportation of goods and commodities across borders with no inspection or passport examination; and (6) a common currency.

The EU has made tremendous progress toward its stated goals. The member states have been afforded more efficient, large-scale production because of larger markets within the EU, permitting them to achieve scale economies and lower costs per unit, something that pre-EU conditions denied them. The union allowed northern European capital to flow into poorer southern European states, and workers from Greece, Spain, and Italy to hold jobs in northern ones.

On New Year's Day 1999, many (but not all) members of the EU adopted a unified currency, the Euro. Britain, Denmark, and Sweden opted to retain their own currencies. The exchange rate of participating countries' currencies is locked in against one another and against the Euro. However, the Euro floats against non-EU currencies such as the US dollar. When most (but not all) members of the European Union adopted a single currency, they were well aware that each country effectively surrendered the right to manage its own money supply, which greatly limits its ability to depreciate its currency during a crisis. The Euro has the advantage of eliminating exchange rate fluctuations among member states, which can wreak havoc with trade. However, the downside is that it deprives countries of the ability to devalue their currency during times of crisis, which makes their exports cheaper and improves the trade balance. Greece, for example, was greatly hampered by its inability to depreciate during its severe debt crisis of 2007–2008.

4.9b The North American Free Trade Agreement

Regional economic integration in North America began with the US–Canadian Free Trade Agreement, which phased out trade barriers between the world's two largest trading partners. The North American Free Trade Agreement (**NAFTA**), which included Mexico, went into effect on January 1, 1994. Today, NAFTA encompasses 330 million Americans, 35 million Canadians, and 120 million Mexicans. Unlike the EU, NAFTA includes a developing country.

NAFTA was not well received by all parties in North America. The critics' main argument was that it would lead to the migration of lower-skilled assembly and manufacturing jobs from the United States to Mexico, where labor costs are one-fifth as high. In addition, companies may flee America's more stringent climate regarding environmental pollution and workplace safety controls. Critics of NAFTA also suggest that Japan, Korea, Taiwan, and other East Asian countries would build plants in Mexico and import goods, duty free, to the American and Canadian markets, which would hurt US firms and workers but benefit consumers.

The principal argument in favor of NAFTA was that free trade would enhance the US, Canadian, and Mexican comparative advantages. It was a project to raise incomes in Mexico and increase Mexican demand for goods from the United States and Canada. Another argument held that higher living standards in Mexico would help control the flow of undocumented immigrants crossing the US border, which is now estimated at one million per year.

NAFTA accelerated trade with its elimination of performance standards on foreign producers in Mexico, who are often required to export a certain proportion of their Mexican output. NAFTA also frees up financial investments by reducing restrictions on the establishment of US and Canadian financial service subsidiaries, including banking, securities brokering, and insurance operations in Mexico. Trucking companies are allowed free access to the market in each of the three states.

Before NAFTA, a free trade zone that stretched 100 miles south of the border was in operation; NAFTA, in effect, extended the maquiladora zone to all of Mexico, allowing imports from manufacturing plants to America with duties imposed only on the value added by manufacture in the maquiladora plants.

NAFTA expanded trade and investment with Mexico, but American imports have grown more quickly than exports, turning what was once a US trade surplus with Mexico into a deficit. Mexico saw its US-bound exports jump in the years since l994 by 241 percent, while imports from the United States climbed by 170 percent. Government-subsidized US corn has flooded Mexico and bankrupted two million farmers there.

Mexico's burgeoning maquiladora industry has grown by 150 percent since NAFTA began, but it has depressed wages in US border states. Roughly 800,000 US jobs moved to Mexico as a result of NAFTA, many of them in manufacturing—a small fraction of the 165 million jobs in the US labor force. About an equal number of jobs have been created on the US side of the border but are concentrated in lower-paying service occupations. Meanwhile, the United States continues to lure Mexican workers, many of whom were uprooted from rural communities when Mexico opened its markets to subsidized US agricultural goods.

4.9c The Organization of Petroleum Exporting Countries

A **cartel** is an agreement among producers that seeks to increase prices by controlling supplies. The most successful commodity cartel is the Organization of Petroleum Exporting Countries (OPEC). Founded in 1960, OPEC consists today of fourteen countries—Venezuela, Ecuador, Saudi Arabia, Iran, Iraq, Kuwait, Qatar, the United Arab Emirates, Libya, Algeria, Nigeria, Angola, Gabon, and Congo (Figure 4.10); former members include Indonesia, Ecuador, and Gabon. The success of OPEC in raising oil prices encouraged other underdeveloped countries to create new nonpetroleum cartels.

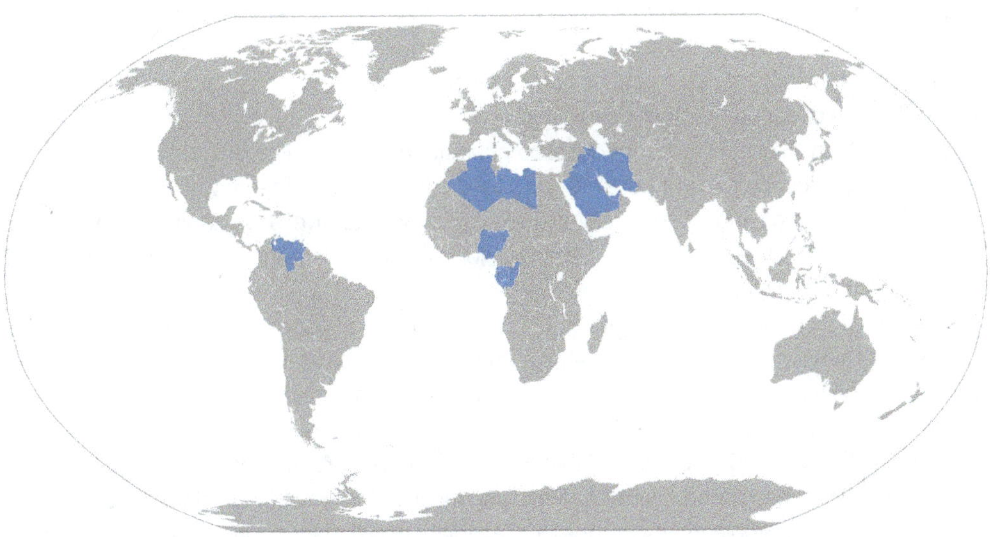

Figure 4.10 Members states of the Organization of Petroleum Exporting Countries.

OPEC became well known worldwide during the oil shocks of 1973 and 1979. The shocks led to high inflation and unemployment in the West, accelerating the deindustrialization that had already begun. It also generated a huge pool of **petrodollars** that were deposited in Western banks. As the price of oil soared, major new exploration by American and European oil companies commenced. In North America, this led to the growth of oil drilling in Louisiana, Texas, Oklahoma, and Alaska, as well as offshore. In Canada, it turned Alberta into a major oil producer. As world oil prices eventually declined, these new oil operations and oil explorations likewise declined, which sent shock waves through the oil industry and depreciated home and business values throughout Houston, Texas, and other oil-dependent cities.

World oil prices depend on the resumption of economic growth not only in the developed nations, but also in the less developed nations, and the demand for crude petroleum. The resumption of production and refining in Russia and Central Asia will also affect world petroleum prices. The future demand for OPEC oil during the twenty-first century is likely to be substantial and increasing because of the expected recovery of world economic output and OPEC's large reserves. At present, OPEC produces less than 50 percent of the world's output but controls 80 percent of the proven reserves. Despite its tremendous oil reserves, generating enough funding to maintain its production and increased capacity will be a problem for OPEC nations because, surprisingly, most have serious financial debt problems.

4.10 Summary

This chapter examined some dimensions of the global economy. International business is any form of economic activity that crosses a national border, such as movements of goods and services (trade) and capital (investment).

The analysis of trade theory began with the notion of *comparative advantage*. The original Ricardian model concluded that specialization within a larger international division of labor led to higher standards of living for all parties involved, even less-efficient trading partners. However, classical trade theory overlooks too many issues to be realistic, including innovation, demand, government, and economies of scale. Free trade was established in the nineteenth century within a colonial framework of inequality among countries. Many developing countries are primary goods producers heavily dependent on foreign demand and, therefore, vulnerable to the business cycle of expansion and contraction in developed countries. Thus, some critics argue that it is a facade that disguises the ability of rich countries to take advantage of less developed ones. One country's free trade may appear like another's exploitation.

A major extension of the original theory of comparative advantage, Porter's notion of competitive advantage, includes the roles of demand, labor skills, knowledge and innovation, agglomeration economies, financing, and government policy. Unlike comparative advantage, which is static, the theory of competitive advantage is dynamic, noting that regional and national advantages will always change over time.

International trade—the movements of outputs—is only one aspect of globalization. Another is investment. In some respects, international movements of capital, technology, and managerial skills are now more important than international trade. There are many types of investment, including, for example, purchases of government debt, but the most important is foreign direct investment (FDI). Most FDI is organized through multinational corporations (TNCs) and originates in the economically developed countries. However, most TNCs invest in developed, not developing, countries. For example, many European, Canadian, and Japanese firms invest heavily in the United States, where they generate jobs, wages, output, and profits.

Money must be exchanged on international markets for goods and services, and so exchange rates play a critical role in influencing the prices of imports and exports among countries. Explaining why exchange rates fluctuate is no easy matter but is related to levels of real output, inflation rates, demand factors, and currency speculation.

Although the vast majority of economists agree that free trade is optimal, many states practice protectionism. Protectionist measures include tariffs and nontariff barriers such as quotas. Arguments in favor of protectionism hold that it saves jobs, protects infant industries, and guarantees supplies of strategically significant minerals and oil. None of the arguments in favor of protectionism is viable; invariably, protectionism protects inefficient firms that are unable to compete internationally at the expense of consumers, who pay higher prices.

Most countries participate in multinational organizations and agreements designed to enhance trade. Major organizations that can be important to international business are *international financial institutions* and groups that promote regional integration. A common way of shaping global output and trade in a given sector is through commodity cartels. The most famous cartel, OPEC, has played a fundamental role in affecting the price of petroleum worldwide.

After the Second World War, the General Agreement on Tariffs and Trade (GATT) reduced trade barriers worldwide. In 1995, GATT transitioned into the World Trade Organization, a permanent institution designed to minimize trade barriers and arbitrate disputes among countries. Unlike GATT, the WTO covers services. The WTO is plagued, however, by the huge subsidies paid by the governments of countries such as the United States and Western European states to their agribusinesses, resulting in low prices that cripple farmers in less developed countries.

The most successful example of regional integration is the European Union, which includes twenty-seven states. The EU involves the most complete economic integration in world history, including the absence of trade and investment barriers among its member states, as well as flows of labor. A common currency, the Euro, has been accepted by most but not all of its members. The EU includes most Eastern and Western European nations in a powerful trade bloc with more than 500 million people, and it manages almost 50 percent of worldwide trade. Starting in 1994, the US–Canadian free trade agreement included Mexico in the North American Free Trade Agreement (NAFTA).

Glossary

Absolute advantage: the cost savings in production that a firm or region enjoys

Cartel: a group of producers who conspire to raise prices by limiting the supply of a given product

Common market: a form of economic union in which labor and capital are free to move among member states

Comparative advantage: the relative savings in production costs that a firm or region enjoys compared to other regions

Economic integration: the degree of economic union in a trade area can create, including similar external policies to the rest of the world

Economic union: the highest degree of economic integration, including harmonization of government policies

European Union (EU): the twenty-seven members of the largest trading bloc in the world

Exchange rate: the value of one currency relative to another

Export restraint agreement: policies by a government to limit its country's exports to avoid trade frictions with other states

Export subsidies: government support for exporters, which may include training programs and subsidized research

Foreign direct investment (FDI): capital invested in buildings, offices, machinery, and equipment located in one state by firms located in a different one

Free-trade area: the lightest form of economic integration that eliminates trade barriers among member states

General Agreement on Tariffs and Trade (GATT): the series of trade agreements that reduced protectionism worldwide after the Second World War; it became the World Trade Organization in 1992

Infant industry: groups of firms that are starting a new industry in a country and often receive government support

Intellectual property rights: the laws that protect intellectual property, including patents, regulations against plagiarism, copyright infringement, and theft of ideas

International currency markets: the global market that trades in national currencies

International Monetary Fund (IMF): one of two organizations founded by the Bretton Woods Agreement, it is largely concerned today with the management of foreign debt among developing countries

Intra-corporate trade: trade among divisions of the same transnational corporation

NAFTA: the North American Free Trade Agreement, including Canada, Mexico, and the United States, that began in 1994

Nontariff barriers: limitations on imports into a country excluding tariffs, such as quotas and licensing requirements

Organization of Petroleum Exporting Countries (OPEC): a cartel of countries designed to regulate the global supply of oil

Petrodollars: revenues from oil sales priced in US dollars

Protectionism: limitations on imports designed to protect domestic producers, such as quotas and tariffs

Quota: a limit on the volume of imports of a good from a foreign country

Regional integration: various levels of economic cooperation and harmonization among member states

Tariff: a tax on imports of a good from a given country

Tax-haven country: countries with exceptionally low corporate taxes designed to attract financial investments

Terms of trade: the relative value of a country's exports and imports

Trade deficit: the difference between a country's exports and imports

Transnational corporations (TNCs): firms that operate in two or more countries

Unequal exchange: the flow of surplus value from one country or region to another, part of the reproduction of uneven development

World Bank: one of two organizations set up by the Bretton Woods Agreement, designed primarily to facilitate long-term development projects

World Trade Organization (WTO): the successor to the GATT, a permanent organization designed to reduce protectionism and manage trade disputes

Study Questions

1. Why does international trade occur?
2. What is comparative advantage?
3. What is competitive advantage, and how does it differ from comparative advantage?
4. What are the inadequacies of existing trade theory?
5. What is meant by the terms of trade? Why have they declined for many Third World nations?
6. What forces have driven the internationalization of banking?

7 How do exchange rates affect trade?

8 What caused the huge US trade deficits?

9 What is foreign direct investment? What are its impacts?

10 How has the US FDI balance changed over time? Why?

11 What are the principal barriers to international business?

12 What are the major arguments in favor of protectionism?

13 What are tariffs, quotas, and nontariff barriers?

14 What was the GATT, and what is the WTO?

15 What is regional economic integration, and why does it occur?

16 Compare and contrast the EU and the NAFTA.

Suggested Readings

Coe, N., and H.W.C. Yeung,. 2015. *Global Production Networks: Theorizing Economic Development in an Interconnected World*. Oxford University Press.

Coe, N., P. Dicken, M. Hess, and H.W.C. Yeung. 2010. Making connections: Global production networks and world city networks. *Global Networks* 10(1):138–49.

Derudder, B., and P. Taylor. 2020. Three globalizations shaping the twenty-first century: Understanding the new world geography through its cities. *Annals of the American Association of Geographers* 110(6): 1831–54.

Dicken, P. 2015. *Global Shift: The Internationalization of Economic Activity*, 7th ed. Guilford.

Donaldson, D. 2015. The gains from market integration. Annual Review of Economics 7(1): 619–47.

Friedman, T. 2005. *The World Is Flat: A Brief History of the Twenty-First Century*. Farrar, Straus and Giroux

Hugill, P. 1993. *World Trade since 1431: Geography, Technology, and Capitalism*. Johns Hopkins University Press.

Jeanne, L., S. Bourdin, F. Nadou, and G. Noiret. 2023. Economic globalization and the COVID-19 pandemic: Global spread and inequalities. *Geojournal* 88(1):1181–8.

Krugman, P. 1991. *Geography and Trade*. Cambridge University Press.

Porter, M. 1990. *The Competitive Advantage of Nations*. Free Press.

Ricci, A. 2019. Unequal exchange in the age of globalization. *Review of Radical Political Economics* 51(2):225–45.

Rickard, S. 2022. Economic geography, politics, and the world trade regime. *World Trade Review* 21(3):367–79.

Rodrik, D. 2021. Why does globalization fuel populism? Economics, culture, and the rise of right-wing populism. *Annual Review of Economics* 13(1):133–70.

Sachs, J. 2019. *The Ages of Globalization: Geography, Technology, and Institutions*. Columbia University Press.

Steger, M. 2020. *Globalization: A Very Short Introduction*. Oxford University Press.

Steger, M., and P. James. 2019. *Globalization Matters: Engaging the Global in Unsettled Times*. Cambridge University Press.

Stiglitz, J. 2002. *Globalization and Its Discontents*. W.W. Norton.

5 Resources and the Environment

The capitalist world economy and economic growth depend heavily on the availability of natural **resources**. There is mounting concern that the consumption of inputs and goods in developed countries, and increasingly in developing countries, is depleting the world's stock of resources and irreparably degrading the planet's environment.

Optimists argue that economic growth in a market economy can continue indefinitely, minimize the limits in raw materials, and emphasize gains in productivity. In contrast, pessimists hold that there are inherent ***limits to growth*** imposed by the earth—by the fact that air, water, **minerals**, and **energy** sources can be exhausted or ecosystems overloaded. This view maintains that the world system is nearing these limits and points to existing food, mineral, and energy shortages and to areas now beset by **deforestation** and soil erosion.

This chapter deals with the complex issue of population and resources. It opens with a brief overview of concepts such as **carrying capacity**. Next, it describes types of resources, including renewable and nonrenewable ones such as minerals. It delves into the issue of food production and security. The latter part of the chapter is largely concerned with energy resources such as petroleum, coal, natural gas, and **nuclear energy**. Throughout, it is concerned with the environmental issues surrounding resource extraction.

5.1 Resources and Population

Many people suffer from a view that resources are limitless and do not appreciate that our rapid consumption of them ultimately threatens our affluent way of life. The developed countries are, in short, annihilating the resources on which high standards of living are built. The growth of some developing countries aggravates the situation. Their growing populations put increasing pressure on resources and the environment, and many aspire to Western-style affluence that is contingent on the intensive use of resources. Poor countries generally do not have the means for running the high-energy consumption and transportation systems prevalent in the industrial West. Resource constraints prevent the large-scale production of consumer goods for the growing populations of the developing countries.

However, by the most common measures of material well-being (per capita incomes, calories consumed, life expectancy), people in most, but not all, countries are better off today than their parents were. But this assessment is not without problems. These gains in standards of living are based on averages, and do not reflect the distribution of wealth and associated standards of living. Moreover, the world's population may be achieving improvements in well-being at the expense of future generations.

5.1a Carrying Capacity and Overpopulation

The population–resources relationship has been extensively debated, particularly during periods of economic downturns. Neo-Malthusian pessimists (Chapter 7) assert that the world will eventually enter a stationary state at carrying capacity that sustains the maximum population with available resources. They often point to recurring food crises and famines in Africa as the result of **overpopulation**, in which demand exceeds supply. However, this view is simplistic because it ignores the historical, geographic, political, and technological contexts in which the production and consumption of goods occur. Human beings are capable of modifying their environment and altering the constraints and opportunities it presents. Conversely, optimists stress the role of modern technology. Technological advances in the past two hundred years have greatly raised the world's carrying capacity, and future technical innovations offer the promise of raising carrying capacity still further. The question of how many people the planet can sustain also depends on the standard of living deemed acceptable. To give large numbers of people a quality of life resembling the American middle class would require vast quantities of additional resources, likely more than the planet has available.

5.2 Types of Resources and Their Limits

What are resources and what are their limits? Natural resources acquire meaning only in relation to historically specific technical and social contexts and are defined in relation to a particular level of development. Resources include all materials of the biological and physical environment that may be used under particular technological and socioeconomic conditions. Because these conditions are always subject to change, the determination of what is useful also changes. For example, petroleum was not considered a resource until the mid-nineteenth century, when the Industrial Revolution led to rising demand for **fossil fuels**. Uranium, once a waste product of radium mining in the 1930s, is now a valuable ore.

5.2a Renewable and Nonrenewable Resources

A major distinction is frequently made between nonrenewable and **renewable resources**. ***Nonrenewable resources*** consist of finite masses of materials, such as fossil fuels and metals and minerals, which cannot be used without eventual depletion. They are fixed in amount, or in some cases, such as soils, form very slowly over time. Large populations with high rates of per capita consumption of goods deplete these resources quickly. Many nonrenewable resources are completely destroyed by use; petroleum is an example. Others, such as iron, may be recycled, which expands the limits to the use of a nonrenewable resource.

Renewable resources are capable of yielding output indefinitely without impairing their productivity. Examples include water, wind, sunlight, vegetation, fish, and animals. As the world gradually shifts to a post-carbon economy, renewable energy has become increasingly important (Figure 5.1). However, even renewable resources can be depleted and permanently reduced by overuse. For example, productive fishing grounds can be destroyed by overfishing. Fertile topsoil, destroyed by erosion, can be impossible to replace. The future of agriculture is guaranteed only when production does not exceed its **maximum sustainable yield**, or maximum production consistent with maintaining future productivity of a renewable resource.

The overuse of resources is often described as the ***tragedy of the commons***, a term coined by biologist Garrett Hardin in 1968. This metaphor refers to the way public resources are ruined

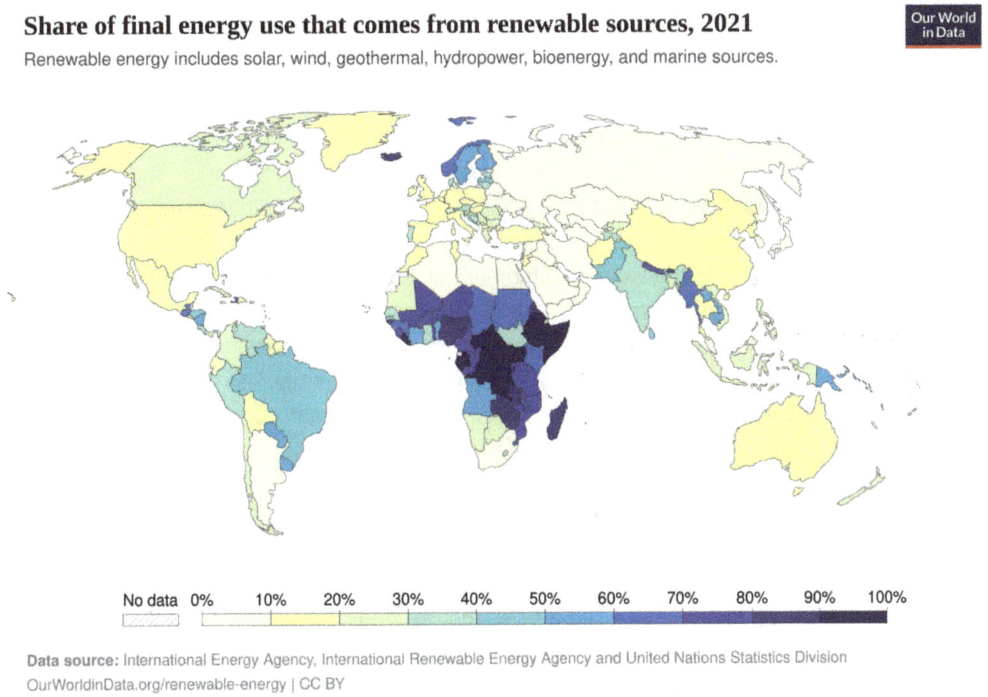

Figure 5.1 Energy consumption from renewable resources. Because poor countries tend to rely more heavily on animate energy resources such as wood, they use much more renewable energy than do wealthy, industrialized countries.

by the isolated actions of individuals: when each individual maximizes their well-being, the collective result is a degradation of the environment. This happens when the costs of individual actions are not captured in market prices. Originally, the term referred to the tendency of shepherds to use common grazing land; as each one sought as much of the commons as possible, it became overgrazed. Similarly, people who fish are likely to try to catch as many fish as they can, reasoning that if they do not, others will. Thus, the tragedy of the commons illustrates a market failure, a problem generated by individual actors who behave rationally but collectively create an irrational and self-destructive outcome. Similarly, dumping waste and pollutants in public waters and land or into the air is the cheapest way to dispose of worthless products. Firms are generally unwilling to dispose of these materials by more expensive, environmentally friendly means unless mandated by law.

5.2b Food Resources

Due to wave upon wave of innovation in farming, global food production has been increasing faster than the population for decades. While there is sufficient food to feed everyone in the world, there are enormous spatial variations in access to a sufficient number and quality of calories. Average daily per capita calorie consumption is 3,300 in developed countries and 2,650 in developing countries. Even with a sufficient calorie intake, people may suffer from *chronic* ***malnutrition***—a lack of enough protein, vitamins, and essential nutrients.

The populations of the developed world are typically well-fed; indeed, in the United States, the major dietary problem is an overabundance of calories and widespread obesity. In the developing world, in contrast, hundreds of millions of people worldwide still go hungry. The growth of the world's population, coupled with widespread poverty, threatens the environmental base on which agriculture depends, such as topsoil. Even worse, environmental degradation reproduces poverty as degraded ecosystems diminish agricultural returns to poor people.

These international patterns are compounded by those within countries. The sharpest nutritional differences are not among countries, but between rich and poor people. The poor of the earth are the hungry, and those with the least political power often suffer in terms of an insufficient food supply.

Hunger among the poor of the world is often attributed to deforestation, soil erosion, groundwater depletion, droughts, and hurricanes. Although environmental problems do affect the food supply, it has limited significance compared to the role of social conditions such as war and a world economy whose rules are tilted against the impoverished. Hunger is a social problem, and reflects prevailing patterns of power: everywhere, hungry people are poor people with the least access to resources.

Population Growth, Hunger, and Poverty

Population growth is one of many causes of the global food problem, and Malthusian views often influence the public's view of this issue (Chapter 7). However, worldwide, there is no food shortage. In fact, world food production grew steadily from 1961 to 2021. Over the next several decades, productivity increases, assuming continuing high investments in agricultural research, are likely to be sufficient to meet effective demand and rising world population. However, some argue that food production will be constrained by the limits to the biological productivities of rangelands, the fragility of tropical and subtropical environments, massive overfishing of the world's oceans, the increasing scarcity of fresh water, the declining effectiveness of fertilizer applications, and social disintegration in many developing countries.

The successes of global agriculture have not been shared equally. In sub-Saharan Africa, per capita food production has not been able to keep up with population growth. In contrast, East Asia, and to a lesser extent Latin America, have experienced tremendous gains. The reasons for this are complex and have to do with the egalitarian patterns of land ownership, government policies toward farmers (e.g., price ceilings on agricultural crops), the existence of transportation and water infrastructures, government credit to small farmers, and the role of different states in the world economy (i.e., trade relations, investments, and debt).

The food and hunger problem is most severe in sub-Saharan Africa, a region that has long suffered from centuries of colonial abuse, corrupt governments, tribal conflicts and secessionist wars, rapid population growth, and lack of foreign investment. Indeed, famine, the most extreme expression of poverty, is now mainly restricted to Africa. Worldwide, famines have been decreasing for decades, and almost always follow severe drought or wars. Africa's multiple civil wars, which have killed millions, have left millions of other people hungry. Such conflicts divert resources from civilian use, disrupt the production of crops, terrorize populations, destroy the infrastructure, destabilize markets, complicate the stability of the governments, and inhibit the flow of development aid. But peace alone is not a sufficient condition for eliminating acute hunger. Appropriate policies and investments are needed to stimulate rural economic growth that generates food security. Rural infrastructure development, credit to farmers, and land redistribution are also necessary steps in this regard. Price controls on food crops create disincentives to produce, and heavily subsidized food imports from the developed world, especially the United States, bankrupt local farmers.

The unequal allocation of food is the direct result of poverty, the single greatest cause of hunger. Hungry people are inevitably poor people who lack the financial means to feed themselves. During famines, the prices of food rise dramatically, with calamitous results for the poor. Because so many local food systems are linked to the world market, even in the midst of hunger, food may be exported for profit. Since the populations of the developed world can afford to pay much more for food than people in less developed countries can, the global food market fails to include the poor. Solving the world food problem is ultimately a matter of overcoming poverty in developing nations. While some parts of the developing world have made great economic strides over the past forty years (e.g., East Asia), much of Africa and parts of India and Latin America remain mired in persistent poverty and hunger.

Maldistribution of Resources

The problem of world food distribution has several dimensions. First, there is the problem of transporting food from food-surplus to food-deficit regions. Although transport systems in developing countries usually lack the speed and efficiency of those in developed countries, they are not serious impediments under normal circumstances. The problem is acute either when massive quantities of food aid must be moved quickly or when the distribution of food is disrupted by military conflict. Second, developing countries often suffer serious problems with food and storage. Food may be hoarded by merchants until prices rise and then sold for a higher rate of profit. Pests such as rats consume considerable quantities, and investments in concrete storage containers can help to minimize this loss. A third aspect of the distribution problem is the inequitable allocation of food. Only a few countries have large grain surpluses. But grain is not always given when it is most needed.

Closely associated with poverty as the leading cause of hunger is the structure of land ownership. Land is frequently concentrated in the hands of a small elite. Sometimes, this can be attributed to the legacy of colonialism, as in Latin America. Many rural residents own no land at all. They are landless laborers who depend on extremely low wages for their livelihoods.

Civil Unrest and War

Political conflict is also an important cause of hunger. On occasion, governments withhold food to punish rebellious populations. For example, the government in Nigeria starved one million Biafrans in the 1970s, and the government in Ethiopia starved Eritreans into submission. In Sudan, the Arab government's genocidal attacks against the African population in Darfur led to the starvation of millions. Civil wars, which are frequent in developing countries whose political geographies were shaped by colonialism (Chapter 2), devastate farming systems. Without a stable political environment, the social mechanisms necessary to produce and distribute food to the hungry cannot operate effectively.

Environmental Decline

As population growth increases pressure on a given land area, the need for food pushes farmers into marginal lands that are too steep, too hot, too cold, or too dry to raise crops. One result of farming in areas ill-suited to it is ***desertification*** and ***deforestation*** (Figure 5.2). Overgrazing and cutting of trees encourage deserts to advance because they destroy windbreaks that prevent soil erosion. Removing the vegetation cover also reduces soil moisture, which amplifies the risk of drought.

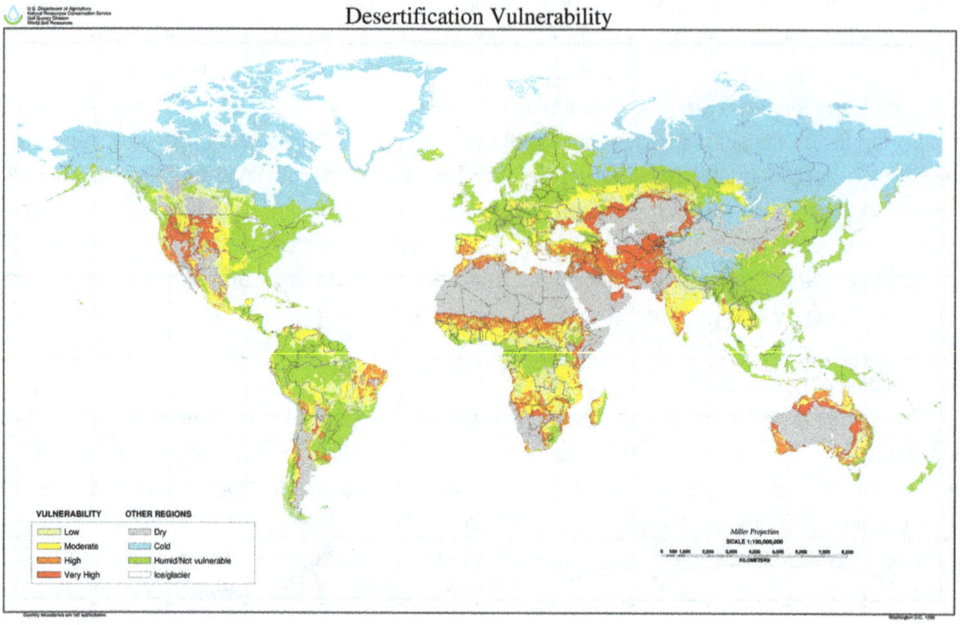

Figure 5.2 Global desertification vulnerability.

Government Policy and Debt

In many developing countries, governments have invested heavily in their militaries and cities at the expense of improving agricultural production. In addition, some governments provide food at artificially low prices in order to make food affordable in cities, a practice that keeps labor affordable for multinational corporations and placates the middle class. However, cheap urban food lowers prices and robs farmers of the incentive to farm.

Debt compounds this problem. The average debt of many developing countries runs into the billions. In 2020, the combined foreign debt of African countries stood at more than one trillion dollars. Simply put, African governments have little surplus capital to invest in their infrastructure or food production systems. Instead, they enforce austerity, cutting government services (often at the behest of the IMF). Debt repayments absorb a large share of foreign revenues, decreasing funds available for agricultural investment.

In recent decades, agriculture in developing countries has expanded. Unfortunately, this expansion is in the export sector, not in the domestic food-producing sector. Governments and domestic elites have opted for the promotion of export-oriented agriculture. The result is the growth of an agricultural economy based on profitable export products and the neglect of small farmers producing food for local populations.

Imports from the developed world, particularly the United States, also exacerbate food problems. For example, after the passage of the North American Free Trade Agreement (NAFTA) in 1994, US exports of government-subsidized corn caused the price of corn in Mexico to fall by 70 percent, bankrupting two million Mexican farmers. All over the world, farmers protest subsidized US grain exports for undermining local food-producing systems. The issue has become a major obstacle in WTO negotiations.

5.3 Increasing Food Production

Yield increases will be the major source of future food production growth. These can be achieved in two ways, that is, through the expansion of arable land and increased productivity.

5.3a Expanding Cultivated Areas

Much of the world's land is already given over to agriculture (Figure 5.3). Large parts of the remaining earth's surface, however, are too dry, too hot, or too cold for farming. Indeed, the supply of arable land may be decreasing because of environmental degradation. About half of the world's potentially arable land lies within the tropics, especially in sub-Saharan Africa and Latin America. Much of this area is under forest in protected areas, and most of it suffers from soil and terrain constraints as well as insufficient rainfall. In Asia, two-thirds of the potentially arable land is already under cultivation; South Asia's agricultural land is almost totally developed.

The expansion of tropical agriculture into forest and desert environments contributes to global deforestation and desertification. Since the Second World War, half of the world's rainforests in Africa, Asia, and Latin America have disappeared, and they continue to shrink at an alarming rate. Conversion of this land to agriculture has led to the loss of biodiversity, increased carbon dioxide emissions, and decreased carbon storage capacity. Desertification—the growth of deserts due to humanly caused factors, typically on the periphery of natural deserts—threatens about one-third of the world's land surface and the livelihoods of a billion people. Many of the world's major cattle rangelands are at risk. The main factor responsible for desertification is overgrazing, but deforestation (particularly the cutting of fuel wood), overuse of marginal soils, and soil salinization caused by poorly managed irrigation systems are also important.

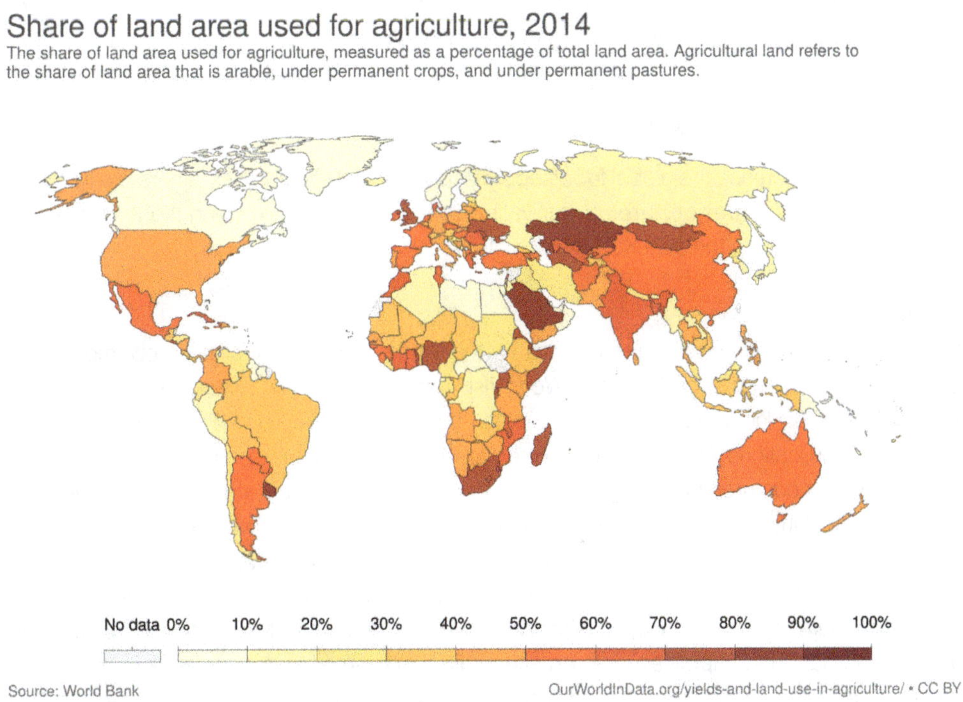

Figure 5.3 Share of land use for agriculture.

5.3b Raising the Productivity of Existing Cropland

The quickest way to increase food supply is to raise the productivity of land currently under cultivation. Remarkable increases in agricultural yields have been achieved in developed countries through the widespread adoption of new technologies, including genetically modified crops, hybrid varieties, herbicides, and fertilizers.

The most famous approach for increasing yields in developed countries is known as the ***Green Revolution***, which was started by American agronomist Norman Borlaug in the 1960s. Borlaug and associates developed, in which new high-yielding varieties of wheat, rice, and corn through plant genetics, including crops that grow more quickly, yielding several harvests per year, are more pest- and drought-resistant, and have higher protein content than traditional crops. The Green Revolution has had enormous impacts in Asia and Mexico, increasing the food supply, but it also has problems. It depends heavily on farm machinery, which the poor lack sufficient capital to buy. It relies on seeds of new hybrid crops, which poor farmers cannot afford. It also uses chemical fertilizers, pesticides, and herbicides, which have contaminated water supplies. Green Revolution farming is a monoculture, which is ecologically unstable because of its susceptibility to pestilence. It is confined largely to a group of eighteen heavily populated countries, extending across the tropics and subtropics from South Korea to Mexico. It is also benefiting countries that include half of the world's population.

Politically, the Green Revolution promises more than it can deliver. Even in areas where the Green Revolution has been technologically successful, it has not always benefited hungry people without the means to buy the newly produced food. It has benefited mainly wealthy farmers who can afford to adopt its technical inputs and management practices. Some effects of labor-displacing machinery include agricultural unemployment, increased landlessness, rural-to-urban migration, and increased malnutrition for the unemployed who are unable to purchase the food produced by the Green Revolution. The Green Revolution has helped to create more and larger commercial farms alongside fewer and smaller peasant plots.

5.3c Creating New Food Sources

A third method to increase the global food supply is to increase food sources. There are three main ways of doing this: (1) **mariculture**, or cultivating the oceans; (2) developing high-protein cereal crops; and (3) increasing the acceptability of inefficiently used present foods.

Mariculture

Although oceans cover three-fourths of the earth, fish provide a very small proportion—about 1 percent—of the world's food supply. World fish consumption has increased steadily. Today, because fishing trawlers are so large and efficient, the oceans have been heavily overfished, with catastrophic implications for marine ecosystems. Over the past thirty years, the total tonnage of fish caught by commercial fishing fleets has leveled off as a result of overfishing. Overfishing has been particularly acute in the North Atlantic and Pacific Oceans. Countries such as Iceland, Japan, Norway, and Peru, whose economies rely on fishing, are sensitive to the overfishing problem. Indeed, if current fish consumption levels continue, the ocean's ecosystems are likely to experience severe stress, or collapse, with dire consequences not only for the world environment but for the human food supply as well. An alternative to commercial fishing fleets is fish farming. *Mariculture* is now expanding rapidly and accounts for 40 percent of the world's fish caught yearly.

High-Protein Cereals

Another source of future food production rests in higher-protein cereal crops. Agronomists have developed high-yield, high-protein cereal crops in the hope that they will be able to help the protein deficiency of people in developing countries who cannot afford meat. Fortification of rice, wheat, barley, and other cereals with minerals, vitamins, and amino acids is an approach that also deserves attention. But developing countries rely on unprocessed, unfortified foods for the vast bulk of their food intake.

More Efficient Use of Foods

In many developing countries, foods to satisfy consumer preferences as well as religious taboos (e.g., concerning pork or beef) and cultural values limit the available supply. One way around this obstacle is to increase the palatability of existing foods that are plentiful. Fish meal is a good example: currently, one-third of the world's fish output is turned into fodder for animals and fertilizer. Another underused food resource is soybeans, a legume rich in both protein and amino acids. Most of the world's soybeans are processed into animal feed or fertilizer and into nondigestible industrial materials.

A Solution to the World Food Supply Situation

The overpopulation diagnosis of the food problem is simplistic and incorrect (Chapter 7). Current world grain output can more than adequately feed every person on earth. Rather than blaming the poor for being hungry, we should examine the mechanisms that produce hunger, including the dynamics of the global economy, trade relations, debt, and corrupt governments. Governments affect the infrastructure, the availability of credit to small farmers, and the efficiency of marketing and distributing food on a broad scale. Subsidized grain exports from the United States depress prices and hurt farmers.

The industrialization of agriculture in many ways is hurting, not helping, the world's hungry population. Small, carefully farmed plots are more productive per unit area than are large estates because the families that tend to them invest as much labor as necessary to feed themselves when they can. Yet, despite considerable evidence from around the world, government policies in many developing countries ignore small farmers.

5.4 Nonrenewable Mineral Resources

A ***mineral*** refers to a naturally occurring inorganic substance in the earth's crust. For example, iron is a mineral, whereas petroleum is not, since the latter is organic in nature. Although minerals abound on the earth's surface, they are usually not sufficiently concentrated to be economically recoverable. Moreover, the richest deposits are unevenly distributed and are being rapidly depleted. Most nonmetallic elements are consumed at much greater rates than metallic ones. Industrial societies use nonmetallic minerals, which are plentiful and often widespread, including nitrogen, phosphorus, potash, or sulfur for chemical fertilizer, or sand, gravel, or clay for building purposes.

5.4a Strategic Minerals

Only five countries—Australia, Canada, South Africa, the United States, and Russia—are significant producers of six ***strategic minerals*** vital to modern technology. Of the major mineral-producing

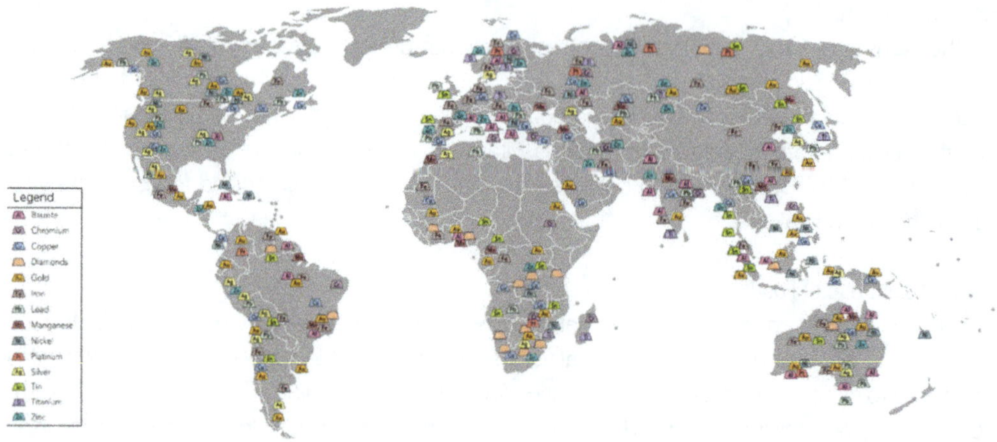

Figure 5.4 Locations of the world's major mineral deposits and mines.

countries, only the United States and Russia are also major processors and consumers. The other major consumer countries—Japan and Western Europe—are deficient in strategic minerals.

Most key minerals will be exhausted within a hundred years, and some will be depleted within a few years at current rates of consumption, assuming no new **reserves** are discovered. The United States' dependence on imports has grown steadily over the last seventy years. When measured in terms of percentage imported, US dependency increased from 50 percent in 1960 to over 82 percent in 2020. Minerals projected to be future needs are unevenly distributed around the world. Many of them, such as manganese, nickel, bauxite, copper, and tin (Figure 5.4) are concentrated in Russia and Canada and in developing countries, notably Africa. Whether these critical substances will be available for consumption may depend less on economic scarcity and more on international politics.

The oceans, which cover 71 percent of the Earth, contain large quantities of dissolved minerals such as salt, magnesium, sulfur, calcium, and potassium. More valuable marine minerals include copper, zinc, tin, silver, and magnesium. More feasible than mining the oceans is improving mining technology, especially to reduce waste in the extraction and processing of minerals. Also feasible is to utilize technologies that allow minerals to be used more efficiently in manufacturing.

Reusing minerals is another solution for the mineral problem. Affluent countries deposit huge quantities of household and industrial waste in landfills. These sites are sometimes called "urban ores" because they can be recovered and used again. For years, developed countries have been **recycling** scarce and valuable metals such as iron, lead, copper, silver, gold, and platinum, but large amounts of scrap metals are still being wasted.

Mineral extraction has numerous impacts on the environment, depending on the size of the operation, mining procedures, and local hydrology. Mining, such as mountain top removal, leads to soil erosion, deforestation, water pollution, and calamitous local ecological changes. As the demand for minerals rises, ever-lower quality minerals will be extracted, and even with the best of methods, environmental degradation will continue.

5.5 Energy Resources

Energy sources are crucial for economic development and the lifeblood of modern economies. Indeed, energy-related products are the biggest single item in international trade; petroleum alone accounts for about one-quarter of the volume (but not value) of world trade. The US economy consumes vast amounts of energy, overwhelmingly consisting of fossil fuels. Energy forms the fuel for the economic machine that feeds, houses, and moves the population. The largest uses of petroleum are transportation and manufacturing, whereas the major use of coal is electrical power generation.

In the late twentieth century, rapid rises in the cost of petroleum initiated a new regime of energy production and consumption. Oil-consuming countries, including developing countries strapped by debt, were crippled; in developed countries, deindustrialization accelerated rapidly after the oil shocks of the 1970s. Oil prices regularly rise and fall, pointing to the cyclical nature of the petroleum industry, which has huge impacts on other sectors of the world's economy.

5.5a Energy Production and Consumption

Most of the world's energy is produced from nonrenewable resources such as petroleum, natural gas, and coal. Renewable energy sources, particularly wood and charcoal, are used primarily by poor people in developing countries. Although renewable energy sources such as solar and wind power have declined in cost and risen in popularity, fossil fuels remain the largest source of energy. Only a few countries produce much more energy than they consume, including Saudi Arabia, Iraq, Mexico, Iran, Venezuela, Indonesia, Algeria, Kuwait, Libya, Qatar, Nigeria, and the United Arab Emirates. Saudi Arabia is by far the largest exporter of petroleum and has the largest reserves. The world's leading industrial powers—most notably Japan, many Western European countries, and the United States—consume much more energy than they produce, making them heavily reliant on imported oil, largely from the Middle East.

The United States is the world's largest energy producer but consumes almost all that it generates. With roughly 5 percent of the world's population, the United States consumes 25 percent of its oil, largely for transportation, which consumes 40 percent of American energy inputs. The automobile, for all the convenience it offers, is a very energy-inefficient way to move people. In contrast, developing countries consume about 30 percent of the world's energy but contain about 80 percent of the population. There is, therefore, a huge imbalance between per capita energy consumption and the level of development.

5.5b Petroleum

The world's fossil fuels are highly unevenly distributed around the globe. Most of the world's petroleum reserves are concentrated in a few countries, many of which are politically unstable. Two-thirds of the world's petroleum resources are located in the Middle East (Figure 5.5); other large reserves are found in northern Africa, Latin America (especially Mexico and Venezuela), Russia, Nigeria, and Indonesia. Offshore drilling, such as in the North Sea and South China Sea, forms another supply. The United States has tapped its petroleum deposits in Alaska and the Gulf of Mexico, making it the single largest oil producer (but not exporter) in the world.

Global trade in petroleum reflects the geographies of supply and demand. Because developed countries rely heavily on oil for transportation, heating, and energy production, they consume the bulk of the world's petroleum. Wealthy lifestyles are energy-intensive lifestyles. The spatial distributions of the world's supply and demand for petroleum create a distinct pattern of trade flows of petroleum, the most heavily traded commodity (by volume) in the world. Primarily, these flows consist of exports from the Middle East to Europe, East Asia, and North America, although the United States also imports considerable quantities from South America and Nigeria.

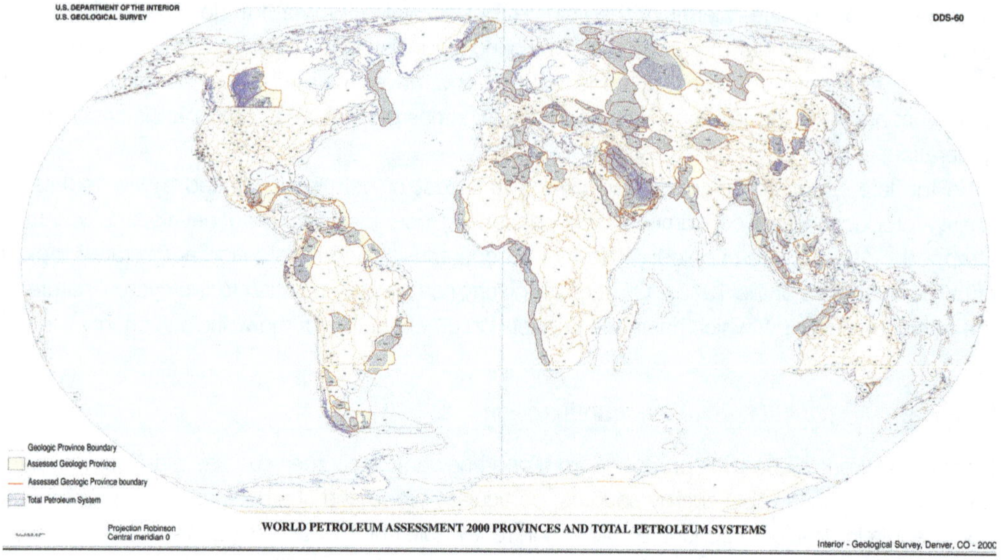

Figure 5.5 Locations of the world's major petroleum producing regions.

Although oil reserves have increased over time, most of this is due to new discoveries in the Middle East, Venezuela, and the South China Sea. Regionally, however, reserves have been declining in important consuming countries. For example, reserves in Russia declined by 9 percent between 1990 and 2020, and they declined by 9 percent in the United States during the same period. Europe's reserves are likely to be depleted by 2050. Moreover, exports of oil from Africa and Latin America will probably start to fall by 2050. The Middle East will then be the only major exporter of oil, but political turmoil there could cause interruptions in oil supplies, creating problems for the import-reliant regions of North America, Western Europe, Japan, and East Asia.

5.5c Natural Gas

A significant substitute for petroleum is natural gas, the fossil fuel experiencing the fastest growth in consumption. Natural gas has declined in price over the last several decades, causing economic problems for regions reliant on coal. Like petroleum, natural gas is also unevenly distributed, with nearly 40 percent in Russia and central Asia and one-third in the Middle East (Figure 5.6). Natural gas production is increasing rapidly, and so are estimates of gas reserves due to major finds in Siberia and large discoveries in China, South Africa, and Australia. The pattern of production is similar to that of consumption. Much natural gas is shipped via pipelines; Russia is a major supplier of Europe in this way, making Europe reliant on Russia and vulnerable to its foreign policies, such as the invasion of Ukraine in 2022.

5.5d Coal

Coal is the most abundant, and dirtiest, fossil fuel, and is a major contributor to **acid rain** and climate change. Most coal is consumed in the country in which it is produced (Figure 5.7). Coal is not a cost-effective fuel for mobile energy units such as trains and automobiles, although it can be adapted through gasification techniques. Coal-powered power plants are a leading generator of carbon dioxide emissions, the most important driver of climate change. For this reason, many

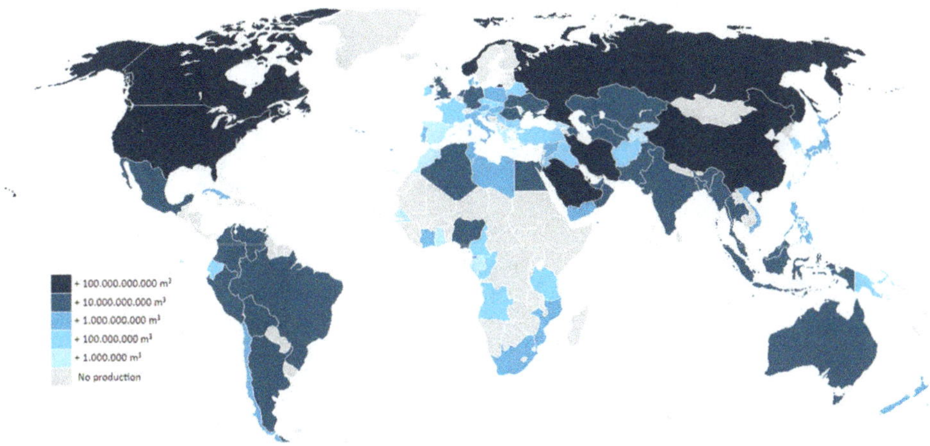

Figure 5.6 World natural gas production.

countries have sought to minimize their coal production; for example, Germany is phasing out coal altogether. Coal-producing regions have also been threatened by cheaper substitutes, such as natural gas.

5.6 Energy Options

The age of cheap fossil fuels will eventually come to an end. As societies prepare for a post-carbon future, they must find alternatives to fossil fuels. The options in this regard include:

5.6a Conservation

One way to reduce the gap between production and consumption is for consumers to restrict their consumption. This is easier said than done. Energy **conservation** lengthens the time in which finite fuel resources can be used and reduces environmental stress, and helps to buy time for the development of more acceptable sources of energy. It is possible to enjoy economic progress while simultaneously consuming less energy per unit of output. In the United States, GDP per capita has increased enormously in the twentieth century, whereas energy use per capita only slightly more than doubled. Energy efficiency increased steadily, partly as a result of more fuel-efficient cars and industries installing better equipment.

5.6b Nuclear Energy

The form of nuclear energy currently in use—**nuclear fission**—involves splitting large uranium atoms to release the energy within them. But fission causes serious problems, including the disposal of radioactive waste and the dangers of accidental explosions, such as occurred at the Three Mile Island plant in Pennsylvania in 1979 and Chernobyl in Ukraine in 1986. Many places near nuclear power plants throughout the United States have contaminated groundwater supplies and leak radioactive wastes. Another problem is its high costs: each nuclear plant costs billions of dollars to build.

Nuclear power is less widely used in North America than in some European countries and Japan (Figure 5.8). In France, one-half of energy comes from nuclear power; in Japan, 25 percent;

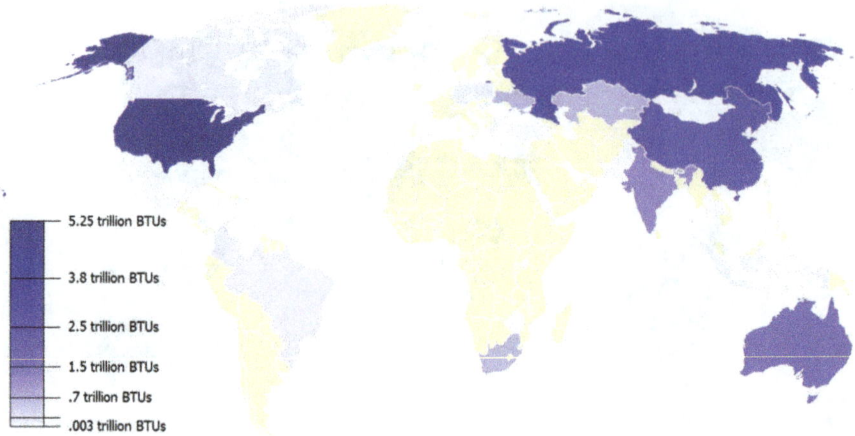

Figure 5.7 Distribution of the world's coal reserves: China, Russia, and the United States are the leading producers and consumers.

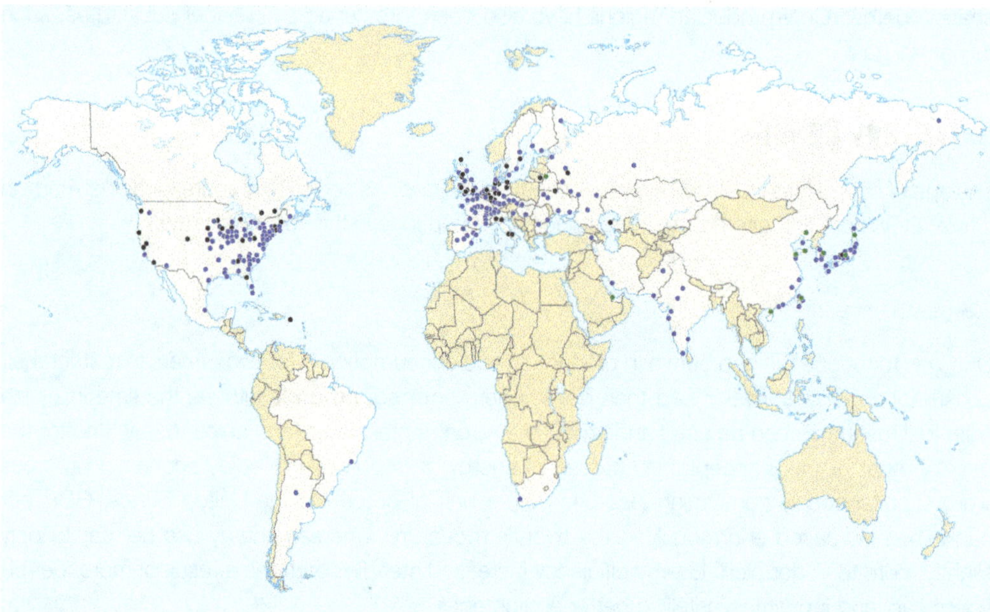

Figure 5.8 Locations of the world's nuclear power plants.

Belgium, France, Hungary, and Sweden produce more than one-half of their energy from nuclear power plants. Interestingly, some countries have decided to phase out nuclear power generation because of high risks and high costs. For example, Sweden began phasing out its nuclear plants in 1995, a process completed in 2010.

Nuclear fusion, the combining of smaller atoms to release their energy (the process that fuels the sun), has the potential to be a solution to the environmental concerns of nuclear fission because it does not release radioactive waste. The raw material for nuclear fusion is hydrogen, which is abundant. Fusion can be made to occur artificially, but it is not yet commercially viable, although many experts expect that it will soon be. If this technology is ever commercially successful, fusion would provide limitless amounts of cheap energy and pose no radiation dangers.

5.6c Geothermal Power

Geothermal power is a potential energy source in several countries that have hot springs, geysers, and other underground supplies of hot water. This renewable resource is highly localized, however. New Zealand obtains about 10 percent of its electricity from geothermal power, and lower amounts are utilized by countries such as Italy, Japan, Iceland, and the United States. Iceland is the world's leader in this regard. In the United States, Wyoming and California use considerable quantities.

5.6d Hydropower

Another renewable source of electric power is hydropower, or energy from rivers through hydroelectric dams. In developed countries, the expansion of hydropower is limited mainly by environmental concerns; dams do significant damage to riverine ecosystems, particularly fish species. Behind the dam, water creates a reservoir; below the dam, the river may be reduced to a trickle. Moreover, reservoirs increase the rate of evaporation and the salinity of the water. In tropical areas, reservoirs harbor hosts that cause diseases such as malaria and schistosomiasis. For example, since the construction of the Aswan High Dam in the 1970s, schistosomiasis has become endemic in lower Egypt, infecting up to one-half of the population. In addition, reservoirs can become silted over time, reducing their potential to produce electricity. When silt trapped in reservoirs cannot proceed downstream, it fails to enrich agricultural land. This has had significant impacts on agricultural productivity downstream from the Aswan High Dam. China completed the Three Gorges Dam on the Yangtze River in 2006, the largest hydroelectric dam on the planet, and is experiencing similar problems. Turkey's Southeast Anatolia Project is planning for the construction of twenty-two dams and nineteen hydroelectric power plants on the Tigris and Euphrates rivers, which will generate problems for downstream users in Syria and Iraq.

5.6e Solar Energy

Solar energy is trapped by large solar panels. The other way of capturing solar energy is through photovoltaic cells. Like hydropower, solar energy is inexhaustible. It does not have the risks associated with nuclear energy, and it is free of pollution. Europe, China, and the United States are the world's leaders in using this source (Figure 5.9). In the United States, solar energy and incoming solar radiation are highest in the Southwest and lowest in the Northeast.

The main problem with solar energy traditionally has been cost. Currently, solar energy production is more expensive than other sources of fuel, but declining rapidly, and it is becoming competitive with coal and petroleum. To promote the development of innovative energy supplies, many governments offer tax deductions for solar units mounted on housetops.

5.6f Wind Power

Wind power also provides an inexhaustible, renewable resource where there are frequent surface winds of 15 mph or more. However, high-tech wind mills are expensive, and most **wind farms** meet with local opposition. Like solar energy, wind power has steadily declined in price and

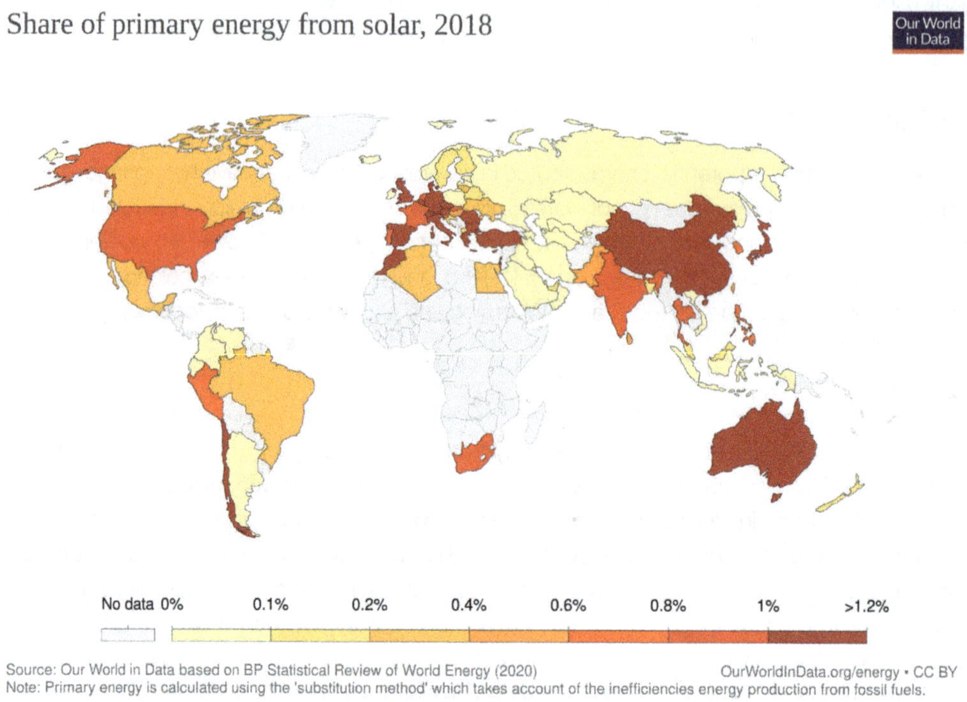

Figure 5.9 Percentage of energy consumption derived from solar energy, 2018.

is becoming competitive with fossil fuels. As wind farms have proliferated, they have become increasingly important to energy production in much of the Midwest as well.

5.7 Resource Extraction and Environmental Degradation

The extraction of natural resources, ranging from growing food to minerals to energy production, inevitably imposes costs on the environment. These may be divided into three overlapping categories: (1) pollution, (2) habitat preservation, and (3) environmental equity.

5.7a Pollution

Pollution is the discharge of waste liquids, gases, and chemicals into the air, land, and water. At sufficiently high levels, pollution creates health hazards to plants, animals, and humans, and degrades the environment. Although ecosystems have the capacity to regenerate and cleanse themselves, when large amounts of gases and solids are released from industrial activity, recycling and purification needs are overwhelmed. At that point, the quality of the environment is reduced as pollutants defoliate forests, poison waters, contaminate soils, reduce fish stocks, and burden wildlife habitats.

Air Pollution

Air pollutants are normally carried high into the atmosphere, but occasionally become concentrated at ground level, particularly when a temperature inversion prevents the escape of pollutants to higher altitudes. These conditions lead to the creation of smog that has several effects: it causes respiratory problems; stings the eyes; and creates a haze over large, congested cities. Air pollution at the local scale is a major concern in cities because of the release of carbon monoxide, hydrocarbons, and tiny particulates that infiltrate the lungs. When sulfur from air pollution combines with water in the atmosphere, it forms acid rain, which has destroyed countless trees in eastern North America and Europe. Around the world, air pollution kills several million people annually, notably in India and China.

At the global scale, air pollution may damage the atmosphere's **ozone layer** and contribute to **global warming**. The Earth's protective ozone layer is threatened by **chlorofluorocarbons** (CFCs) such as freon, which leak from air conditioners and refrigerators; carried into the stratosphere, they cause ozone depletion. In 1987, the world's developed countries signed the Montreal Protocol, which began phasing out CFCs; it has been widely heralded as a very successful example of global cooperation to address an environmental crisis.

Concerns about global warming center on the burning of fossil fuels in enormous quantities, which increases the amount of carbon dioxide in the air (Figure 5.10); carbon dioxide makes the atmosphere more opaque, reducing thermal emissions to outer space, producing a **greenhouse effect**. The vast majority of these gases emanate from the economically developed world (Figure 5.11), although China recently surpassed the United States as the world's largest contributor. Since the 1890s, the average temperature of the Earth's surface has increased by 2 degrees Fahrenheit, which has had serious environmental impacts around the globe: droughts, forest fires, melting glaciers, changing distributions of plant and animal species, warming oceans, rising sea levels, declines in coral reefs, and hurricanes. The overwhelming majority of the world's atmospheric scientists agree that climate change is primarily caused by human beings. Most scientists agree that it is socially irresponsible to delay actions to slow down the growth of gases such as carbon dioxide and methane.

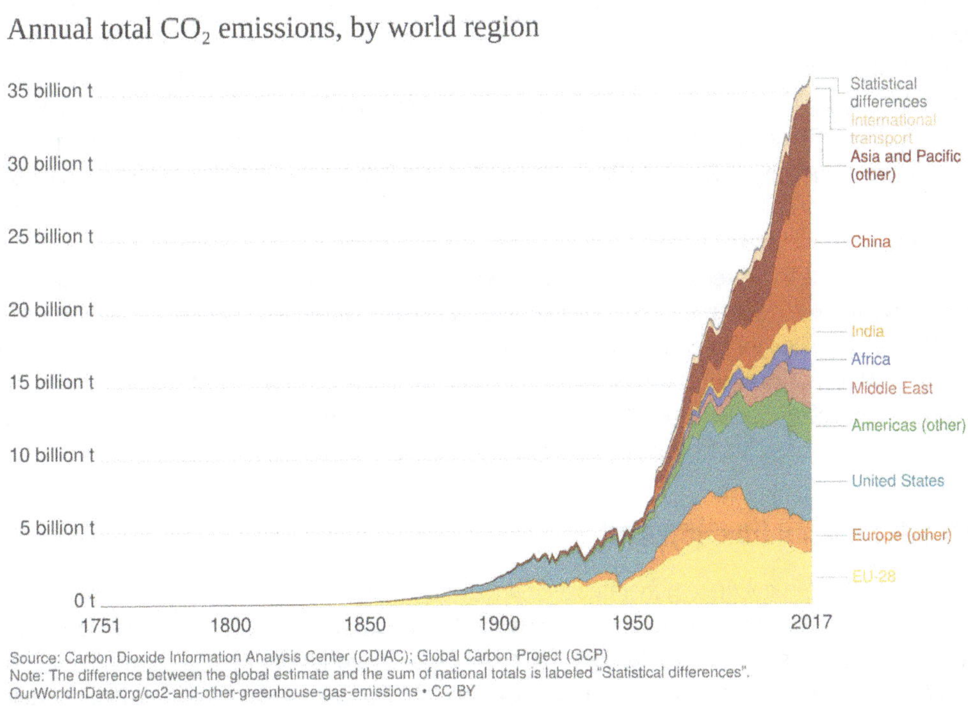

Figure 5.10 Total world annual carbon dioxide emissions, 1751–2017.

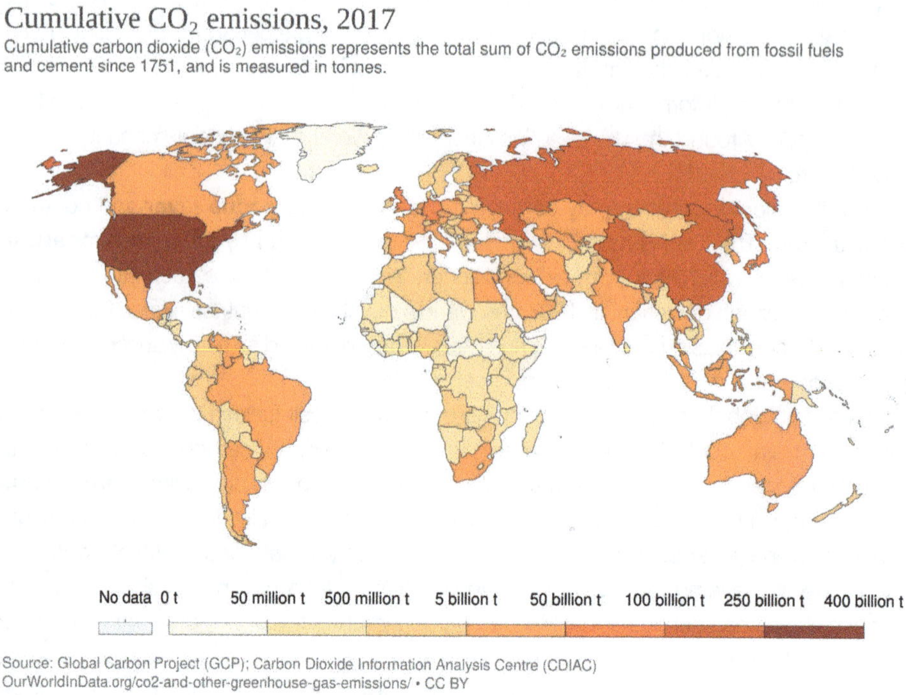

Figure 5.11 Distribution of the sources of the gases that cause climate change.

Water Pollution

When societies use large amounts of water, it invariably becomes contaminated. The largest wastewater sources include urban runoff, mining, industrial discharges, and agricultural runoff. As populations and standards of living rise, the demand for water increases. Water has become a contentious political issue because it has become increasingly in short supply, such as in the western United States and the Middle East. Water shortages are most acute in developing countries, where roughly one billion people already find it difficult to obtain drinking water. In the United States, the major water management problem throughout the twentieth century focused on acquiring additional water supplies to meet the needs of expanding populations and economic activities. The Clean Water Act of 1972 resulted in improvements in the water quality of water bodies that receive discharges from point sources such as municipal waste-treatment plants and industrial plants. Recent efforts to improve water quality have focused on the reduction of pollution from diffuse or **nonpoint sources** such as agricultural and urban runoff and contaminated groundwater discharges. These efforts to control pollution are often costly and meet with resistance from entrenched agribusiness interests.

5.8 Sustainable Development

Economic development often generates as many costs as benefits. For example, a dam to generate hydroelectric power may drown fertile river bottomlands. Mining leaves despoiled lands and water pollution in its wake. New housing tracts cover over productive farmland. In the long run, the Western lifestyle that has spread throughout the globe has increasingly threatened the viability of the planet and its ability to sustain life. From climate change to deforestation, soil erosion to desertification, ocean acidification and overfishing to habitat destruction, environmental problems have become increasingly more severe. There is something fundamentally wrong with treating the Earth as if it were disposable. This trend has given rise to a quest to create a sustainable society that will provide a high quality of life not only today, but for future generations.

Sustainable development is defined as economic growth that meets the needs of the present generation without compromising the ability of future generations to meet their own needs. It typically involves long-range planning. It promotes a systematic shift in policies to allow the market system to internalize environmental costs such as air pollution and climate change. The environmental costs of cars, for example, should include acid rain, primarily in the form of more expensive gasoline. An extension of this sentiment is a desire to protect nature rather than dominate it. Sustainable development invites us to recognize that future growth will be constrained by resources that are finite. A sustainable future will be the result of innumerable social and policy changes at the local and global levels.

Many critics in the developing world point to the West's excessive consumption of resources such as meat, wood, and fossil fuels; sustainability thus involves the critical question of the maldistribution of the world's resources. Building a sustainable society involves putting environmental concerns on a par with development. Because the futures of the planet's rich and poor peoples are inextricably intertwined, sustainability involves a more just and equitable distribution of the world's resources. Sustainability is thus an environmental and social justice issue often at odds with the imperatives of uneven spatial development that lie at the heart of capitalism.

The **growth-oriented lifestyle** characteristic of Western, urbanized societies continues to place increasingly severe stress on the planet. The world's climate and ecosystems are changing rapidly; species are going extinct in large numbers; diverse ecosystems are replaced by monocultures; glaciers are melting; and coral reefs are dying. Endless growth is simply unsustainable and will ultimately lead to self-destruction. Sustainable development is thus the only feasible long-term solution to the planet's myriad environmental crises; it demands that we transform our present growth-oriented, economic system into a balance-oriented one that recognizes that natural resources are exhaustible, promotes large-scale recycling, and avoids permanently damaging the world's environment. A balance-oriented economy does not mean an end to growth, but a type of society in which only low-energy growth occurs.

5.9 Summary

This chapter concludes by restating the resources-population problem. Following a review of renewable and nonrenewable resources, it explored the matter of food resources. World hunger is essentially a consequence of social relations, including war and disruptions of agricultural systems. Food production is increasing faster than population growth, yet malnutrition persists. Famine, like poverty, is a social construction, not a natural event, and its origins, like its solutions, must be found in the uneven distribution of resources among and within countries. Nonetheless, famine has declined steadily over the last several decades.

Unlike food, which is replenishable, nonrenewable minerals and fossil fuels, once used, are gone forever. The rise of the industrialized West was predicated on harnessing enormous amounts of cheap energy in the form of fossil fuels, which still power the global economy. The chapter discussed alternatives to fossil fuels such as solar, wind, and **geothermal energy**. Finally, it noted the role of sustainability in shaping the long-term trajectory of the world's future. Growth and inequality are closely linked parts of social change and environmental protection.

Glossary

Acid rain: rain that has absorbed sufficient CO_2 to make carbonic acid, which is highly destructive to vegetation
Aquaculture: raising fish in pens in the ocean
Biomass: the amount of organic material in a region
Carrying capacity: the ability of a region to sustain a population of a given species
Chlorofluorocarbons (CFCs): gases released from industrial goods such as propellants that are destructive to the atmospheric ozone layer
Conservation: the protection of ecosystems and resources of a region
Deforestation: the annihilation of forest cover
Desertification: the expansion of deserts, often due to human activities
Energy: the capacity to do work
Flow resources: resources that are not fixed to one place, such as river water
Fossil fuels: the family of fuels formed in the geological past, such as coal, natural gas, and petroleum
Geothermal energy: energy obtained from underground geothermal vents
Global warming: a popular term to describe heating caused by anthropogenic climate change
Green Revolution: the spread of genetically engineered crops that raised agricultural productivity in many countries
Greenhouse effect: a term that compares heating due to anthropogenic climate change to the higher temperatures found inside greenhouses designed to grow plants
Growth-oriented lifestyle: the typical Western way of living that maximizes resource consumption and pollution
Limits to growth: a popular term introduced by the Club of Rome to delimit the uppermost bounds of world population and economic growth
Malnutrition: the lack of adequate calories resulting in poor health
Mariculture: the production of seafood
Maximum sustainable yield: the largest yield that can be generated in a given area over the long term
Minerals: resources located underground that provide metals and related useful products
Nonpoint sources: pollution that does not emanate from a single place, such as agricultural runoff
Nonrenewable resources: resources with limited supplies, such as minerals, petroleum, and topsoil
Nuclear energy: energy obtained from nuclear power plants, all of which currently use fission
Nuclear fission: energy obtained by splitting uranium atoms, which fuels all nuclear power plants
Nuclear fusion: energy obtained by fusing hydrogen atoms; while this occurs naturally in the sun, it is not yet commercially feasible
Organization of Petroleum Exporting Countries (OPEC): a cartel of oil-exporting countries
Overpopulation: the notion that a given population has exceeded the carrying capacity of a region
Ozone layer: the atmospheric layer of ozone that protects life from ultraviolet radiation
Passive solar energy: energy captured by solar panels
Projected reserves: estimate volumes of a resource in a given area, including those that have not been exploited and those likely to be discovered in the future
Recycling: the reuse of waste products such as glass, metals, plastics, paper, and cardboard
Renewable resources: resources that are naturally replenished over time, such as water
Reserves: the known quantity of a given resource
Resource: materials useful for human needs
Solar energy: energy obtained from the sun
Stock resources: resources with a finite reserve, such as minerals
Strategic minerals: minerals that are critical to the economy and national security
Sustainable development: development designed to minimize long-term impacts on the environment, including renewable energy and recycling

Tragedy of the commons: a frequent term used to describe over-exploitation of communally held resources such as land or the oceans
Undernutrition: inadequate supply of calories
Wind farm: collections of modern wind mills designed to produce energy

Study Questions

1. What is meant by carrying capacity?
2. Differentiate renewable from nonrenewable resources.
3. What are the major causes of Third World hunger?
4. What are three methods of expanding world food production?
5. What was the Green Revolution? Where was it largely located?
6. Summarize the major flows of oil on the world market.
7. Where are the major world coal deposits located?
8. What are some alternative energy options to fossil fuels?
9. What are some environmental consequences of high energy use? Be specific.
10. What is sustainable development?

Suggested Readings

Alexandersson, G., and B. Klevebring. 2019. *World Resources-Energy, Metals, Minerals: Studies in Economic and Political Geography*. Walter de Gruyter.

Bridge, G., and L. Gailing. 2020. New energy spaces: Towards a geographical political economy of energy transition. *Environment and Planning A: Economy and Space* 52(6):1037–50.

Castree, N., and B. Braun (eds.) 2001. *Social Nature: Theory, Practice, and Politics*. Blackwell.

Cruz, J., and E. Rossi-Hansberg. 2024. The economic geography of global warming. *Review of Economic Studies* 91(2):899–939.

Ellis, R. 2003. *The Empty Ocean: Plundering the World's Marine Life*. Shearwear Books.

Falola, T., and A. Genova. 2005. *The Politics of the Global Oil Industry*. Praeger.

He, C., S. He, E. Mu, and J. Peng. 2022. Environmental economic geography: Recent advances and innovative development. *Geography and Sustainability* 3(2):152–63.

Heffron, R.J. 2020. The role of justice in developing critical minerals. *The Extractive Industries and Society* 7(3):855–63.

Hine, A., C. Gibson, and R. Mayes. 2023. Critical minerals: Rethinking extractivism? *Australian Geographer* 54(3):233–50.

Huber, M. 2019. Resource geography II: What makes resources political? *Progress in Human Geography* 43(3):553–64.

Klare, M. 2002. *Resource Wars: The New Landscape of Global Conflict*. Owl Books.

Molotoks, A., P. Smith, and T. Dawson. 2021. Impacts of land use, population, and climate change on global food security. *Food and Energy Security* 10(1):p.e261.

Petrova, S., and N. Simcock. 2021. Gender and energy: Domestic inequities reconsidered. *Social & Cultural Geography* 22(6):849–67.

Robbins, P. 2019. *Political Ecology: A Critical Introduction*. John Wiley & Sons.

Scholten, D., M. Bazilian, I. Overland, and K. Westphal. 2020. The geopolitics of renewables: New board, new game. *Energy Policy* 138:111059.

Ziegler, S. 2019. The Anthropocene in geography. *Geographical Review* 109(2):271–80.

Zimmerer, K. 2006. *Globalization and New Geographies of Conservation*. University of Chicago Press.

6 Development and Underdevelopment in the Developing World

The modern world originated in Western Europe in the late fifteenth and early sixteenth centuries, when capitalism began to displace feudalism (Chapter 2). This event was marked by the growth of markets, the wealthy ruling class, ideological and cultural changes, and uneven spatial **development**. On the heels of centuries of colonialism, the capitalist world economy became steadily divided between rich and poor regions. During and after the Industrial Revolution, this division witnessed an international system in which a small, wealthy minority of countries in Western Europe, North America, and Japan rose to prominence using primary sector resources produced by the impoverished peoples of their colonies. The creation of today's world, with a rich core and a poor periphery, was the outcome of a systemic process by which the world's economy reproduces uneven spatial development. In short, the birth of the modern world system and the schism between rich and poor countries are two sides of the same coin.

This chapter deals with the implications of global uneven development. It begins by discussing the notion of *development*, which reflects many different concepts and measures. Next, it examines the dynamics of urbanization in the developing world, where most urban growth occurs. Third, it summarizes the major problems of development that poor countries confront around the world. The fourth part turns to several major theories of development, such as modernization, dependency, world systems, and neoliberalism. Fifth, it points to the issue of uneven development within developing countries. In the sixth section, it examines development strategies. The seventh focuses on industrialization in the developing world. Eight, it turns to the economic growth of East Asia, which has been the fastest-growing part of the world for decades. Finally, it addresses the question of **sustainable development**.

6.1 What Is Development?

If Europe, North America, Japan, South Korea, Australia, and New Zealand are developed countries, then what label describes the poorer countries of the world? A variety of terms have been used in this regard, including "primitive," "backward," "undeveloped," "underdeveloped," "less developed," and "developing." The term "underdeveloped" was formerly used to describe situations in which resources were not yet developed, implying that development was sure to come, eventually. The problem with this view is that it assumes poverty to be natural rather than a social product and hides the historical processes that *make* people poor. The adjective "underdeveloped" describes not an initial state but rather a condition arrived at through colonialism, which established the inequality between rich and poor countries.

While economic development may seem straightforward, it involves multiple complex goals and concepts. Broadly speaking, development entails the growth of incomes and the reduction of poverty. However, some countries have experienced rapid growth of incomes and rising poverty and inequality simultaneously. Other measures of development include nutrition, health care, infant mortality, education, housing, and civil liberties. Yet other measures of development include capital inflows, the capacity to produce capital goods, and trade balances.

6.1a How Development Is Measured

Economic development is measured through several social, economic, and demographic indices. The principal ones are GDP per capita, the composition of the labor force by economic sector, the ability to produce consumer goods, educational and literacy, the health of a population, and its degree of urbanization.

GDP per Capita

By far the most common measure of wealth and poverty internationally is gross domestic product (GDP) per capita, i.e., the total value of the goods and services produced by a national economy in a year divided by its population. As Figure 6.1 illustrates, GDP per capita is more than $50,000 in most developed countries. At the same time, three billion people live on less than $2 a day, or $750 per capita per year, and one billion people live on one dollar per day or less. Japan, North America, Western Europe, Saudi Arabia, Australia, and New Zealand have the highest per capita incomes in the world. South Asia, Southeast Asia, and sub-Saharan Africa have the lowest. In between lie most of Latin America and China.

However, GDP is a flawed measure in several ways: it does not capture nonmarket, noncommodified economic activity (e.g., barter and subsistence production) and it is vulnerable to fluctuations in exchange rates among countries.

Per capita purchasing power, which includes the cost of living, is a more meaningful measure of actual incomes than GDP per capita. The purchasing power in developed countries is more than $20,000 per capita per year, whereas in Africa it is less than $1,000 per year. The United States is surpassed by Japan, Scandinavia, Switzerland, and Germany in per capita income. However, it surpasses almost all countries in per capita purchasing power because goods and services are relatively inexpensive in the United States compared with other industrialized countries. In other respects, however, including poverty rates, income inequality, and access to health care, the United States lags behind much of Europe.

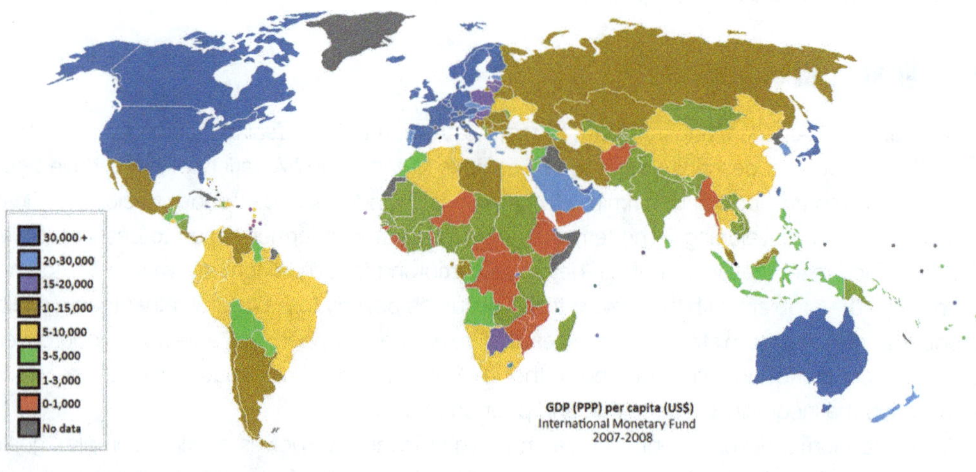

Figure 6.1 GDP per capita, 2019.

Economic Structure of the Labor Force

The sectoral distribution of employment in a country also reflects its economic development. Employment is typically divided into three major categories:

1. The *primary sector* involves activities that extract materials from the earth—mining, lumbering, agriculture, and fishing (Chapters 5, 8).
2. The *secondary sector* consists of manufacturing, the transformation of raw materials into finished products (Chapter 9).
3. The *tertiary sector* includes the provision of services, such as finance and business services, wholesaling and retailing, personal services, health care and entertainment, tourism, transportation, and communications (Chapter 10).

In **less developed countries**, it is common that a large share of the labor force works in the primary sector, primarily as peasants and farmers (Figure 6.2). The need for many farm workers reflects the reliance on animate sources of energy. In the United States, only 2 percent of the labor force is employed in agriculture (Chapter 8), whereas in some African countries, more than 75 percent of the laborers work in fields or farms.

Education and Literacy of a Population

Development can also be measured by the extent of education in a country, such as the proportion of children who attend school. The literacy rate is the proportion of people who can read and write. In sub-Saharan Africa, often the majority of the population is illiterate (Figure 6.3). Other measures of education include the average number of students per teacher. Developed societies generally have low student-teacher ratios, whereas poor countries have high ones. In many

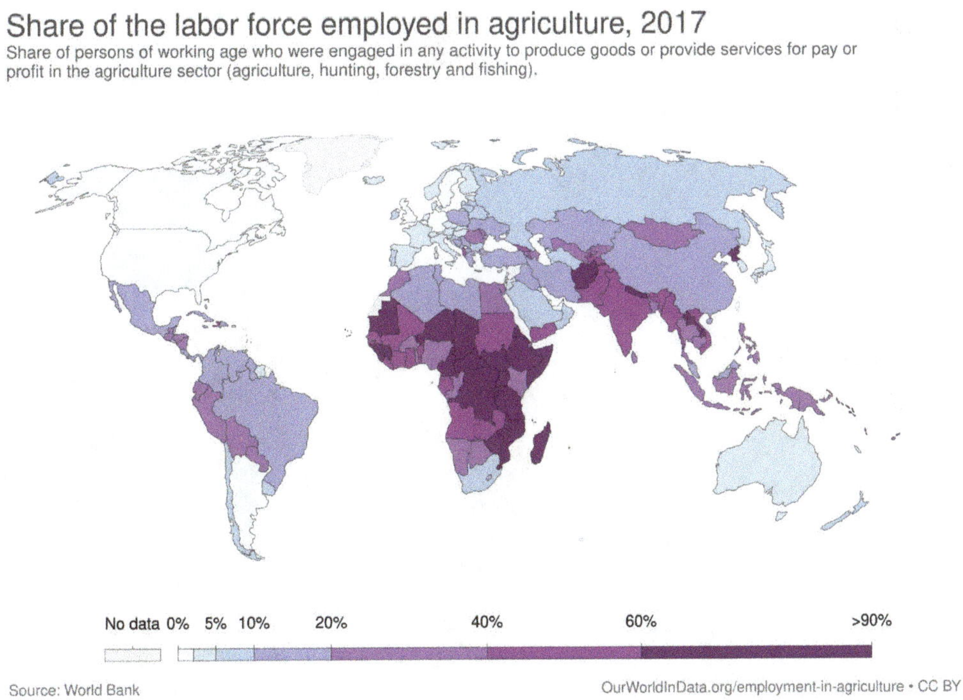

Figure 6.2 Percentage of labor force in agriculture.

developing countries, teachers' salaries are low, the buildings are decrepit, and there are insufficient funds for textbooks, uniforms, or equipment such as computers. Notably, many countries have ample funds for their militaries, indicating that having poor educational systems is a policy choice, not a "natural" problem.

There are frequently large gender differences in literacy rates within developing countries. In many poor societies, rural families can often only afford to send one child to school (the others are working the fields), and that child is most likely to be male, given the entrenched patriarchy of much of the world. In some countries, women's literacy rates are much lower than men's. The Middle East and South Asia show the greatest disparities. However, in the highly developed world, the literacy rates of men and women are identical. Countries with low percentages of women attending school also usually have poor social and economic conditions for women, including job opportunities. Raising women's literacy rates is a key means of lowering fertility rates and empowering women economically and politically.

Health of a Population

The population's health status is almost inevitably much better in developed countries than in less developed countries. One measure is diet, measured as caloric consumption per capita (Figure 6.4).

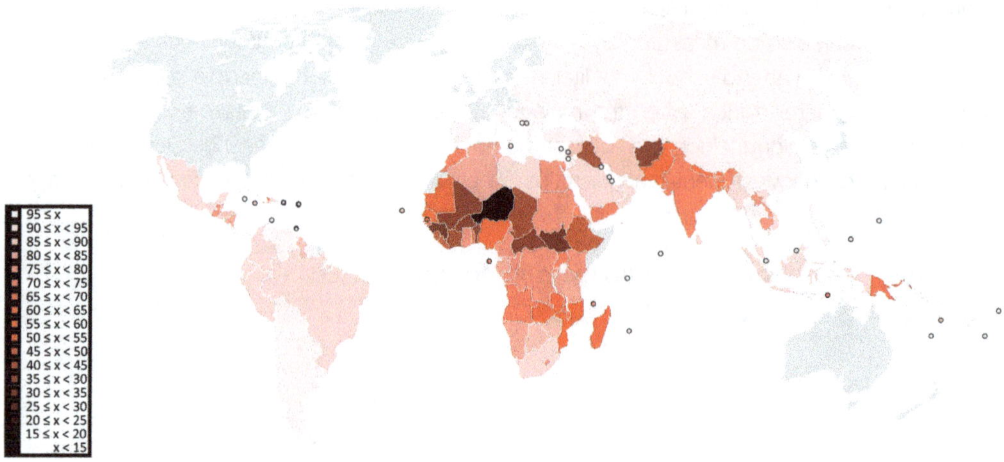

Figure 6.3 Global literacy rates.

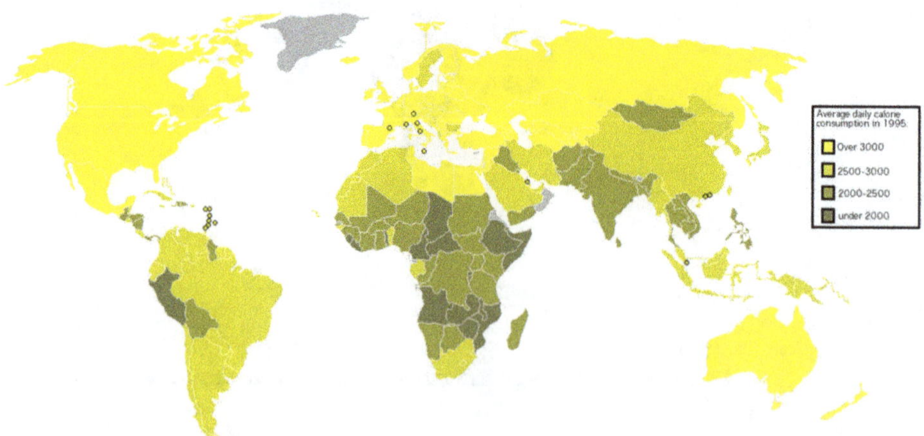

Figure 6.4 Daily average caloric consumption.

In developed countries, most people, on average, consume one-third more than the minimum daily requirement and maintain a high level of health. However, even in developed countries such as the United States, pockets of hunger and malnutrition exist. Conversely, an overabundance of cheap food and inadequate exercise have led to an obesity epidemic. Obesity has become an epidemic worldwide; there are today more obese people worldwide than hungry ones.

People in developed countries typically have better access to doctors, hospitals, health care providers, and specialized care. Whereas in developed countries, there is one doctor per 1,000 people, in some developing countries, each person shares a doctor with many thousands of others. Africa, by far, has the worst access to health care; in some countries, there are more than 15,000 people per doctor, which essentially means that most people *never* see one. Within countries, there are also discrepancies, usually between urban and rural areas.

Infants and children are the most vulnerable members of any society because their immune systems are not well developed, and they lack effective political power. This status is reflected in the infant mortality rate, the proportion of babies who die in their first year. In developed countries, on average, fewer than 10 babies in 1,000 die within their first year; in many less developed countries, more than 100 babies die per 1,000 births, the result of poor prenatal care, lack of clean water, poor sanitation, malnutrition, and infectious diseases. The geography of infant mortality (Figure 6.5) is thus perhaps the best measure of economic development, or the lack of it.

AIDS is a serious disease worldwide. More than thirty million people have died of this disease, and an additional fifty million are infected with HIV. The worst region in the AIDS epidemic is sub-Saharan Africa, particularly the southern parts, where in some countries 40 percent of the adult population is infected. The sub-Saharan region accounts for more than 60 percent of the people living with HIV worldwide. The disease is spread by prostitutes and truck drivers, and

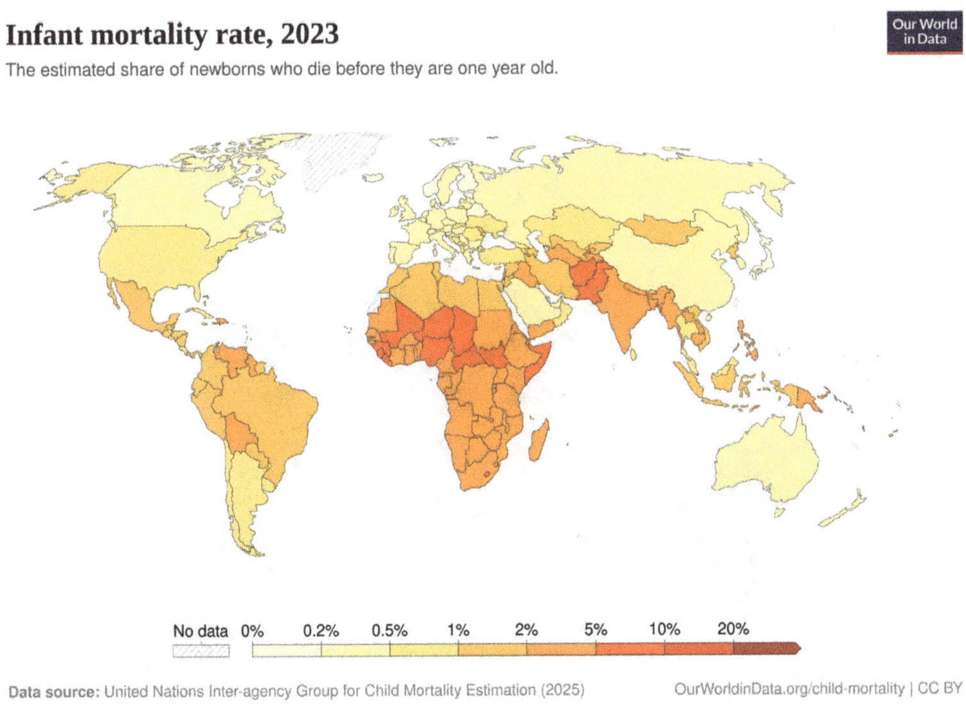

Figure 6.5 Infant mortality rates around the world.

into rural areas, as migrant workers bring the virus home. The social consequences of this die-off are catastrophic, including millions of children who have lost their parents to the disease.

The reliability and quality of the food supply, access to clean drinking water, public health measures, ability to control infectious diseases, and access to health services all shape life expectancy. People in developed countries typically live a relatively long time—often more than seventy-seven years, on average—while those in underdeveloped ones live considerably less (Figure 6.6). In parts of Africa, few adults can realistically expect to live past their fiftieth birthday. In almost all countries, life expectancy is closely tied to income and lifestyle choices such as smoking.

Availability of Consumer Goods

The quantity and quality of consumer goods purchased in a society is another measure of the level of economic development. Easy availability of consumer goods means that the economy has fulfilled the basic human needs of shelter, clothing, and food, and that disposable incomes allow purchases of other goods such as automobiles, textiles, electronics, jewelry, cell phones, refrigerators, and washing machines.

6.2 Urbanization in Developing Countries

In the developed world, urbanization occurred on the heels of the Industrial Revolution; that is, it was synonymous with industrialization (Chapter 2). While large parts of the developing world are urbanized, typically underdeveloped countries lag behind the economically advanced countries in this regard. However, cities throughout the developing world are growing quickly, much more so than those in Europe, North America, Japan, or Australia. Today, about 56 percent of the world's people live in cities (Figure 6.7): for the first time in human history, the majority of the planet resides in urban areas.

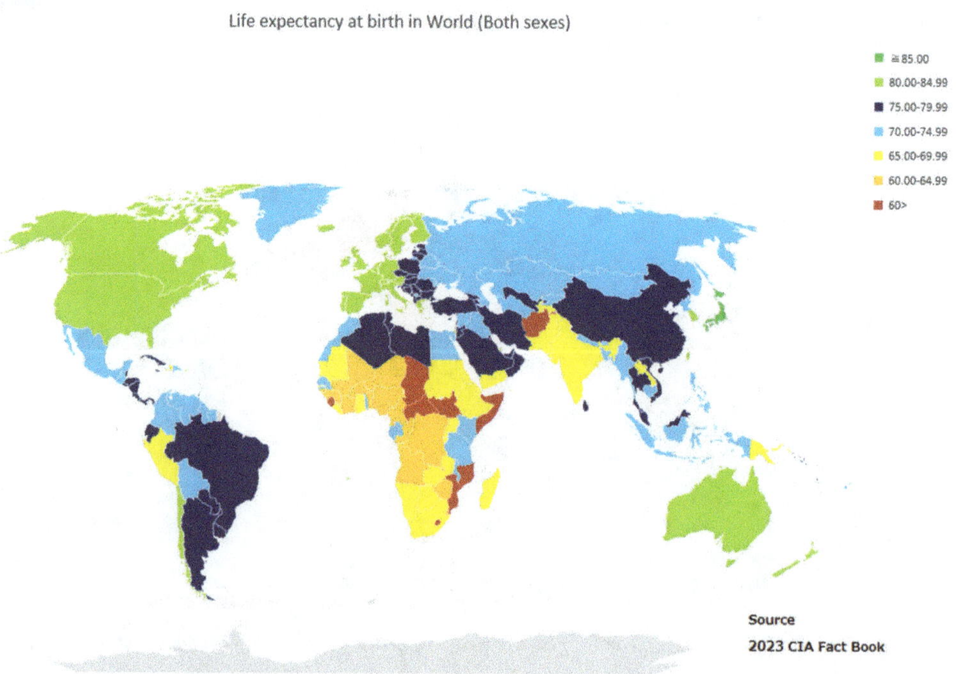

Figure 6.6 Life expectancy at birth, 2021.

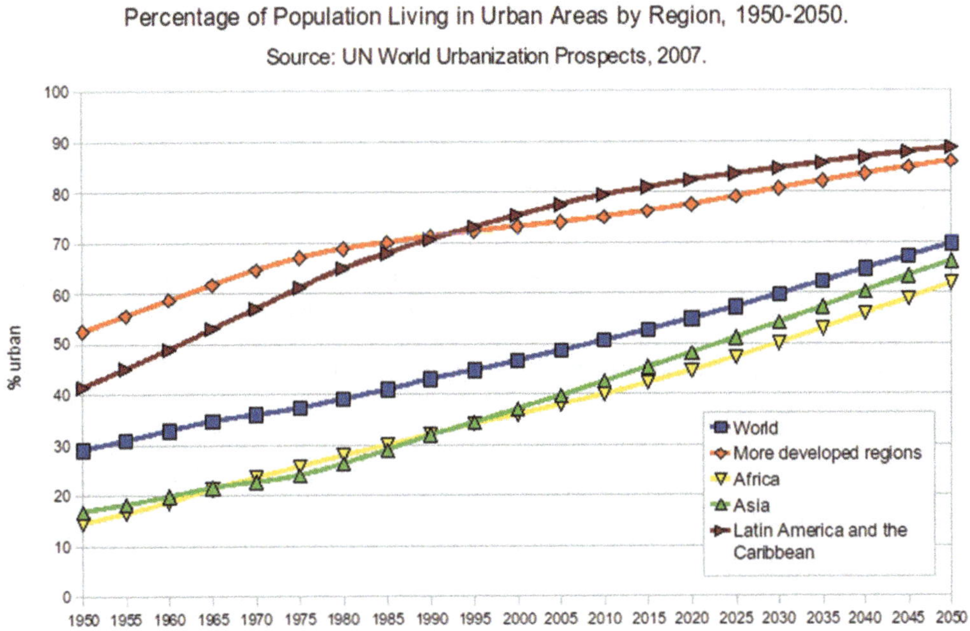

Figure 6.7 Urbanization rates over time.

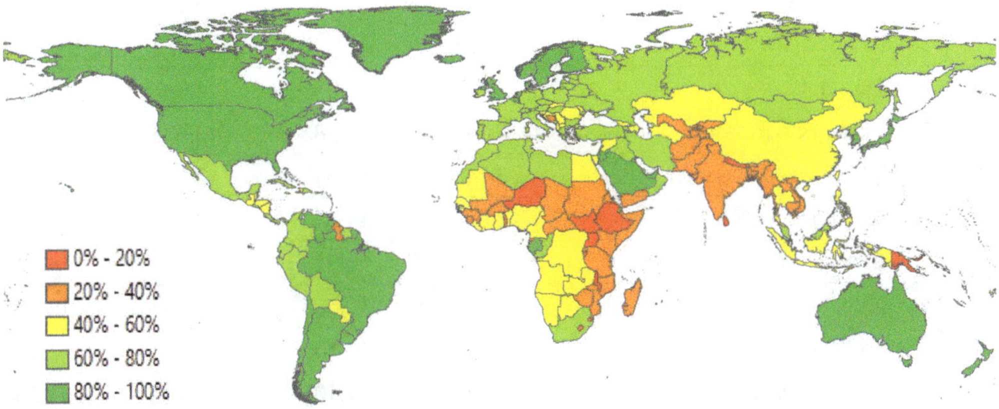

Figure 6.8 World urbanization rates.

Not surprisingly, the proportion of each country's people who live in cities varies widely around the world (Figure 6.8). In parts of Latin America, urbanization rates resemble those of North America or Australia, where 75 percent or more of the population lives in cities. In Asia and Africa, however, the proportions are generally lower. Only half of China lives in cities, and in wide swaths of Africa, less than 20 percent do so.

Crucially, urbanization in the developing world differs significantly from that in the developed world. Because the historical contexts of less developed countries differ from those of the West, particularly through the impacts of colonialism, the patterns of urban growth are also very different. Some countries in the developing world had well-established traditions of urbanization before the Europeans came, such as the Arab world and China. In others, colonial powers such as Britain played a major role, constructing cities such as Calcutta in India, Rangoon in Myanmar (formerly

Burma), or Singapore; similarly, the Dutch founded Batavia, Indonesia, which later became Jakarta (Chapter 2).

Today, the overwhelming bulk of the world's urban growth is in the developing world. Most of the world's largest cities, for example, are found in Latin America and Asia (Table 6.1). In 2020, the greater Tokyo metropolitan region, with thirty-six million, was the world's most populous, while New York, Mexico City, Mumbai, and São Paolo, Brazil vied for second place, with roughly twenty million inhabitants each. Many of the others, with populations over five million, are located in China and India (Figure 6.9).

Moreover, cities in the developing world are growing much more rapidly than their counterparts in developed countries. While natural growth rates in cities in the developing world are higher than

Table 6.1 **World's Largest Urban Areas, 1970–2020 (millions)**

Rank	Agglomeration, Country	Pop. 1970	Pop. 2000	Pop. 2010	Pop. 2020
1	Tokyo, Japan	26,615	34,450	35,676	36,371
2	New York, USA	15,880	17,846	19,040	19,974
3	Mexico City, Mexico	10,690	18,022	19,028	20,189
4	Mumbai, India	7,082	16,086	18,978	21,946
5	São Paulo, Brazil	9,614	17,099	18,845	20,544
6	Delhi, India	4,426	12,441	15,926	18,669
7	Shanghai, China	7,326	13,243	14,987	17,214
8	Kolkata, India	7,888	13,058	14,787	17,039
9	Dhaka, Bangladesh	2,221	10,285	13,485	17,015
10	Buenos Aires, Argentina	8,745	11,847	12,795	13,432
11	Los Angeles, USA	8,926	11,814	12,500	13,160
12	Karachi, Pakistan	3,989	10,019	12,130	14,855
13	Cairo, Egypt	6,450	10,534	11,893	13,465
14	Rio de Janeiro, Brazil	7,557	10,803	11,748	12,775
15	Osaka-Kobe, Japan	9,844	11,165	11,294	11,365
16	Beijing, China	6,034	9,782	11,106	12,842
17	Manila, Philippines	4,999	9,958	11,100	12,786
18	Moscow, Russia	7,623	10,016	10,452	10,524
19	Istanbul, Turkey	3,600	8,744	10,061	11,177
20	Paris, France	8,558	9,692	9,904	10,007
21	Seoul, South Korea	6,808	9,917	9,796	9,740
22	Lagos, Nigeria	1,890	7,233	9,466	12,403
23	Jakarta, Indonesia	4,813	8,390	9,125	10,792
24	Chicago, USA	7,160	8,333	8,990	9,516
25	Guangzhou, China	2,673	7,388	8,829	10,414

Figure 6.9 Distribution of the world's largest metropolitan areas.

that in the developed one (Chapter 7), the growth of cities in less developed countries is primarily due to the massive number of rural-to-urban migrants, many of whom are displaced by agricultural mechanization, unequal land distribution, high unemployment, lack of public investment and services, low farm prices, and war in rural areas. This pattern was also true of cities in the developing West during the Industrial Revolution. However, because few developing countries have generated the industrial job growth that the West did, the labor markets and employment conditions in the developing world are fundamentally different. In short, urbanization in the developing world and the crisis of rural areas are two sides of one coin.

Because many migrants move to the cities on the basis of their perceptions that there are greater opportunities to be found there (views that may be erroneous due to imperfect information), they often find themselves plunged into desperate circumstances. Less developed countries' urban labor markets rarely generate sufficient employment opportunities, leading to high unemployment rates or **underemployment**, in which migrants do not utilize their skills. Many find work in low-paying service jobs in the informal economy, which includes a wide array of jobs that are generally oppressive, low paying, and offer little financial security (Figures 6.10, 6.11). In many poor countries, the informal sector constitutes the bulk of employment opportunities. In Manila, the Philippines, for example, thirty thousand people earn their living recycling garbage on the Payatas garbage dump, as adults and children work together, stepping over debris to scavenge bits of plastic or tin cans. Others find marginal incomes selling trinkets and food on the streets, prostitution, selling illegal drugs, the black market, illicit currency exchange, or as casual day laborers doing construction. In contrast, formal sector, export-oriented jobs that tie developing countries' cities to the global economy, such as the garment industry or electronic assembly plants—which, however exploitative they appear to Western ideas, still offer higher wages than most local opportunities—tend to be the exception, not the norm.

Residential patterns in developing countries' cities also differ from those in the developed world. Whereas in the developed countries the poor tend to be a minority, often living in the city center, in the developing world, the poor may be a majority of the city's inhabitants, depending on the overall level of economic development. Often, the relatively wealthy command the city centers, whereas

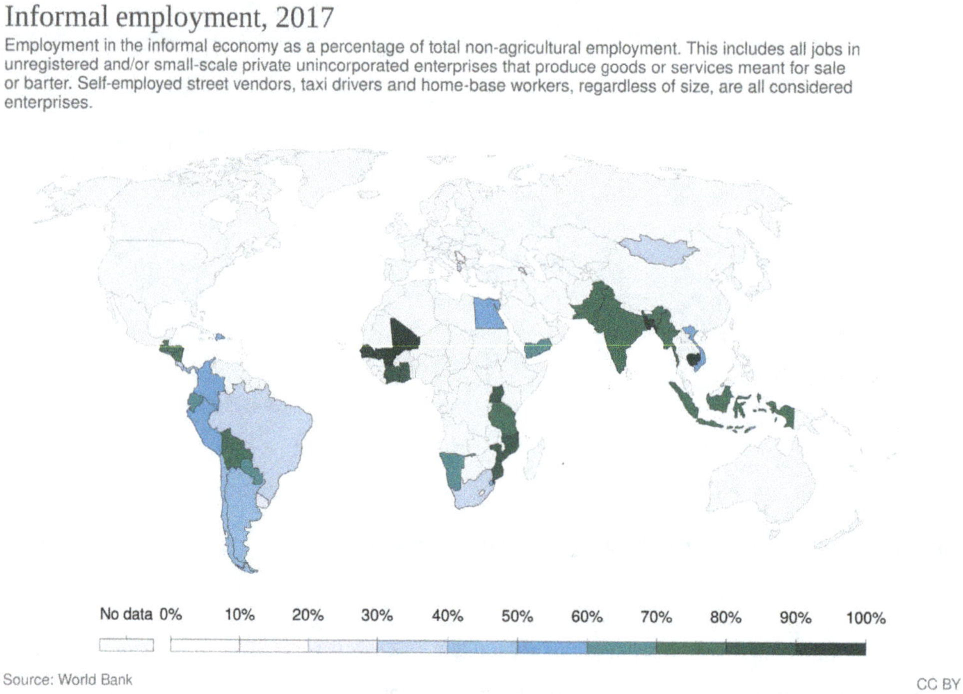

Figure 6.10 The extent of the informal economy across the globe.

Figure 6.11 This street vendor in Mexico is typical of employment in the informal economy.

the poor live in the urban periphery. In many Latin American cities, for example, housing near the old colonial plaza in the center is relatively expensive, while many live in **slums** on the periphery.

Billions of people today live in slums (Figure 6.12). In small, dilapidated houses, often self-made from cheap materials such as cinder blocks or sheets of corrugated tin, many urbanites inhabit squalid neighborhoods. These go by a variety of names: slums, Indian *chawls* or *bustees*, Brazilian *favelas*, Indonesian *kampong*, Turkish *gacekondu*, South African townships, and West African *bidonvilles* (or "tin can cities") (Figure 6.13). Countless numbers of people live in conditions that lack adequate housing, electricity, roads, transportation systems, schools, clean water, or sewers. High unemployment, lack of access to health care, inadequate education, and road systems all conspire to produce vast pools of suffering. Often, densities in such places are far higher than in any developed country's city. When migrants seize buildings or land that belongs to wealthy land-owners, the government may rule such **squatter settlements** to be illegal and tear them down with bulldozers.

6.3 Development Problems of Less Developed Countries

The developing world consists of an enormously diverse set of societies with profound cultural and economic differences. Nonetheless, they share a common set of problems, to varying degrees, that inhibit economic development.

6.3a Population Growth

Population growth is one of several issues that confront most developing countries. We must avoid the Malthusian error of blaming all of the world's problems on population growth (Chapter 7). Focusing exclusively on population growth detracts from other important factors, such as the long-term effects of colonialism, the dynamics of the world system, and government corruption and indifference to their populations. However, it is nonetheless true that in many countries the rate of population growth exceeds productivity gains, reducing the average standard of living.

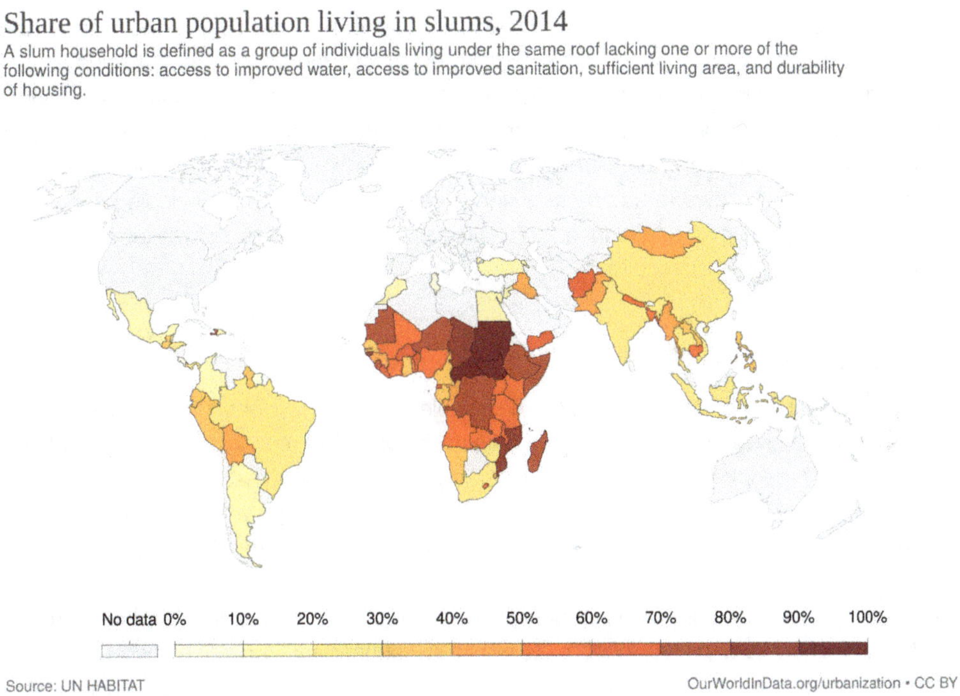

Figure 6.12 Proportion of urban population living in slums.

Figure 6.13 Mathare slum in Nairobi, Kenya, illustrates the housing of many poor people in the developing world.

In many poorer countries, particularly in Africa, rapid population growth rates strain food supplies, creating food insecurity. Rapid growth also reduces the ability of households to save, reducing the supply of investment capital. High population growth rates in agriculturally dependent countries mean that farmland must be used more intensively. It is also associated with overgrazing, overplanting, and soil erosion. Rural population growth, which is generally higher than that of cities, induces large flows to urban areas and greater urban problems, including congestion, pollution, housing shortages, and reduced supply to public services. Rapid increases in population also intensify unemployment.

6.3b Unemployment and Underemployment

Unemployment and underemployment are major problems in less developed countries. Put bluntly, their economies rarely generate enough jobs for all, for a variety of reasons. Unemployment is the condition in which people cannot find jobs; underemployment means that people cannot work as much as they would like, and that their skills are underutilized. Unemployment in many developing countries often exceeds 20 percent, a level comparable to that during the Great Depression in the United States in the 1930s.

Most cities in less developed countries experience large flows of migrants from rural areas as a result of low agricultural prices, inadequate public investment, rural unemployment, and lack of land reform. Many migrants are attracted to cities by the expectation of jobs and higher incomes, expectations that may not be met in reality. Once in the cities, many migrants cannot find work and contribute to the crisis of unemployment. Other migrants find limited employment as handicraft peddlers or street vendors, forming part of the unregulated, low-wage, informal economy.

Unemployment and underemployment are not the sole reasons for the problems of less developed countries. Jobs are generated only when there is sufficient investment, which is insufficient in most less developed countries. Because the domestic supply of investment capital is limited, poor countries rely on **foreign direct investment** (FDI) supplied by transnational corporations, which can generate social, economic, and environmental costs. Thus, unemployment involves a complex, globalized framework in which public and private resources are allocated, jobs are generated, occupations and skills are created, and investment capital is channeled.

6.3c Low Labor Productivity

One reason for low levels of productivity in developing countries is the small scale of farming operations often found there, which are subsistence-based and generate few economies of scale (Chapter 3). Another is that investments in farm machinery are often very low, interest rates are relatively high, and most capital is provided by foreign-owned firms, whose major incentive is exports and profits, not job generation. Most developing countries lack the buildings, machinery, power networks, and factories that enable people to produce efficiently. In addition, less developed countries typically have low levels of human capital, or skills. Investments in human capital—such as education, health, and social services—make unskilled workers into skilled, productive workers. Schools in developing countries are often inadequate, literacy rates may be comparatively low, and technical skills are limited. Many highly skilled professionals, such as engineers and doctors, emigrate to developed countries, where they are more highly paid. The United States and Europe, for example, attract some of the best-trained labor from less developed countries, including doctors, engineers, and scientists. This immigration has led to a brain drain, whereby less developed countries lose talented people to the developed ones.

6.3d Lack of Capital and Investment

Most less developed countries suffer from a lack of productive capital, such as machinery, equipment, factories, public utilities, and infrastructure. Because there is a close relationship between output per worker and investment, if a country hopes to increase its output, it must find ways to increase capital investment.

Capital accumulation for less developed countries must come from savings and investment. When a less developed country generates savings and revenues from exports, capital is released for investment. But the barriers to saving and investing are often high in less developed countries, including high interest rates, political unrest, corruption, and a lack of good regulations. The two main problems with investment in less developed countries are the lack of investment opportunities and the lack of incentives to invest locally. Usually, less developed countries have low levels of domestic spending per person, so aggregate demand is low, and their markets are relatively small compared to developed countries. Without domestic industries, consumers must turn to imports, especially for high-value-added products.

Many less developed countries suffer from **capital flight**, which occurs when wealthy individuals and firms deposit their monies in overseas banks in developed countries. Capital flight occurs when the wealthy confront politically unstable governments, unfavorable exchange rates, and high rates of inflation and taxation. Foreign investment in developing countries worldwide is almost completely offset by capital flight to banks in Europe and North America. In 2020, for example,

Mexican firms were estimated to have held about $160 billion in assets abroad, which is roughly equal to their international debt.

6.3e Inadequate and Insufficient Technology

Typically, less developed countries lack the technologies necessary to increase productivity, accumulate wealth, and raise standards of living. Some countries acquire new techniques through technology transfer that may accompany foreign investment by transnational corporations. Unfortunately, to put this technology to use, countries must have a minimal level of human capital for the installation, repair, and maintenance of equipment.

In developed countries, where labor has historically been relatively scarce, technology was developed primarily to save labor costs and to increase output. However, in less developed countries, capital intensification displaces workers, eliminating critically needed jobs. In agriculture, much of the midlatitude technology of the developed countries (e.g., tractors and combines) is unsuited for agricultural systems with tropical or subtropical climates and soils. Thus, even the same technology can have different impacts geographically. In short, to understand the impacts of technology, we must examine it within its social and economic context.

6.3f Unequal Land Distribution

In most developing countries, a large part of the population lives in rural areas, and land is critical to survival. Unfortunately, land is often highly unequally distributed, and often a small minority of wealthy landowners controls the bulk of arable (farmable) land. In Brazil, for example, 2 percent of the population owns 60 percent of the arable land; in Columbia, 4 percent own 56 percent; in El Salvador, 1 percent own 41 percent; in Guatemala, 1 percent owns 36 percent of the land; and in Paraguay, 1 percent owns an astounding 80 percent of all farm land. These numbers reflect the historical legacy of the Spanish land grant system (*encomiendas*) imposed over centuries of colonialism (Chapter 2). Wealthy rural elites often own vast estates and plantations designed primarily for growing export crops, not for sale in domestic markets. The majority of the rural population, consequently, is landless and must sell its labor, often at very low (below subsistence) wages, and live in serf-like conditions. Throughout much of the developing world, land reform is lacking because the government is too corrupt to redistribute the land owned by a few wealthy families. For some countries, such as the Philippines or El Salvador, land reform is the single most pressing problem deterring them from economic development. Shortages of farmland are accentuated by agricultural mechanization and high population growth. The unsurprising result is frequent political unrest, including demands for land redistribution; in much of Latin America, for example, peasant social movements to regain control of the land are active, and violent confrontations with landowners are frequent. Further, rural areas with high natural growth rates and widespread poverty become the source of rural-to-urban migrants, who swamp cities with desperately poor people seeking jobs.

6.3g Poor Terms of Trade

Terms of trade is a term that refers to a country's relative values of its exports and imports (Chapter 4). When the terms of trade are good, a country sells relatively high-priced goods (e.g., manufactured goods) and imports relatively low-priced ones (e.g., mineral ores, agricultural products). This situation generates foreign revenues that allow the country to pay off debts or to invest in infrastructure, human capital, or new technologies. Unfortunately, many less developed countries, with economies shaped by centuries of colonialism in which they became suppliers of minerals and foods, export low-priced primary sector goods and import expensive, high-priced, manufactured ones. Many less developed countries rely heavily on exports of cheap goods such as petroleum, copper, tin and iron ore, timber, coffee, and fruits and must import expensive ones such as automobiles, pharmaceuticals, office and telecommunications equipment, and machine tools. The persistent global glut in raw materials, including petroleum, agricultural goods, and mineral ores, has depressed the prices for many less developed countries' exports. Without local manufacturing firms, many are forced to sell low-priced goods such as bananas for high-valued ones such as computers and pharmaceuticals. This situation makes it difficult to generate foreign revenues and exacerbates a country's debt situation. Further, poor terms of trade perpetuate a country's cycle of poverty: low foreign revenues yield little to reinvest, which helps to create a shortage of capital, resulting in low rates of productivity.

6.3h Foreign Debt

Much of the developing world is deeply in debt to foreign governments and banks. In 2022, total world debt amounted to more than $180 trillion. In absolute terms, the largest debtors are in the developed world (Figure 6.14), led by the United States, which owes the world more than $22 trillion (Table 6.2). However, when we examine relative debt, or debt as a percent of GDP, a different picture emerges: measured in terms of national income, it is developing countries that owe the most. Having borrowed hundreds of billions of dollars, they find themselves unable to repay either the interest or the principal. Debt repayments—annual interest and amortization payments—often consume a large proportion of their export revenues that might otherwise be used for development (Figure 6.15).

The origins of the debt crisis lay in the oil crises of the 1970s and 1980s, when many less developed countries took out large loans with the expectation of paying them off when their economies improved in the future. Western banks had a huge influx of petrodollars invested by OPEC countries and were eager for borrowers; developing countries took advantage of this access to cheap loans. Even oil-exporting countries such as Mexico, Libya, and Nigeria overborrowed based on their expectations of rapidly rising oil prices. When oil prices fell in the 1980s, these countries found themselves saddled with debts that they could not afford. In addition, many developing countries faced long-term, declining terms of trade and governments that spent heavily on their

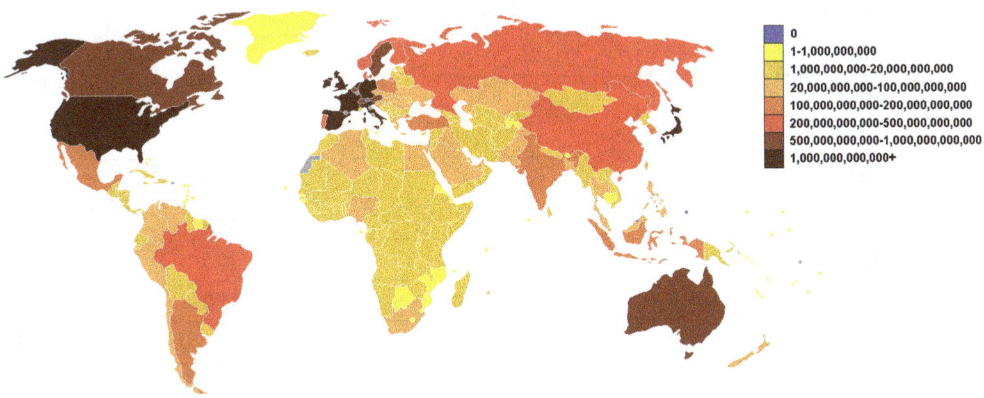

Figure 6.14 Map of absolute external debt.

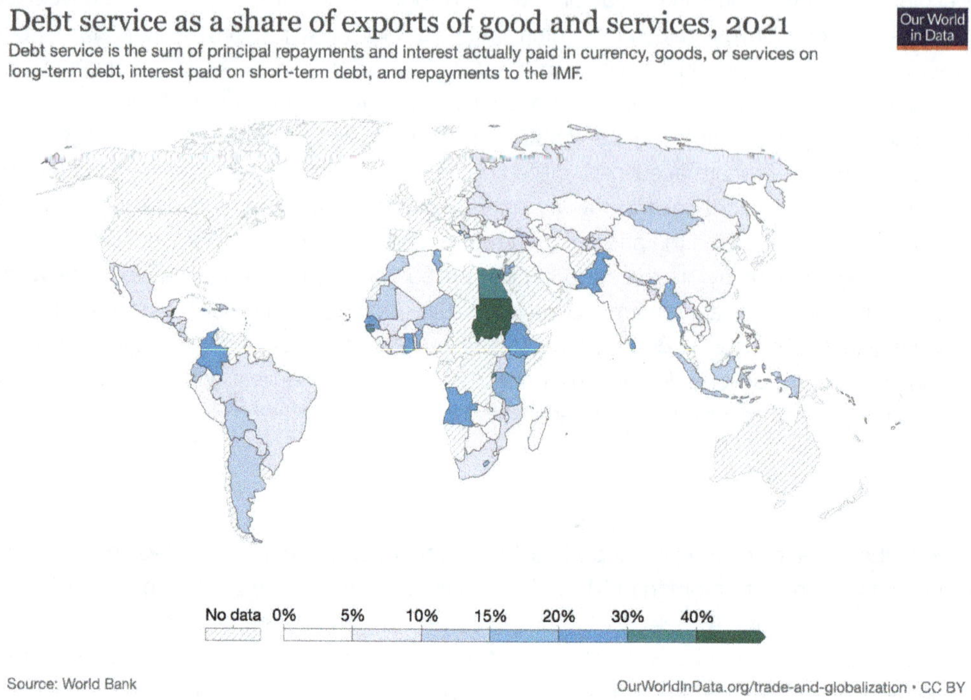

Figure 6.15 Debt service (repayments as percentage of export revenues), 2017.

Table 6.2 Countries with Greatest External Debt, 2024 ($ trillions)

	$ trillions	% GDP
United States	22.9	102
United Kingdom	9.1	345
France	7.3	230
Germany	5.7	165
Japan	4.8	96
Netherlands	4.6	453
Italy	2.5	141
Spain	2.3	170
China	2.1	29
Canada	1.9	143
Switzerland	1.8	285
Australia	1.8	130
India	0.6	20
Singapore	0.2	471

militaries as well as overly ambitious, unrealistic development schemes such as highways and hydroelectric dams.

The mounting debt caused concerns about the stability of the international monetary system. In 1982, Mexico ran into severe difficulties meeting interest and capital payments on its debts; other Latin American countries also appeared ready to default. A collapse of the financial system was forestalled by a series of emergency measures designed to prevent large debtor countries from defaulting. The International Monetary Fund (IMF), the World Bank, and the governments of lending countries collectively engaged in massive bailout exercises that accompanied debt reschedulings, or changes in the repayments of debt. Debt restructuring policies imposed by the IMF typically tie debt relief to "**structural adjustment**," which usually includes reductions in government subsidies to the poor (such as for kerosene, cooking oil, or mass transit), a rise in interest rates to attract foreign investors, and devaluations of currencies (which drive up the costs of imports and make exports cheaper). Such policies lower the quality of life for hundreds of millions, if not billions, of people, and have made the IMF a much-hated institution in many countries. For example, a poor mother in a less developed country with a sick infant relies on imported pharmaceuticals; IMF-imposed currency devaluations drive up the costs of these drugs, often making them out of reach for the poor, who need them the most. For these reasons, many scholars and politicians call for a debt moratorium to allow less developed countries a respite from what are often crushing debt burdens.

As the debt crisis grew in the late twentieth and early twenty-first centuries, some less developed countries required foreign banks to rewrite their loans and cancel a portion of the principal and interest. The result was a loss of confidence in their future ability to repay. At the same time, the United States began to generate enormous federal budget deficits; to finance these, it borrowed a large portion of the world's investment, driving up interest rates worldwide. Higher oil prices, declining prices for raw material exports, higher interest rates, and a decline in public and private lending all contributed to the debt crisis that continues today.

6.3i Restrictive Gender Roles

Deeply entrenched social stratification systems work against disadvantaged people, especially women, in many developing countries. Gender refers to the socially reproduced differences between women and men, and shapes every society's allocation of resources and people's life chances. Typically, gender roles work to the advantage of men and the disadvantage of women.

The economic, political, and social status of women around the world is geographically variable. Worldwide, nowhere do women claim the same rights and opportunities as men. Women account for most of the world's 1.5 billion people living in extreme poverty (under $1.90 per day). In most countries, women do most of the field work while still being responsible for household chores such as cooking, cleaning, carrying water and firewood, and raising children. Women still spend more time working than men in all regions except North America and Australia, and their wages are lower everywhere. In Muslim countries, women are usually not economically active outside of the home because of religious prohibitions. In Latin America, labor force participation of women in the economy is increasing, but mostly outside of the agricultural sector. Households in sub-Saharan Africa, India, and Southeast Asia are highly dependent on female farm and market income, as well as commodified labor in the labor market. At a world scale, women generally garner only 60–70 percent of what men earn, often for the same work.

6.3j Corrupt and Inefficient Governments

Often, less developed countries' governments are controlled by bureaucrats whose primary interest is catering to the wealthy elite and foreign investors, a situation that creates governments that are at best indifferent and often outright hostile to the needs of their own populations. Many less developed countries' public policies are ineffectual or counterproductive, and typically ignore the rural areas in favor of cities. Government jobs are frequently allocated through patronage, not a merit system. In Africa, public sector jobs often go disproportionately to members of the same tribe as that of the president. The military is often the most well-funded and well-organized institution and may be a source of political instability, as during military takeovers of the government in coups d'état or takeovers of the state. Corruption often is endemic and may become a way of life, generating inefficiency and inequality (Figure 6.16). Corruption erodes trust, misallocates public resources, and breeds resentment among the public. It is one of the most serious development problems across the world.

Corruption often accompanies political repression. In many poor countries, tyrannical, dictatorial governments curtail their citizens' civil liberties, censoring the media and routinely imprisoning or executing dissidents. Attacks on journalists, labor union leaders, student movements, religious groups, and ethnic minorities are often common under such situations. China, Myanmar, Iran, and Sudan are cases in point. Poor countries often have oppressive governments; widespread and well-protected civil liberties are more common with economic development. Understandably, many people feel alienated from their own governments under these conditions and may sympathize with various resistance movements, contributing to political instability.

Less developed countries' governments, often with limited budgets, frequently have great difficulty building or maintaining their infrastructures. Roads, dams, ports, bridges, airports, tunnels, and highways may be neglected and fall into disrepair, driving up transportation costs. Electrical power stations may not be maintained correctly, leading to frequent brownouts and blackouts. Telecommunications networks may be missing or unreliable, especially in rural areas. Unsanitary water supplies become major sources of infectious diseases such as cholera and malaria; indeed, unclean drinking water is one of the world's major public health hazards. More than a billion people

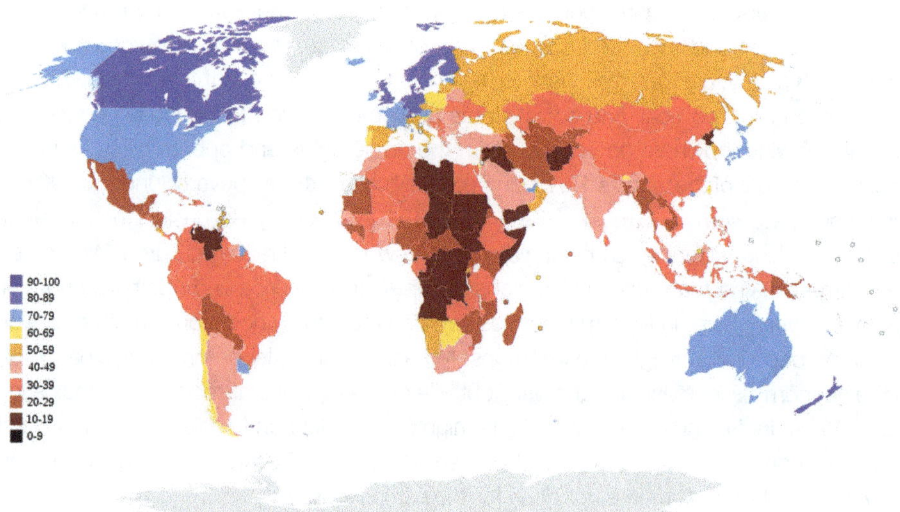

Figure 6.16 Map of world corruption scores, 2019.

worldwide do not have access to safe drinking water. Because the infrastructure is essential to economic development, governments that do not reinvest in their country's transportation, water, and communication lines do not facilitate economic growth.

Similarly, public services in many less developed countries are often underfunded and inadequate. Public schools—the major avenue of upward mobility for many—are often crowded and understaffed, and may fail to educate the country's young people, leading to high illiteracy rates, especially for girls. Without a minimum level of literacy, the labor force is unattractive to foreign investors. Salaries for teachers are often far too low for them to support themselves. This situation creates dire shortages of skills and human capital. Inadequate public health care systems, including shortages of physicians and nurses, reduce the health of the labor force and lower its productivity. Underfunded, overcrowded public transportation systems make circulation within and among cities difficult. Images of terribly overcrowded buses and trains in developing countries testify to government priorities as much as population growth. Only the military tends to be a well-funded public service in much of the world.

6.3k Solutions

There is clearly no one-size-fits-all approach to ending poverty in poor countries. If ending poverty were easy, it would have been ended long ago. Poverty-reducing strategies must include the following: investments in health care and education; land redistribution and protecting land tenure rights for the rural poor; agricultural development; better access to clean water; establishing political stability; reducing corruption; democratizing decision making through representative government; debt relief; improving terms of trade; ending developed countries' trade restrictions against imports from less developed countries; and increasing women's decision making power in the household. Obviously, these steps are difficult to implement, often because powerful, entrenched interests benefit from the status quo. Perhaps the best hope lies in a smoothly functioning world economy that can pull billions of people out of poverty, as has happened in much of East Asia.

6.4 Major Conceptual Perspectives on Development

Theories of development have existed for many years. Three perspectives on development in economic geography hold widely varying assumptions and conclusions, including **modernization theory**, **dependency theory**, and **world-systems theory**. These arose from very different philosophical foundations and have varied political emphases. All of these grapple with the complex question of how the global economy shapes patterns of opportunities and constraints to development in countries.

6.4a Modernization Theory

The first and most widely accepted theory of development is modernization theory. This perspective starts with the question: Does the development of the less developed world follow the same historical trajectory as the West? The intellectual origins of this line of thought lie with the famous sociologist Max Weber, who attributed the economic success of northern Europe to the Protestant ethic (Chapter 2). Weber held that although capitalism originated in religion, it led to a process of rationalization that gradually undermined the role of religion in public life. Weber's work established a precedent that viewed economic development and social change as a function of people's ideas, culture, and beliefs rather than their social relations and historical context.

Weber's ideas were enormously popular in the United States following the Second World War. In the United States, where they were translated and elaborated by Talcott Parsons, it equated modernity with being Western, denying the possibility of modern, non-Western cultures. Writing during the Cold War, a period of intense competition between the United States and the Soviet Union for influence in the developing world, modernization theory explicitly advocated capitalism as the best possible path any country could follow, that is, the only one that leads to modernity, wealth, and democracy. This theory argued that if less developed countries adopted Western values, market relations, and political systems, they would eventually become affluent, democratic societies. Tradition, in this view, is equated with irrationality and superstition, an obstacle to development. Hence, progress is held to be a unidirectional path from the traditional to the modern. In this view, rich industrial countries are modern, whereas poor countries must undergo the modernization process to acquire it.

This theory argued that less developed countries could eventually achieve the same prosperity that Europe and North America enjoyed if they gave up the ways of their premodern past. Modernization theory advocated stability and gradual political and economic changes, not revolutionary leaps. It is worth noting that whereas Weber held that capitalism was a unique institution to Europe, modernization theorists maintained that the triumph of capitalism was inevitable worldwide.

Modernization theory held that economic development occurred through a series of stages. One version of these stages is the widespread view that all economies pass through successive eras in which the labor force is employed first in agriculture, then manufacturing (i.e., the Industrial Revolution), then services (a postindustrial world). Developed by two economists, this became known as the Fisher-Clark model of structural change. However, while this notion reflected the history of the Western world, many developing countries leapfrog directly from agriculture into services. The fact that the developing world does not automatically follow in the footsteps of the West points to the dangers of generalizing from the experience of the industrialized to the developing world, which has very different historical contexts and cultures. Stages models imply that development is always linear, which is not necessarily the case.

Modernization theory placed great emphasis on the diffusion of progress: new ideas, technologies, and institutions were held to flow from the core to the periphery. Core areas, with abundant human capital, were seen as rational and creative, with much to offer less developed peripheries. Cores and peripheries exist at both the international and national scales. Internationally, therefore, diffusion occurred from the world's core, i.e., the developed countries, to the global periphery as multinational corporations invested in less developed countries. Investment was held to realize a country's comparative advantage and allow it to carve a niche in the world economy. Thus, modernization theory advocated unfettered free trade and no protectionism. Within countries, diffusion occurred from the urban cores, i.e., cities, which were held to be the foci of modernity, to the rural periphery, i.e., rural areas. Cities offered agglomeration economies, new technologies, and possibilities of life that rural areas did not. Thus, modernization theory held that urbanization was inherently a good thing, that rural areas were trapped in cycles of tradition and stagnation. The more people who moved to cities, the better. Rural-to-urban migration was held to equalize wages and standards of living, ironing out uneven development.

Poverty, in the modernization view, is held to be the result of the incomplete formation and diffusion of markets. Markets promote an equalization of standards of living in poor and rich regions through the free flow of capital and labor. Thus, modernization theory stands in direct contrast to Marxist claims that capitalism inevitably generates uneven development. Economic development in this conception is likened to a race, in which the rural areas must catch up with the cities, much as less developed countries must catch up with the developed world.

Modernization theory also advocated major social, cultural, and political changes. Population growth, for example, must be brought under control lest the demand for resources exceed productivity gains. Thus, modernization theorists advocated family planning programs, a line of thought similar to the demographic transition (Chapter 7). During the Cold War, population control policies were advocated as a means to raise standards of living and reduce the appeal of the communist bloc's alternative to capitalism. Thus, in the 1960s, when modernization theory was at its height, the US founded programs such as the Peace Corps and the Agency for International Development, which put population control front and center of their agendas.

Culturally, modernization theory argued that tradition was synonymous with irrationality and fear of change, whereas modernity holds that change is good and necessary for the sake of progress. Thus, cities were alleged to be inherently modern, and rural areas were to be transformed and brought into the modern age. The culture of modernity was to be diffused through mass education, the media, and growing literacy rates.

Politically, modernization theory equated capitalism with democracy, arguing that only market-based societies could protect civil liberties. This view was widely popularized by Nobel Prize-winning economist Milton Friedman. In contrast to the centralized power structures of precapitalist societies, markets create multiple centers of power, which limits the chances of tyrannical states. The growth of a middle class was often held to be a central event in protecting civil liberties: autocratic societies tend to be poor, with a small authoritarian elite in control. By generating an educated, informed citizenry, markets were held to be antithetical to dictatorship. Modernization theorists point to East Asian countries such as South Korea, Taiwan, and Singapore as evidence of this thesis.

A major advocate of this approach was Walter Rostow, an influential economist and presidential policy advisor in the 1960s who proposed a very famous five-stage model of development. Rostow's model viewed economic growth as an airplane taking off. He proposed five stages in this process:

1. Traditional society—A society that has not yet started the process of development and is mired in poverty. Such societies tend to be highly traditional, with a high percentage of people engaged in agriculture. Contemporary examples include Mali or Bhutan.
2. Preconditions for take-off—A society in which the process of development begins, usually because an elite group initiates economic activities. The country starts to invest in its infrastructure, such as water supplies and transportation systems. A culture of growth begins to take root. Examples include Thailand, Mexico, and Indonesia.
3. The take-off—A society that sees rapid economic growth in light manufacturing, such as textiles. Public education is introduced and literacy grows. As a result, productivity rises, and so do wages and standards of living. Urban growth accelerates. Examples today include China and Chile.
4. The drive to maturity—A society that enjoys a healthy and diverse economy, and rapid productivity growth. The middle class mushrooms and literacy is universal. Heavy manufacturing flourishes, and services grow quickly. Workers become more skilled and specialized. Examples include Singapore, Taiwan, and South Korea.
5. The age of mass consumption—A society that has become highly urbanized, wealthy, and service based. Disposable incomes are high and the consumer economy dominates. Examples include the United States, Canada, Japan, and Germany.

Modernization theory grouped every country into one of these stages. It is assumed that less developed countries will follow the steps of the developed countries.

The policy implications of modernization theory are straightforward. Overall, it advocates the creation of markets with appropriate state intervention as needed. Thus, barriers to trade and investment should be removed; foreign capital and investments from multinational corporations should be welcomed. Urban development should be emphasized at the expense of rural areas. Internationally, modernization theory was central in the switch from import substitution to export promotion that occurred in much of East Asia in the 1970s and 1980s. More recently, it has been used to justify neoliberal structural adjustment policies of the International Monetary Fund, including currency devaluations and reductions in government subsidies.

Modernization theory has been soundly criticized on a number of grounds. Critics note that it is ethnocentric: it posits the history of the West as an ideal to be imitated by everyone else, implying that non-Western cultures are inherently inferior. It offers a simplistic view of history that culminates only in the experience of Europe and North America. Less developed countries are reduced to little more than backward versions of the West, not as entities with their own distinct cultures, geographies, and histories. Modernization theory is silent about colonialism and its long-term impacts. Finally, by uncritically celebrating capitalism, modernization theory only focuses on the benefits but ignores the costs: markets always look fine to the winners, who preach free trade, but appear less rosy to those who have not benefited from them. Finally, modernization theory focuses only on the internal dynamics within countries (i.e., their cultures) and ignores the external context, such as colonial legacies and the countries' position in the global division of labor. In this way, it ends up blaming the victim (i.e., attributing poverty to poor people), which is politically easy but not accurate.

6.4b Dependency Theory

In the 1960s, scholars from the developing world, particularly Latin America, began to question the assumptions, conclusions, and promises of modernization theory. What appeared as comparative advantage and interdependence to the West, for example, appeared to less developed countries as exploitation. Drawing from the intellectual heritage of Marxism, dependency theorists reversed modernization theory's assumptions. They claimed that the development of the core countries was intrinsically dependent on the **underdevelopment** of the world's periphery. The theory suggests that unequal development of the world economy directly reflects the historical experience of colonialism (Figure 6.17). The development of Europe and North America, especially during the nineteenth and early twentieth centuries, relied on the systematic exploitation of underdeveloped areas by means of unequal terms of trade and the abuse of low-paid labor. Dependency theory is thus thoroughly historical, whereas modernization theory is not.

In this view, poverty did not simply happen; it is a historic product. Like commodities such as shirts or cell phones, poverty is actively made by the dynamics of capitalism, not just something that happens naturally. The process of creating poverty is called the "development of underdevelopment," which holds that underdevelopment is not a state but an active process. Dependency theorists thus argue that the less developed countries were made to be poor by the West. Exploitation of the world's periphery occurs through the process of uneven exchange, in which less developed countries produce low-valued goods in the primary sector (e.g., foods and mineral ores) and purchase high-valued goods from the core. Unequal exchange is a market transaction that appears on the surface like an exchange of opposites but conceals the appropriation of surplus value that low-income workers in developing countries produce.

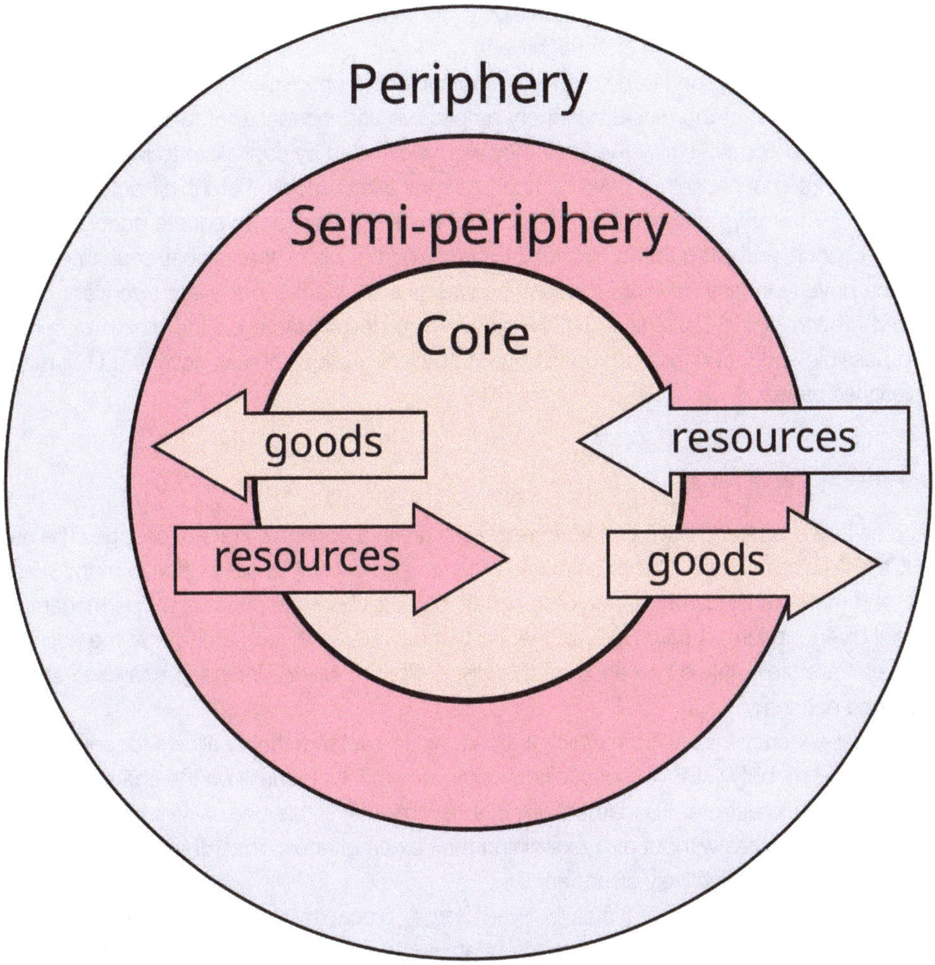

Figure 6.17 Mechanics of dependency theory.

Unlike modernization theory, dependency theory analytically prioritizes external, not internal, causes of poverty. Because developed and underdeveloped countries had very different histories, they occupy very different roles in the capitalist world economy. The West, unlike its colonies, was *undeveloped*, but not *underdeveloped*, that is, ruled by a colonial power that deformed its economies. Less developed countries do not simply "lag" behind the West but are mired in poverty produced by the world economy, a situation portrayed as inevitable by the economics of comparative advantage. Dependency theorists argue that development and underdevelopment are two sides of global capital accumulation: the wealth of developed countries is derived from the labor and resources of the less developed countries. Whereas modernization theory holds that markets eliminate uneven development, dependency theorists maintain that capitalism inevitably perpetuates it. Development and underdevelopment are therefore two sides of the same coin, a zero-sum game: development somewhere (i.e., the West) requires underdevelopment somewhere else (i.e., its colonies).

The policy implications of dependency theory, which are directly opposite to those espoused by modernization theorists, emphasize self-reliance. In this view, countries should exclude transnational corporations because they are predatory and practice import substitution to promote domestic production. Thus, it calls for shielding domestic industries behind tariffs and quotas. Some even advocate defaulting on foreign debt.

Like modernization theory, dependency is also subject to criticisms. Dependency theory tends to view the countries of the global periphery as passive and incapable of taking action; it lumps all less developed countries together as if they were victimized by capitalism to the same degree. It ignores the internal causes of poverty, such as rural aristocracies that inhibit development. Its explanation for the core's wealth is simplistic and does not offer an adequate account of technological change and productivity growth. Empirically, the claim that global capitalism always generates poverty in less developed countries was unsustainable: in the late twentieth century, the rapid growth of the East Asian NICs showed that development on the global periphery is indeed possible and that capitalism does not automatically reduce all less developed countries to impoverished states.

6.4c World-Systems Theory

The forces that shape and drive the world economy are in a continual state of change. The rise of the NICs in Asia, particularly China, deindustrialization and economic stagnation in many Western states, and the increasing importance of transnational firms illustrate capitalism's permanent state of spatial disequilibrium. A third body of theory, world-systems theory, started by the sociologist Immanuel Wallerstein, takes into account this disequilibrium in explaining the changing structure of the world economy.

Unlike dependency theory, from which it arose, world-systems theory allows for some upward and downward mobility within the capitalist world economy. Its focus is on the entire world rather than individual nation-states. Fundamentally, this view maintains that one cannot study the internal dimensions of countries without also examining their external ones. Thus, the boundary between foreign and domestic effectively disappears.

World-systems theory differentiates between large, precapitalist world empires, such as the Romans, Mongols, or Ottomans, which appropriate surplus from their subject populations through the state rather than the market, and the capitalist world system, in which markets reign supreme. Capitalism arose during the "long sixteenth century" (1450–1650), which saw colonial empires but not the militarized states such as tsarist Russia or Austria-Hungary. Under capitalism, there is no single political entity to rule the world, but many different nation-states. This means that there is a single market but multiple political centers, so one power exerts control over the market. The political geography of capitalism is thus not the nation-state but the interstate system. Occasional attempts to reassert a world empire included the Habsburgs, Napoleon, and Germany in the First and Second World Wars.

World-systems theorists maintain that capitalism takes many forms in time and space. Because its labor markets vary so much, labor is exploited in different ways in different regions. In the core, labor is waged, that is, organized through labor markets, while in developing countries, there is considerable use of unfree labor, including slavery, indentured workers, and landless peasants working on plantations. The world economy shapes places in such a way that high-valued goods are produced in the core and low-valued ones in the periphery. It is the incessant search for profits through low-cost labor that drives the world system forward to expand.

Unlike the simple bifurcation between developed and less developed countries found in both modernization and dependency theories, world-systems theory maintains that there is a tripartite division among core, periphery, and semiperiphery (Figure 6.18). The world's core and periphery are the developed and undeveloped countries, respectively: one is wealthy, urbanized, industrialized, and democratic, the other rural, impoverished, agriculturally based, and dominated by authoritarian governments. The semiperiphery lies in between: it has characteristics of both core and periphery and includes relatively well developed less developed countries, such as the NICs, Brazil, Mexico, Thailand, and Saudi Arabia. In the core, the world system generates high wages, high levels of urbanization, industrialism and postindustrialism, the service economy, advanced technology, and a diversified product mix. In the world periphery, there are low wages, low levels of urbanization, preindustrial and industrial technology, and a simple production mix. Consumption is low. In between are semiperipheral countries where both sets of processes coexist. The theory suggests that the semiperipheral countries are exploited by the core countries through uneven exchange, while at the same time exploiting peripheral countries.

World-systems theory emphasizes the role of a single hegemon that dominates the global political and economic system. Hegemony exists when one core power enjoys supremacy in production, commerce, and finance and occupies a position of global political leadership. The hegemonic power owns and controls the largest share of the world's economy. It is the leading trading and investment country, its currency is the universal medium of exchange, and its city of primacy is the financial center of the world. Because of its political and military superiority, the dominant core country maintains order in the world system and imposes solutions to international conflicts that serve its self-interests. Early in the history of capitalism, Spain was the world hegemon. Britain played this role in the nineteenth century, and the United States has done so since the Second World War. Consequently, hegemonic situations are characterized by periods of relative peace (e.g., the Pax Britannica of the nineteenth century).

The hegemon "sets the rules of the game," the ideological and legal framework within which everyone else operates. During the period of Spanish dominance in the sixteenth and seventeenth centuries, for example, mercantilism was the dominant ideology. Under the Pax Britannica of the nineteenth century, free trade became the norm. And since the rise to dominance of the United States since the Second World War, **neocolonialism** has been the dominant pattern.

When powers in the core have conflicts among themselves, they open opportunities both for new hegemons and for countries on the periphery. For example, the Napoleonic Wars of the early nineteenth century and the Second World War in the twentieth were openings for nationalist anticolonial movements worldwide, first in Latin America and later in Asia and Africa. During a

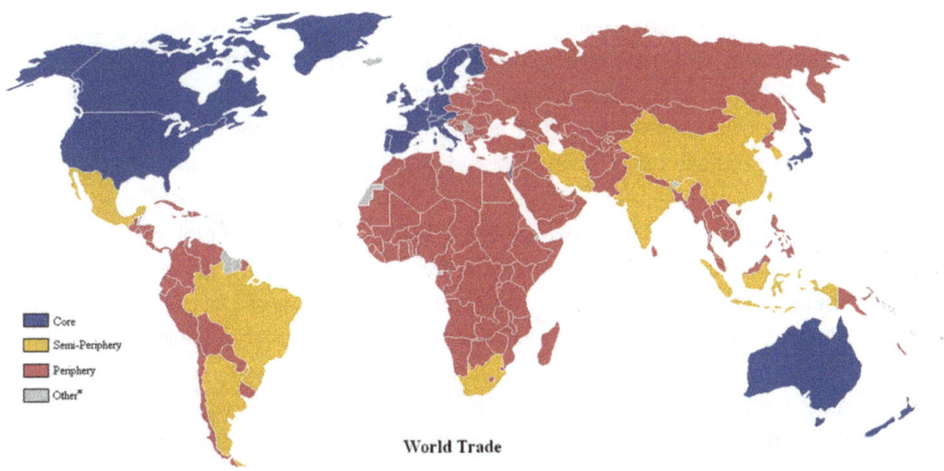

Figure 6.18 The core, semiperiphery, and periphery of world-systems theory.

hegemonic power's decline, rival states, which can focus on capital accumulation without the burden of maintaining the political and military apparatus of supremacy, catch up and challenge the hegemonic power. Thus, in the early twentieth century, Germany challenged Britain for global leadership, with catastrophically violent results. The growing Chinese challenge to the United States can also be seen in this light.

6.4d Neoliberalism

Neoliberalism is an enormously influential, contemporary global political and development ideology that has had dramatic effects on the world economy. It is called neoliberalism because it is based on eighteenth-century liberal economic theory (e.g., Adam Smith and David Ricardo), which idealized free markets and minimal government regulation.

The basic tenet of both liberal and neoliberal economic theory is the strong belief in free markets that should be allowed to self-regulate without state intervention. Anything that interferes with "free markets" is deemed counterproductive. For example, state measures such as minimum wages and unemployment benefits would be considered to be detrimental to the market because they would impede the labor market's ability to self-regulate. In this view, only markets create wealth; governments simply redistribute it. Markets are held to reward risk, innovation, and hard work. Government regulation is therefore seen as unnecessary and counterproductive. For this reason, neoliberalism is sometimes called "market fundamentalism." The neoliberal social theory and policies that accompany neoliberal economics advocate for limited or no social interventions such as state-funded welfare or health care. Neoliberalism is also characterized by an emphasis on individual initiative: people are seen as autonomous actors responsible for their own actions and well-being. Discursively, neoliberalism positions the poor as the cause of poverty, not its outcome. Thus, it is an updated form of economic thought that marries older traditions of political economy and individualism.

Neoliberalism as an economic and political theory started at the University of Chicago with the work of the conservative philosopher and economist Friedrich von Hayek and his student Milton Friedman. This ideology was enthusiastically adopted by the administrations of Ronald Reagan in the United States and Margaret Thatcher in the UK in the 1980s, both of which promoted deregulation, privatization, and free trade. Tax cuts were justified on the basis of "trickle down" economics, which asserted that reducing taxes on the wealthy would stimulate investment and economic growth. Reagan and Thatcher's policies marked the beginning of the end of the "welfare state" that had been entrenched in Europe and North America since the Great Depression of the 1930s. Both administrations initiated the deregulation of the finance, transportation, and telecommunications sectors. Both governments implemented a series of tax cuts, budget reductions, and cuts in government assistance programs for the socially disadvantaged that continue today.

From the United States and Britain, neoliberalism was exported to the global stage. In part, globalization made neoliberal social policies attractive because many people saw the welfare state, with its attendant protections and regulations, as intolerably expensive in light of heightened international competition. Neoliberalism was also justified by the collapse of communism in the 1990s, which offered "proof" that state-run economies were uncompetitive. Shortly thereafter, neoliberal policies were quickly adopted by China, India, Canada, Mexico, many European countries, and numerous countries in the developing world, all of which shifted from import substitution to export promotion, welcoming multinational firms and foreign investment.

Neoliberal policies were adopted and implemented by the International Monetary Fund and the World Bank. Because these organizations are headquartered in Washington, DC, neoliberalism is sometimes called the "Washington Consensus." Internationally, neoliberal economics have had tremendous effects across the developing world. Embodied in the IMF's Structural Adjustment Programs, neoliberal policy is used as a requirement for acquiring loans from both the World Bank and the IMF.

Many critics of neoliberalism argue that neoliberalism creates an ever-widening gap between the rich and the poor and increasingly high levels of global poverty. Tax cuts for the wealthy and corporations, for example, have done little to solve social problems. Cuts in social services around the world have done immeasurable harm to the lives of billions of people, especially the poor. Critics believe that unregulated markets will only make the rich richer and the poor poorer and that to create equality, government intervention is necessary. Neoliberalism is criticized for the manner in which it blames individuals for their economic problems and refuses to examine more systemic issues such as racism.

6.5 Regional Disparities Within Developing Countries

In addition to the bifurcation between the developed and underdeveloped worlds that forms the primary axis of the global economy, there is also regional inequality within developing countries. The geographies of colonialism, including profound rural-urban differences, are one major dimension of this pattern. Major cities of developing countries operate largely as export platforms linking the rich industrial countries and their sources of raw materials in the periphery. Modern large-scale firms tend to concentrate in capital and port cities. Investments of capital into these urban economies attract new migrants from rural areas to cities. International migration is a response to this uneven development: rural-to-urban migrants provide a limitless supply of cheap labor in cities for foreign firms. Transnational corporations and the governments that support them find little reason to change this situation.

Because economic landscapes are produced by social relations, spatial inequalities within developing countries are reflections of social inequalities. The class structure in much of the less developed world is typified by a small, powerful elite and large numbers of poor peasants, with a small middle class. Countries with highly unequal distributions of income, such as Brazil, tend to have highly unequal standards of living among their different regions. Those countries that have achieved more equality economically, such as the NICs of East Asia, tend to have fewer disparities spatially. In this way, economic landscapes mirror and contribute to social relations, revealing how geographies are both socially produced and socially producing.

For modernization theorists, these inequalities reflect poorly developed markets. The promise, but not the reality, is that deregulation and the free flow of capital and labor will eradicate spatial inequalities. For dependency and world-system advocates, uneven spatial development is an integral and inescapable part of capitalism, both among and within countries. They hold it is unlikely to disappear on its own.

The center-periphery within countries resembles that globally, except that in this case the center—usually one or more large metropolitan areas—appropriates the surplus generated by the periphery. Centers at this level, although considerably less powerful than the global core, still have sufficient strength to appropriate for themselves a fraction of the remaining surplus value created in factories, farms, and plantations. Even in remote peripheral areas, local, regional imbalances are likely to exist.

6.6 Development Strategies

Three approaches are often cited to help less developed countries develop economically: (1) expanded trade, (2) investments of private capital, and (3) **foreign aid**.

6.6a Expansion of Trade with Less Developed Countries

Economists agree that expanding trade with less developed countries is one way to help them. Reducing tariffs and trade barriers with less developed countries will improve their situation. Eliminating protectionism levied against developing country producers gives them access to the large, wealthy markets in North America, Europe, and Japan, increasing their export volumes and revenues. After the North American Free Trade Agreement (NAFTA), the United States removed all trade barriers with Mexico, for example. Mexican manufacturing flourished as a result. However, trade liberalization also opened the doors for US imports, particularly heavily subsidized agricultural products, which drove two million Mexican farmers into bankruptcy.

6.6b Private Capital Flows to Less Developed Countries

Less developed countries are also a destination for investments from multinational corporations, private banks, and large corporations. Some of these investments assume the form of foreign direct investment. For example, major US, Japanese, and European automakers have built numerous maquila assembly plants in Mexico. An international trade climate must also be supported by national financial and marketing systems, favorable tax rates, an adequately maintained infrastructure, and a reliable flow of sufficiently skilled labor. Often, however, less developed countries cannot guarantee a politically stable environment, which is difficult when they face tribal conflicts, ethnic strife, religious struggles, and civil wars. These struggles frighten investors and thwart the major capital flows that might otherwise exist. African states in particular have not been able to tap private capital flows from large corporations and commercial banks because of political instability, itself a reflection of the continent's colonial past.

6.6c Foreign Aid

In many cases, especially in emergencies such as droughts, wars, or hurricanes, foreign aid is needed in the form of direct grants, gifts, and public loans to less developed countries. Some aid is provided by charitable groups and human rights organizations. Other actors include governments. Capital is necessary to finance companies, build the infrastructure, generate jobs, increase productivity, and retrain the labor pool.

In absolute terms, the United States has been a major world player in foreign-aid programs. US foreign aid averages $15 billion per year, for example, less than 0.5 percent of GDP, which is the smallest share among all developed countries. Unfortunately, developed country contributions are much too small to make a meaningful difference to most of the developing world. However, for some countries in Africa, aid can form as much as 15 percent of their national output. In failed states such as Afghanistan, foreign aid comprises much more of GDP.

Unfortunately, most foreign aid has stipulations such as purchase requirements that make the less developed countries buy American products and services. Almost 75 percent of US foreign aid is military in nature, and the vast bulk of it goes to close allies such as Israel, Egypt, and Turkey on the basis of political ties as opposed to economic need. These countries are neither the most populous nor the neediest, but they occupy strategic areas of the Middle East where vast oil deposits exist and where the United States struggles against Islamic fundamentalism. In addition, the United States has guaranteed its support of Israel through extensive foreign aid programs.

6.6d Import-Substitution Industrialization

After the Second World War in 1945, many newly independent developing countries sought to break out of their domination by, and dependence on, developed countries. Their goal was to initiate capitalist development through a strategy of **import-substitution industrialization**. This development strategy involved the production of domestically manufactured goods to substitute for imports. Because only the middle classes could support a domestic market, this strategy focused on consumer durables. Argentina, for example, tried to create a domestic automobile industry. These "infant industries" were often subsidized by the state and protected by tariffs in order to reduce imports from developed countries, but local entrepreneurs had neither the capital nor the technology to initiate widespread domestic industrialization. Many development projects were crippled by a lack of capital and expertise. Many countries experienced an initial burst in manufacturing growth and a reduction in imports, but ultimately, their industries lacked the capacity to establish a niche in the world economy.

6.6e Export-Led Industrialization

By the 1960s, it was apparent to many leaders of less developed countries that the import-substitution strategy had failed. It never had the conceptual foundations, such as comparative advantage, that could lead to success. Only countries that had made an early transition to **export-led industrialization**, such as the Asian NICs, were able to sustain high rates of industrial growth. Now, export-oriented development was to be based on the production and export of manufactured goods, emphasizing foreign markets rather than domestic ones. The rise of export-led industrialization in the late twentieth century coincided with massive deindustrialization in Europe and North America. It was, in short, a reflection of the new international division of labor.

Export-oriented industrialization tends to concentrate in export-processing zones, usually cities, which waive foreign exchange controls and export duties, have an adequate infrastructure such as ports, roads, airports, and warehouses, and abundant, low-wage labor. The first export-processing zone was not established in the developing world, but in 1958 in Shannon, Ireland, with the local international airport at its core. In the 1960s, a number of countries in East Asia began to develop export-processing zones, the first being Taiwan's Kaohsiung, set up in 1965, a strategy soon imitated elsewhere. China established a series of them on its Pacific coast. By 2020, at least 220 zones were established in developing countries, mostly in the Caribbean, Central and Latin America, and East Asia.

Much of the employment in export-processing zones is in electronics and electrical assembly or in textiles. Young, unmarried women make up the largest part of the workforce in these industries—nearly 90 percent of zone employment in Sri Lanka, 85 percent in Malaysia and Taiwan, and 75 percent in the Philippines. In this sense, they represent the feminine side of globalization that has become prominent. Thus, the cheap labor so essential to the labor-intensive industries of the global assembly line also reproduces the patriarchal relations that keep many women in particularly low-paying jobs.

Of course, export-led industrialization also has its drawbacks. It makes export-oriented economies vulnerable to shifts in global demand and exchange rate fluctuations. Nonetheless, all of the world's most rapidly growing economies have adopted this model over the last four decades.

6.6f Sweatshops

Many manufacturing firms use sweatshops as a way to exploit cheap, unskilled labor. Guess jeans are made by underpaid Latino immigrants in Los Angeles, working in crowded conditions hidden from public view. Mattel makes tens of millions of toys each year in China using young female workers who have migrated hundreds of miles from home. Textile producers in Bangladesh run enormous factories filled with working women, who sometimes die in tragic accidents. Nike produces its shoes in tough labor conditions in Indonesia. Many of Disney's seasonal products are made by suppliers in Sri Lanka and Haiti, countries with unsavory reputations for labor rights. Soccer balls are sewn together by child laborers in Pakistan, who work up to twelve hours per day.

Sweatshops raise a variety of ethical and moral issues as well as economic ones. In a global economy, it is impossible for consumers to avoid products made under less-than-ideal labor conditions. Moreover, what may appear to be horrific working environments to most citizens in the world's richest nation are attractive to desperate, impoverished workers who live overseas.

6.7 The East Asian Economic Miracle

The world's most successful cases of economic development since the Second World War have been the countries of East Asia, where billions have been lifted out of poverty. Japan pioneered this model starting in the late nineteenth century. After the devastation of the Second World War, Japan's economy was in ruins. Today, Japan's economy is the third largest in the world (after the United States and China). In the 1960s and 1970s, Japan's lead was followed by a second generation of Asian "tigers," South Korea, Taiwan, Hong Kong, and Singapore. South Korea's economy is now the eighth largest in the world, ahead of Sweden, the Netherlands, and Australia. South Korea became the world's largest shipbuilding nation and the world's fifth-largest automobile manufacturer (e.g., Hyundai). Its iron and steel and chemical industries are thriving. South Korea is also a formidable competitor in semiconductors, biotechnology, telecommunications, and nuclear energy. In the 1980s and 1990s, Malaysia, Thailand, Indonesia, and China headed down the same path, starting with textiles and moving quickly into electronics.

There are several characteristics that these countries share that explain their rapid growth. First, East Asian countries exhibit an enormous commitment to education, which in part derives from the traditional Confucian respect for learning and scholarship. East Asian educational systems also often encourage a docile, well-trained workforce, particularly one consisting of easily exploited, low-wage female labor, and they have relatively few days lost to strikes and labor unrest.

Second, these countries have high levels of national savings, which provide capital for investment. Many countries in East Asia save one-third or more of their gross domestic product (GDP; in America, in contrast, the domestic savings rate is between zero and 3 percent of GDP). The result has been the accumulation of large quantities of low-interest capital that allowed them to finance education, infrastructure, manufacturing, and commerce.

Third, East Asia has enjoyed a stable political system within which economic growth is fostered. Governments actively supported rising industries, and at times acted brutally to keep wages low. Industries targeted for growth were given export subsidies, grants, and tariff protection from foreign competitors. Trade unions were restricted and democracy was constrained. Military governments in South Korea and Taiwan dealt harshly with industrial unrest and political dissidents. Authoritarian governments long ruled Singapore. Asian governments did not follow a laissez-faire model; it is a myth that the NICs demonstrate a "free market" in operation.

Fourth, the Asian NICs engaged in widespread land reform in the 1950s, which was made possible by the upheaval that followed the Japanese occupation of Korea and Taiwan (1895–1945) and the Korean War (1950–1953), which removed the rural aristocracies that owned much of the arable land. As a result, rural land ownership in these countries is relatively egalitarian, with high rates of productivity, unlike Latin America. Besides their thriving agricultural sectors, which have dwindled in the face of rapid industrialization, the NICs benefited from low rates of rural-to-urban migration, which kept urban areas from being flooded by impoverished peasants.

Fifth, East Asian NICs carefully cultivated an international niche and focused on exports (export-led industrialization), in contrast to the import-substitution policies that characterized India, Africa, and Latin America. Most import-substituting states depend on imported manufactured goods, whereas their exports chiefly consist of raw materials such as oil, coffee, and soybeans, a condition that creates poor terms of trade.

Finally, the NICs position in the world economy also played a large role in their success. For example, despite the brutality of war, NICs such as Korea, Taiwan, and parts of China benefited from the legacy of their occupation by Japan. Japan initiated the steel, chemicals, and textiles industries in Korea, Taiwan, and Manchuria. The Japanese also built much of the infrastructure in the NICs, including roads, bridges, tunnels, ports, and airports. Japan centralized the government bureaucracies of these countries, which displaced the rural aristocracies that hindered growth. After the Second World War, many NICs, particularly South Korea, established imitations of the Japanese corporate zaibatsu (e.g., the Korean *chaebol*), which were closely linked to banks and obtained easy credit.

In addition to Japan, the United States supported the Asian NICs. Throughout the Cold War (1945–1991), the United States provided generous economic and military aid, freeing resources that could be harnessed for economic development. The United States gave copious grants and subsidies to the NICs and awarded them preferential trade status, such as exemptions from tariffs, which allowed them easy access to the American market. This combination of factors—Japanese occupation and American largesse—indicates that the geopolitical conditions that enabled the take-off of the NICs were a unique constellation of circumstances that could not be duplicated elsewhere.

In the 1990s, the Four Tigers of South Korea, Taiwan, Hong Kong, and Singapore were joined by a new group, including Thailand, Malaysia, and Indonesia (Figure 6.19). Thailand, which long enjoyed close economic and military ties to the United States, saw foreign investments in textiles, automobiles, toys, and tourism, making Bangkok a prosperous but crowded urban center. Malaysia, under an authoritarian government, launched its "Plan 2020" to become an industrialized country; it is now the world's largest exporter of refrigerators and semiconductors and has close ties to banks and firms in Singapore as well. Kuala Lumpur has become a thoroughly modern city. Indonesia, long trapped in poverty, has enjoyed investments in textiles, shoes, electronics, and even aerospace; it also has petroleum revenues.

Finally, there is the case of China. Following decades of isolation under the communists after the founding of the People's Republic in 1949, China opened itself to the world economy in the 1980s. The new policies aimed at creating growth revolutionized the economy and society of the world's most populous nation, encouraging the growth of markets. China enthusiastically welcomed

Figure 6.19 Map of the Association of Southeast Asian Countries, which form the third generation of Asian "tiger" economies.

foreign investments, making the country by far the largest recipient of foreign direct investment in the developing world. The government targeted coastal areas in particular, designating them Special Economic Zones in which investors were given generous subsidies and tax breaks. These regions include Guangdong province, located near and benefiting greatly from its ties to Hong Kong; growth there created new cities such as Shenzhen. Similarly, Fujian province, across the straits from Taiwan, grew rapidly. Shanghai is China's largest city and business hub, and includes Pudong, the financial center. China became a dominant producer in electronics, toys, garments, and other types of light industry. China's economy has grown roughly 8 percent annually for the last three decades, one of the highest rates in the world, and it enjoys large trade surpluses with most of the world. Today, China is the world's second-largest economy, after the United States. This growth has been geographically uneven (i.e., creating uneven spatial development): coastal areas have prospered even as inland ones have not. In response, millions of peasants moved from less densely inhabited, poorer regions in the interior toward the prosperous coasts.

Although India, the world's second most populous country (with 1.4 billion people), lags far behind China in its level of economic development, India too has seen its pace of economic growth recently increase. Much of this growth is located on the western half of the Indian peninsula, including financial centers such as Mumbai (formerly Bombay), one of the world's largest cities, and the famous software complex in Bangalore, India's Silicon Valley. Indeed, India today is the world's largest producer of software, as well as films.

6.8 Sustainable Development

Economic and population growth, resource depletion, environmental degradation, and climate change are the core concerns that have generated growing interest in sustainable development.

Fundamentally, the capitalist world economy is at odds with the sustainability of the Earth's carrying capacity. The majority of the world's commercial fish species are in decline, and given rising temperatures due to climate change, half of the world's people will experience water shortages in the next decades. The ideology of continuous growth has stressed ecosystems, degraded forests, eroded soils, and undermined the planet's capacity to sustain life. In short, unsustainable development has reached the limits of growth.

Sustainable development can be defined as development that meets the needs of the present moment without reducing the ability of future generations to meet their own needs. It emphasizes the crucial need to reduce poverty and acknowledge the world's environmental and technological limits. If the irreversible damage to the world's ecosystems and oceans is to be halted—and it may be too late—profound transformations in the world are needed sooner rather than later. The implication of this view is that sustainable development efforts require a dramatic restructuring of societal relations.

Critics of capitalism have long argued that the present model of continuous growth is unsustainable. Since the costs of global production and trade are not confined within national boundaries but spill over to affect other parts of the world (for example, with greenhouse gas emissions), and since these effects are often more severe in developing countries than in the developed world, environmentalists argue that sustainable development must promote fair trade and fair share principles. Environmental preservation and social justice thus go hand in hand. If capitalist growth is about the limitless transformation of resources into goods, sustainable development is about improving livelihoods and allowing economic activity only insofar as it does not exceed the capacity of the ecosystem. Thus, sustainable development emphasizes biophysical constraints such as resource availability and carrying capacity. It is abundantly clear that the world's poor will bear the brunt of the damage to the world's ecological systems and will carry most of the costs associated with it, while they are hardly to blame or in a position to make any changes themselves. Sustainable development invites us to consume less, reduce energy usage, limit carbon emissions, recycle, and promote environmentally friendly policies. For some people, this will mean a lower standard of living. But as Mahatma Gandhi once said, "The world has enough for everyone's need, but not enough for everyone's greed." Making such changes is not, and will not be, easy, and many people refuse to accept them. But the stakes are high, including the future of life on the planet.

Ecological modernization is the view that it is possible to maintain economic growth *and* protect the environment, and that environmental degradation can be diminished by policies and technological changes that build environmental criteria into economic systems. Its appeal lies in a capacity to create solutions to problems conceived as zero-sum and to implement structural changes once seen as impossible.

6.9 Summary

This chapter began by noting that the term *development* is often vague and problematic, and that there are many different ways to measure it. While GDP per capita is by far the most common measure, other indices use aspects of health and standard of living.

Second, it noted that urbanization has transformed the landscapes and societies of the developing world. The old notion that industrialized countries are urbanized and developing ones are not is rapidly becoming obsolete. Most urban growth is in the developing world, and the ongoing crisis of rural areas drives waves of migrants to cities.

The third part examined a long list of development problems, including high population growth, unemployment, lack of human and investment capital, low productivity, unequally distributed land, poor terms of trade, foreign debt, and corrupt and uncaring governments.

Fourth, it summarized four major conceptual approaches to the question of development, including modernization theory, dependency theory, world-systems theory, and neoliberalism. These approaches offer starkly different policy prescriptions for addressing poverty. Fifth, the chapter noted that regional inequalities unfold not only among countries but also within them. Often this takes the form of urban-rural schisms.

The sixth section examined development strategies such as foreign aid and means of attracting private and public investment capital. It also addressed import substitution and export promotion. While the former has been largely abandoned, the latter is enthusiastically adopted everywhere. East Asia has witnessed remarkable economic growth in the last half-century. The seventh section explored the reasons behind this success. It may be that the region enjoyed a unique combination of circumstances not easily replicable in other parts of the world.

Finally, the chapter took up the matter of sustainable development. Unbridled growth is killing the planet and causing enormous environmental damage. Sustainable development builds the ecological constraints into its analysis of growth, noting the need for major lifestyle changes. As the limitations of the Earth's ecosystems have become more obvious, it has become evident that global capitalism cannot carry on unchanged into the indefinite future. Rather than a growth-at-all-costs strategy, sustainable development promotes ways of sustaining the lifestyles of the present without jeopardizing the opportunities of future generations.

Glossary

Capital flight: the movement of capital out of a country when it suffers economic or political meltdown

Core-periphery: the division of the world into a small group of wealthy countries and a large number of poor ones

Dependency theory: the view of the world economy that the wealth of the developed world is extracted from poor countries

Development: the process of social and economic growth that raises standards of living

Dual economy: the split between the formal and informal sectors

Export-led industrialization: a development strategy that emphasizes maximization of exports and foreign revenues

Foreign aid: the loans and grants that a donor agency or country provides to people in need in another state

Foreign direct investment (FDI): capital invested in buildings, offices, machinery, and equipment located in one state by firms located in a different state

Formal economic sector: the part of the economy in which jobs have regular working hours, relatively high wages, and benefits such as paid vacations

Import-substitution industrialization: a strategy of limiting imports using protectionism to promote domestic suppliers in an industry

Informal economic sector: the part of the economy where jobs are scarce, low-paying, and working hours and opportunities vary daily

Less developed countries: the world's poorer countries, almost all of which were colonies, in Latin America, Africa, and most of Asia excluding Japan, South Korea, and Singapore

Modernization theory: the view of economic development that holds poor countries should follow the historical trajectory of the developed ones

Neocolonialism: the notion that formal colonialism may have ended with independence, but that de facto colonial economic ties are still in operation

Newly industrializing countries: nations that have successfully industrialized and experienced rapid economic growth

Slums or squatter settlements: poor, informal neighborhoods in many cities in the developing world in which the inhabitants do not own the land they occupy

Structural adjustment: the process by which the International Monetary Fund forces debtor countries to adopt measures to make it easier for them to repay their debt, such as currency devaluations and reductions in government subsidies

Sustainable development: development designed to minimize long-term impacts on the environment, including renewable energy and recycling

Terms of trade: the relative value of a country's exports and imports

Underdevelopment: the process by which foreign powers made a country poor through colonial or neocolonial exploitation

Underemployment: labor markets in which workers cannot find sufficient full-time work or jobs that utilize their skills

World-systems theory: a view of the world economy that divides countries into a global core, periphery, and semiperiphery

Study Questions

1. What does GDP per capita mean?
2. What are five common characteristics of less developed countries?
3. What were the origins of the world debt crises?
4. What are the IMF's structural adjustment policies?
5. What are four ways to measure economic development?
6. Why do many developing countries suffer from capital flight?
7. What is the informal economy?
8. How does corruption impede economic development?
9. Why are cities in the developing world growing so quickly?
10. Contrast modernization, dependency, and world systems theories of development.
11. What is neoliberalism, and what impacts does it have?
12. How does development relate to regional disparities within countries?
13. How can the First World meaningfully assist the Third?
14. What is sustainable development, and why is it important?

Suggested Readings

Baker, A. 2021. *Shaping the Developing World: The West, the South, and the Natural World*. CQ Press.

Banerjee, A., P. Niehaus, and T. Suri. 2019. Universal basic income in the developing world. *Annual Review of Economics* 11(1):959–83.

Davis, M. 2007. *Planet of Slums*. Verso.

England, K., and K. Ward (eds.) 2007. *Neoliberalism: States, Networks, Peoples*. Blackwell.

Handelmann, H. 2019. *Challenges of the Developing World*. Rowman and Littlefield.

Harvey, D. 2005. *A Brief History of Neoliberalism*. Oxford University Press.

Horner, R. 2020. Towards a new paradigm of global development? Beyond the limits of international development. *Progress in Human Geography* 44(3):415–36.

Kohli, A. 2021. *Imperialism and the Developing World: How Britain and the United States Shaped the Global Periphery*. Oxford University Press.

MacKinnon, D., S. Dawley, A. Pike, and A. Cumbers. 2019. Rethinking path creation: A geographical political economy approach. *Economic Geography* 95(2):113–35.

Momsen, J. 2019. *Gender and Development*. Routledge.

Peet, R. 2009. *Unholy Trinity: The IMF, World Bank and WTO*. Bloomsbury Publishing.

Peet, R. 2024. *Global Capitalism: Theories of Societal Development*. Routledge.

Potter, R., T. Binns, J. Elliott, E. Nel, and D. Smith. 2017. *Geographies of Development: An Introduction to Development Studies*, 4th ed. Routledge.

Ruzzante, S., R. Labarta, and A. Bilton. 2021. Adoption of agricultural technology in the developing world: A meta-analysis of the empirical literature. *World Development* 146:p.105599.

Selod, H., and F. Shilpi. 2021. Rural-urban migration in developing countries: Lessons from the literature. *Regional Science and Urban Economics* 91:103713.

Trionfetti, F. 2015. Public debt and economic geography. *International Regional Science Review* 38(1):92–113.

Zhou, Y., and Y. Liu. 2022. The geography of poverty: Review and research prospects. *Journal of Rural Studies* 93:408–16.

7 Population

Human beings are what drive the world economy. People are not only the most important factor of production, but their welfare is also the reason for economic growth. People produce and consume goods and services. Thus, population is a major dimension of economic geography and is closely related to issues such as poverty, hunger, unemployment, and sustainability. This chapter examines the world's population distributions, characteristics, and trends. It also reviews competing theories on the causes and consequences of population growth.

7.1 Global Population Distribution

The existence of large numbers of human beings is a relatively recent historical phenomenon (Figure 7.1). For the overwhelming majority of human existence—upward of 95 percent—people lived as hunters and gatherers, collecting food from various wild plants and stalking animals. This mode of production supports only low population densities. Because births equaled deaths, there was virtually no increase in the number of people in the world for thousands of years. Following the Neolithic Revolution of roughly 9000 BCE when agriculture was discovered or invented, the capacity of many societies to support more people increased considerably. However, the really large gains in population only began with the Industrial Revolution of the eighteenth and nineteenth centuries, when the mechanization of agriculture freed millions of people from lives of rural drudgery and allowed for large, dense cities. Thus, exponential population growth is a product of the modern world.

In 2024, the world's population surpassed 8.1 billion. Humans are very unevenly distributed geographically across the globe (Figure 7.2). Most people are concentrated in a few parts of the world, particularly along coastal areas and the floodplains of major river systems. Four major areas of settlement are East Asia, South Asia, Europe, and eastern North America. There are secondary clusters in Southeast Asia, Africa, Latin America, and along the western coast of North America. Six of the top ten most populous countries—China, India, Indonesia, Japan, Bangladesh, and Pakistan—are in East, South, or Southeast Asia. Europe (including Russia) had 750 million residents, about 12 percent; other continents included Africa (1.5 billion, or 14 percent), Latin America (800 million or 8.5 percent), North America (370 million, or 5 percent), and Oceania (40 million, less than 1 percent). The populations of the developing world—Africa, Asia (excluding Japan), and Latin America—accounted for three out of every four people on the planet.

Ten out of the world's nearly two hundred countries account for two-thirds of the world's population (Table 7.1). Five countries—India, China, the United States, Indonesia, and Brazil—contain half of the world's population. With 1.4 billion people, India is the world's most populous country, and growing. China, traditionally the largest, has 1.4 billion, and is second, but due

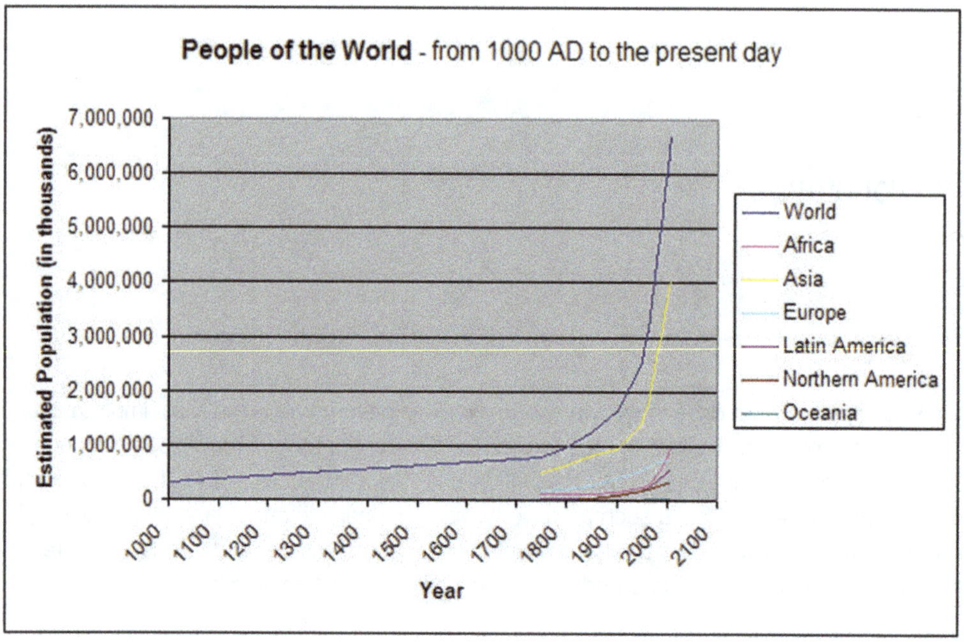

Source of data: Wikipedia http://en.wikipedia.org/wiki/World_population#Population_figures
(BS - Jan 2009)

Figure 7.1 World population growth over time.

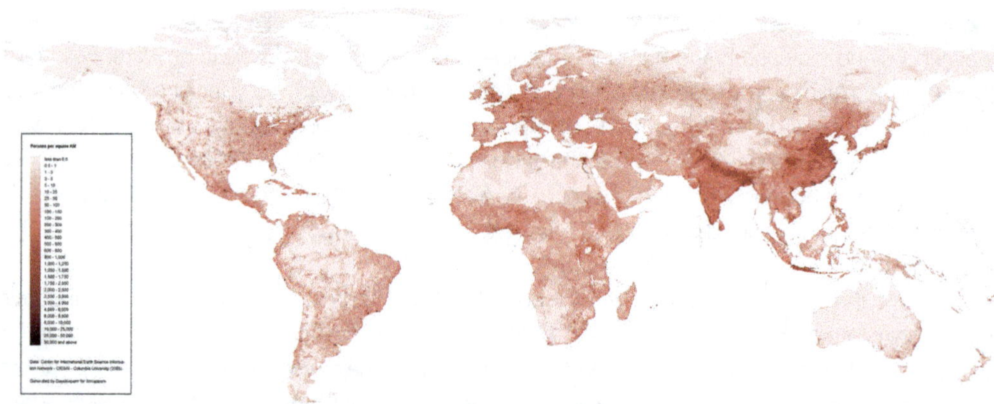

Figure 7.2 Distribution of the world's population.

to low **birth rates**, is losing one million people per year. The United States, with 340 million in 2024, is the third largest and is by far the most populous of the developed nations. Indonesia, the world's largest Muslim nation, is fourth, with 270 million. Other nations with large populations include Brazil (220 million), Pakistan (230 million), Nigeria (200 million), Bangladesh (165 million), Russia (140 million, but declining), Japan (124 million, but declining), and Mexico (120 million). By way of comparison, other countries with significant populations include the Philippines (100 million), Ethiopia (100 million), Vietnam (100 million), Germany (85 million), Egypt (90 million), the Democratic Republic of the Congo and Thailand (80 million each), the UK and France (62 million

Table 7.1 **The World's Ten Most Populous Countries, 2024**

Country	Population 2024 (millions)	% Annual Growth Rate	Estimated Population 2050 (millions)
India	1,430	1.4	1,755
China	1,420	0.6	1,437
United States	335	0.6	438
Indonesia	270	1.4	343
Pakistan	173	2.0	295
Brazil	211	1.0	260
Nigeria	201	2.4	282
Bangladesh	163	2.1	215
Russia	144	-0.4	119
Mexico	127	-0.1	150

Source: World Population Data Sheet

each), and Italy (60 million). Only two of the ten most populous nations are economically developed (the United States and Japan).

7.1a Population Density

Because countries vary so greatly in area, national populations do not reveal the conditions of crowding. The ratio of population to land area is population density—the average number of people per unit area. Several countries with the largest populations have relatively low population densities. For example, the United States is the fourth most populous country, but in 2020, it had a population density of only eighty-four people per square mile. Although they have large, dense metropolitan areas, the United States and Canada form one of the more sparsely populated areas of the world.

Excluding countries with a very small area (such as Singapore), Bangladesh is the world's most crowded nation, where more than 165 million people are crowded into an area the size of Iowa. Three of the ten most densely populated countries—the Netherlands, Japan, and Belgium—are economically developed, whereas another three—South Korea, Taiwan, and Israel—are newly industrializing countries (NICs). These facts remind us that there is no simple relationship between population density and economic development.

7.1b Factors Influencing the World's Population Distribution

The uneven distribution of the world's people is explained by several factors. One of these is the biophysical environment. As noted in Chapter 1, economic geographies are created through the interactions of humans and "nature." Most of the world's people live near the coasts of continents, in river valleys, at low elevations, and in midlatitude and subtropical climates. Areas that are not conducive to agriculture without widespread irrigation, such as the Sahara Desert, are sparsely settled. Few people live in very cold regions, such as northern Canada, arctic Russia, and northern Scandinavia, where growing seasons are short. Many mountainous areas are also low-density habitats. However, it is simplistic to explain the planet's population distribution by looking at the natural environment alone. Clearly, issues such as precipitation and average temperatures are important. But humans are adaptable and can live in many different climates. They use technologies to make hostile environments inhabitable, such as air conditioning, heating, water storage, and irrigation.

As important as the physical environment, if not more so, are the roles of economic and political systems. The geographies of economic activity, such as resource extraction, labor markets, trade, and infrastructures create large, dense populations throughout the world. Population sizes and distributions are influenced by the demographic drivers of fertility, mortality, and **migration**. Policies such as tax policies, zoning, and planning ordinances also shape the population map. All of these forces must be understood through reference to historical context. The current population distribution is explicable only in terms of the past. The location of the world's peoples reflects forces operating at the global, national, and local scales for centuries or longer. For example, the high population densities of Europe or the northeastern United States reflect the earlier geographies of the Industrial Revolution and waves of urbanization. China's large population developed over long periods of growth that depended on fertile agriculture, irrigation, and a social system stretched over vast areas. The distribution of population in the developing world is in part the product of centuries of colonialism, which concentrated growth on coastal areas and gave rise to port cities that were the centers of maritime trade.

7.2 Population Growth over Time and Space

The geography of the world's population is in perpetual change. The world's population is increasing, although at a decreasing rate. Each year, an additional 120 million people are added to the Earth's population, about the population of Mexico, which means the planet adds 300,000 people daily, about three per second. However, many countries, such as in Europe, Russia, and Japan, are losing people as their deaths exceed births. The vast majority of population growth occurs in developing countries, in which more than three-fourths of the human race lives.

7.2a Population Change

The current rapid growth rate of the world population is an historically recent event. It took until 1800 for the world population to reach one billion. The second billion was added 123 years (1850–1927) later, the third billion was added in 33 years (1928–1960), the fourth in 14 years (1961–1975), the fifth in only 12 years (1976–1987), the sixth in 11 years (1988–1999), and the seventh in only 10 years (2000–2020).

One way of measuring population growth is **doubling times**, the number of years that it takes a population to double in size at a particular rate of growth. The doubling time for a population (or savings account) can be estimated by using the rule of 70, which means the number 70 divided by the annual rate of growth. For example, at 1.2 percent, the current rate of world population increase, the doubling time is 58 years, which means that in the year 2080, the world may have 15.8 billion people *if* current rates of growth continue unchanged. With an annual increase of 0.6 percent per year, the doubling time for the US population is 116 years (meaning, if the current rate continues unchanged, the US will have 660 million people in 2122). Malawi, with a population growth rate of 3.2 percent per year, the highest in the world, will see its population double in only 22 years.

Births and deaths are two of the three basic population change processes; the third is migration. In short,

pop. change = births – deaths + in-migration – out-migration, or

$$\Delta P = BR - DR + I - O$$

where ΔP is the rate of population change, *BR* is the birth rate, *DR* is the death rate, *I* is the total in-migration rate (or immigration, if internationally), and *O* is the total out-migration rate (or emigration, if internationally).

The rate of natural increase for a country or a region is the difference between the birth rate and the death rate. The crude birth rate is the number of babies born per 1,000 people per year, and the crude death rate is the number of deaths per 1000 per year. Natural growth (*NGR*), which is the largest component of population change in most societies, is defined as the difference between the birth and death rates:

$$NGR = BR - DR.$$

Similarly, the **net migration rate** (NMR) is the difference between in-migration and out-migration rates:

$$NMR = I - O$$

Thus,

$$\Delta P = NGR + NMR,$$

where *NGR* is the **natural growth rate** and *NMR* is the net migration rate.

For the world as a whole, net migration is obviously zero. However, for any unit smaller than the globe, both natural growth and net migration rates must be included. The US birth rate in 2020 was 12 births per 1,000 people, and the death rate was 8 per 1,000; thus, the natural growth rate was 12 minus 8, or 8 per 1,000, the lowest ever documented. In absolute terms, these rates translated into 3.96 million births and 3.36 million deaths annually.

7.3 Malthusian Theory

The first and most influential social scientist to study population growth was the British reverend Thomas Robert Malthus. Malthus's ideas were formulated during the early days of the Industrial Revolution in the late eighteenth century and had an enormous impact on scholarly and popular understanding of this topic. In his famous 1798 book, *Essay on the Principle of Population Growth*. Malthus laid out his argument that population growth was responsible for the poverty that he witnessed in industrial cities. Thus, Malthus is the inventor of the theory of overpopulation. His dark and pessimistic worldview led economics to be called the "dismal science," and it stood in sharp contrast to the utopian socialism that radiated out of France following the French Revolution of 1789.

Malthus's view began with the idea that human populations, like those of other animal species, grow exponentially (or in the terminology of his times, geometrically). A geometric series of numbers increases at an increasing rate over time. For example, in the sequence 1, 2, 4, 8, 16, 32, ..., the number doubles at each time period, and the increase rises from 1 to 2 to 4 to 8 to 16, and so forth. In the absence of constraints, exponential population growth is commonly observed in bacteria and rodents, among many other species. Note the assumption regarding fertility embedded in Malthus's analysis: he viewed fertility as a biological, not social, phenomenon.

This line of thought reflected the large size of British families at the time and the high rates of fertility during the Industrial Revolution. In short, in Malthus's view, humans, like animals, always reproduce at the biologically highest rate; they were viewed as prisoners of their genetic urges to reproduce. Malthus's argument also carried with it a strong moral dimension, blaming the poor in particular for having large families.

The second part of Malthus's view concerned the growth of food supplies, or resources more generally. He held that food always grew at a much slower rate than did population. In particular, he argued that the food supply grew linearly (or arithmetically). In contrast with a geometric or exponential series, an arithmetic series of numbers grows at a constant rate over time. For example, in the sequence 1, 2, 3, 4, 5, …, the difference between numbers is always the same. Malthus's view that agricultural outputs increased linearly over time reflected the preindustrial agricultural systems of his time. In systems without economies of scale (Chapter 3), increases in output are accomplished only with proportional increases in inputs such as labor and capital. Thus, to double output, farmers had to double labor inputs or land. However, this view of agriculture is actually rather optimistic in Malthus's view. He argued that in the face of limited resources of land and capital, agricultural output would encounter *diminishing marginal returns*. This means that output does not increase as rapidly as inputs. For example, as farmers moved into areas that were only marginally hospitable for agriculture, perhaps because they are too wet, too dry, too cold, or too steep, they would need much larger increases in inputs to achieve larger increases in output (Figure 7.3).

When one graphs the exponential growth of population against the linear growth of food (Figure 7.4), it becomes evident that eventually population must outstrip the food supply, resulting in catastrophe such as famine. Thus, in the Malthusian view, populations will always and inevitably exceed their resources, condemning people to suffering and misery.

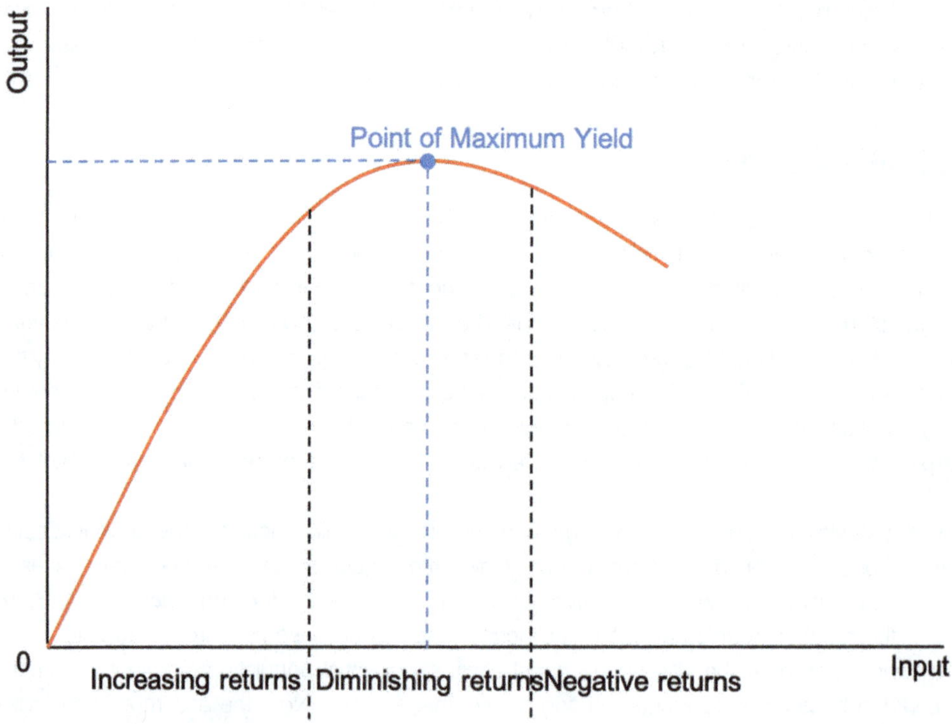

Figure 7.3 Diminishing returns.

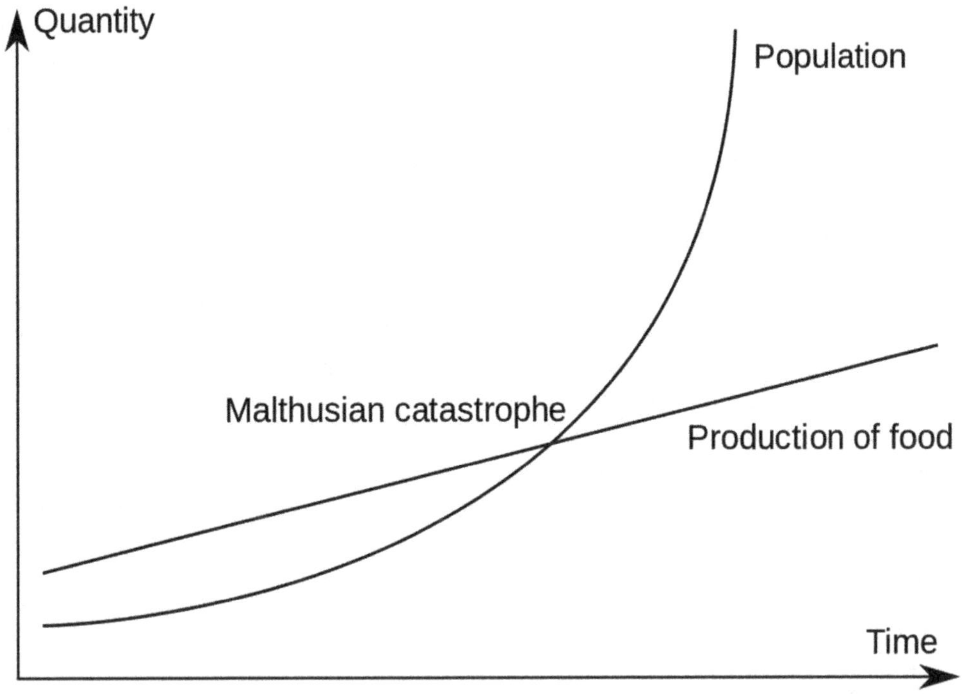

Figure 7.4 Malthus's theory of impending catastrophe.

Malthus blamed many of the world's problems on population growth, and subsequent generations of social scientists and policy makers who were influenced by his ideas have invoked overpopulation to explain everything from famine to deviant social behavior to environmental destruction. Malthus entered into a famous debate with his friend and fellow economist David Ricardo as to whether the British government should subsidize food for the poor. Malthus maintained that such subsidies only encouraged poor people to have more children and thus, they increased poverty in the long run. Indeed, he was explicit in his contempt for the poor, blaming them for their poverty, and even advocating mass death to keep their numbers in check:

> Instead of recommending cleanliness to the poor, we should encourage contrary habits. In our towns we should make the streets narrower, crowd more people into the houses, and court the return of the plague.... The necessary mortality must come.

Malthus added qualifiers to his argument by including checks or factors that limit population growth. Recall that natural population growth is the difference between fertility and **mortality rates**. Malthus's "preventative checks" are factors that reduce the fertility rate. The most obvious preventative check is contraception, which Malthus objected to on religious grounds; instead, he advocated for "moral restraint" or abstinence. Other preventative checks include delayed marriage and prolonged breastfeeding of babies, which inhibits pregnancy. If preventative checks fail, as he predicted, population growth would be curbed by "positive checks" that increased the mortality rate, particularly death, disease, famine, and war.

Malthus's ideas became widely popular in the nineteenth century and were incorporated into the social Darwinism of the time, which viewed social relations as naturally given, not made. Social Darwinism naturalized competition and inequality, holding that the wealthy were simply smarter and harder working than everyone else. However, by the end of the nineteenth century, it became increasingly clear that his predictions of widespread famine were simply wrong. The nineteenth century saw the quantity and quality of available food improve steadily, prices decline, and famine and malnutrition virtually disappear from Europe (except for the Irish potato famine of the 1840s). By the twentieth century, **Malthusianism** was in ill repute.

Malthus's predictions did not materialize due to three major errors that he made. First, he did not and could not foresee the enormous impacts of the Industrial Revolution on agriculture; the mechanization of food production simply rendered the assumption of linear increases untenable. Indeed, the world's supply of food has consistently outpaced population growth for two centuries because productivity growth in agriculture has been higher than the rate of increase in the number of people. This means that there is plenty of food to feed everyone in the world and that hunger is not simply caused by overpopulation but by a variety of other factors, including war, drought, colonialism, and corrupt governments. Second, Malthus did not foresee how the colonial world economy would open up the midlatitude grasslands in much of the world, particularly in North America, Argentina, and Australia, which increased the world's wheat production and facilitated the rise of a global agricultural market. Third, and perhaps most important, Malthus's analysis of fertility was simplistic. During the Industrial Revolution, fertility rates declined, and family sizes decreased, reducing the demand for food. Thus, contrary to his view, humans are not mere prisoners of their genes, and the birth rate is a socially constructed phenomenon, not a biological destiny.

7.3a Neo-Malthusianism

In the mid-twentieth century, when the world saw population growth rates of 2.6 percent annually, Malthusianism underwent a revival called **neo-Malthusianism**. Neo-Malthusians acknowledged Malthus's errors and the failure of his predictions, but argued that while his prognosis was wrong in the short run, it was correct in the long run. In the context of the growing environmental movement of the times, neo-Malthusians also injected an ecological dimension to Malthus's argument, adding environmental destruction as a casualty of population growth. The most famous expression of neo-Malthusian thought was the **Club of Rome**, an international organization of business leaders, policy makers, academics, and others concerned about the future of the planet. The Club of Rome funded a very famous analysis of the planet's future, *The Limits to Growth* (1972), which modeled the earth's population growth, economic expansion, energy and resource consumption, and associated environmental impacts. It concluded that the rapid population and economic growth rates of the post–Second World War era could not be sustained indefinitely, and that there would ensue worldwide economic, environmental, and demographic crises (Figure 7.5). Much of this argument centered on the exhaustion of nonrenewable resources and ecological catastrophe. Unlike Malthus's objections to birth control, neo-Malthusians advocated reducing population growth through birth control policies. They had an important effect on international programs promoting contraceptives and family planning, such as the Peace Corps and the Agency for International Development.

Neo-Malthusianism enjoyed a popularity that the original Malthusian argument did not. However, this view too suffered from a simplistic understanding of how resources are produced. For example, it ignored the effects of rising prices on resource producers, who seek out new supplies. Family planning programs in the developing world often fail for the simple reason that advocating contraception to reduce population growth ignores the reasons why people in poor countries have many children. Due to these failures, alternatives to Malthusianism began to rise, such as the demographic transition.

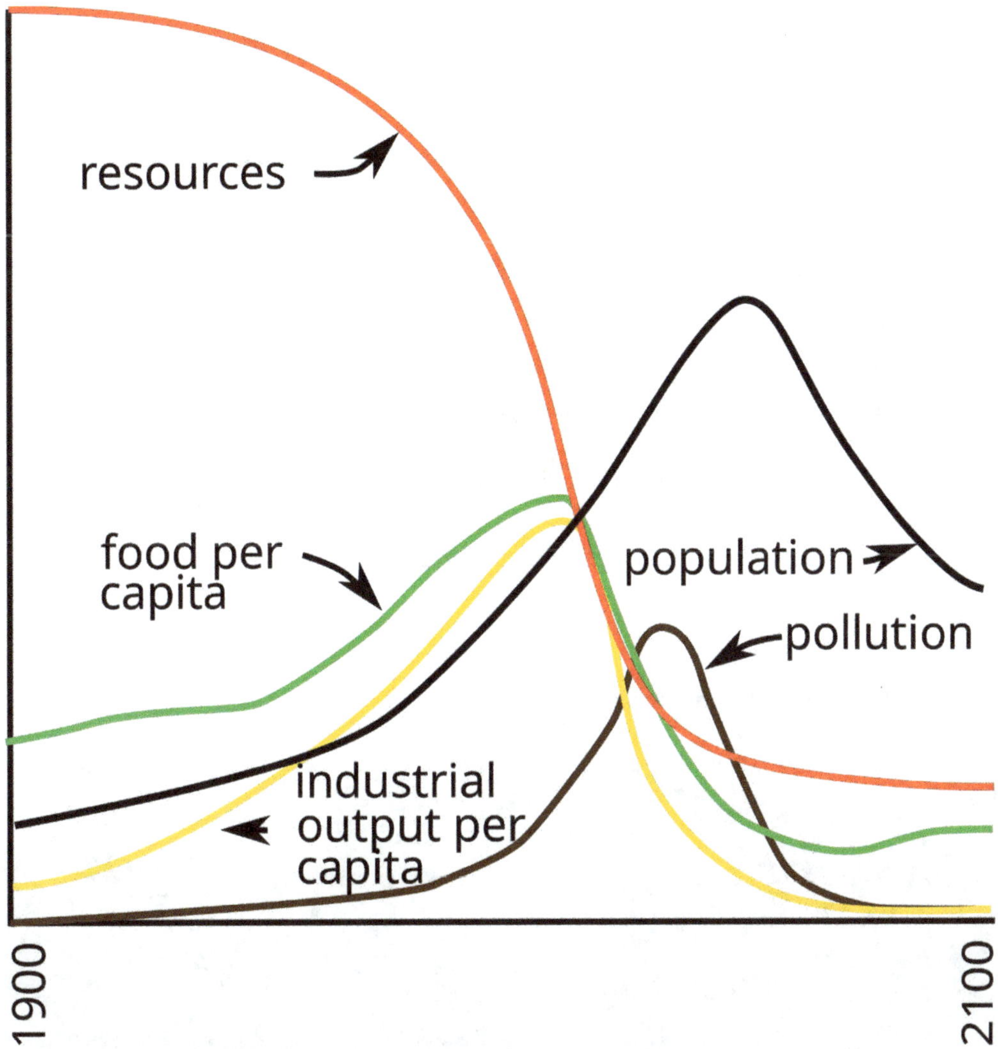

Figure 7.5 A graph illustrating the Club of Rome's predictions of the world's population, food supply, output, and pollution levels in the twentieth and twenty-first centuries.

7.4 Demographic Transition

Developed by American demographers, the demographic transition theory stands as an important alternative to Malthusian notions of population growth. Essentially, this models a society's birth, death, and natural population growth rates over time. Because it is based on the historical experience of Western Europe and North America as they went through the Industrial Revolution, "time" in this model is a proxy for industrialization, urbanization, and rising wealth. Thus, the demographic transition examines how and why birth, death, and growth rates change as a society moves from a rural, impoverished, and traditional state into a wealthier, urbanized, and modern one. This approach is illustrated with a graph of birth, death, and natural growth rates over time that divides societies into four major stages (Figure 7.6).

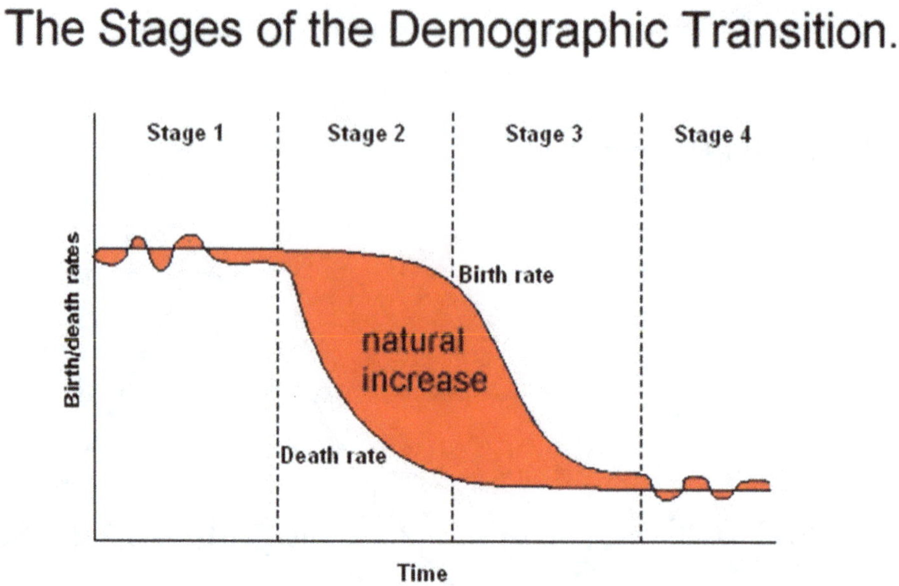

Figure 7.6 The demographic transition.

Figure 7.7 Girls working a field in Africa illustrate agricultural economies' reliance on child labor.

7.4a Stage 1: Preindustrial Society

The first stage represents a traditional, rural, preindustrial society in which fertility is high; families are large and children are numerous. This is exemplified by countries such as Rwanda with large families (7.5 children per mother).

Fertility rates in preindustrial societies have been high for several reasons. In agrarian economies, children are a vital source of farm labor, helping to plant and sow crops, tending to farm animals, performing chores, carrying water, and helping with younger siblings. Even children as young as four can generate more than they consume. For these reasons, scenes of children working in fields are common in the developing world (Figure 7.7).

Families in preindustrial societies are extended ones, with several generations, aunts, uncles, and cousins living in close proximity. In addition, children play an important role in taking care of their elderly parents; in the absence of government programs such as Social Security, the elderly depend on their children for assistance. Finally, in societies with high **infant mortality rates**, where a significant number of babies do not reach their first birthday, having many children is a form of insurance that some will survive until adulthood. In short, there are very rational reasons why poor societies have high fertility rates. In contrast to Malthusianism, which explains high birth rates as given, not made, the demographic transition portrays high fertility as a rational strategic response to poverty.

The relationship between poverty and fertility is illustrated in a map of crude birth rates around the world (Figure 7.8), which reveals that the poorest societies have the highest fertility rates, particularly in Africa. In contrast, birth rates in North America, Europe, Russia, Japan, Australia, and New Zealand are low, and families are relatively small.

However, in preindustrial societies, mortality rates are also high, which means that life expectancy is low. Traditionally, in human existence, few people lived to be more than 40, on average. The primary causes of mortality in poor, rural societies are high infant mortality, inadequate diets, and unsafe drinking water. These contribute to the spread of bacterial diseases. The most common are diarrheal ones, such as cholera and dysentery, which cause dehydration and deplete the body of electrolytes, as well as respiratory ones that cause pneumonia. Many diseases affect poor countries, such as dengue fever, yellow fever, schistosomiasis, influenza, bilharzia, malaria, tuberculosis, plague, smallpox, and measles. Table 7.2 lists the most dangerous infectious diseases in the world in 2024, including respiratory infections brought on by pneumonia and influenza; AIDS, which takes 2.1 million lives annually (especially in Africa); diarrheal diseases; and tuberculosis, which kills 1.7 million people every year.

The global distribution of crude death rates (Figure 7.9) closely mirrors the wealth or poverty of societies. Factors that shape death rates include the amount, quality, and consistency of food supplies; access to health care and antibiotics; the public health infrastructure; care for expectant mothers and babies; and smoking rates. Countries with the highest death rates—and thus lowest

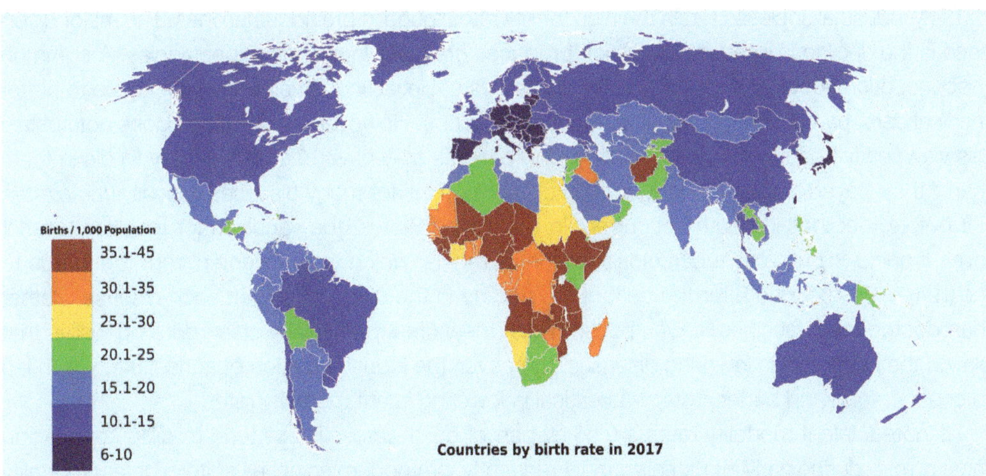

Figure 7.8 Map of world birth rates.

Table 7.2 **World's Most Dangerous Infectious Diseases, 2024 (millions of deaths annually)**

Pneumonia	3.2
AIDS	2.1
Diarrheal diseases	1.9
Tuberculosis	1.7
Malaria	1.0
Measles	0.8

Source: World Health Organization.

life expectancies—are found primarily in sub-Saharan Africa, although Afghanistan, Russia, and several southeast European countries are also relatively high. Conversely, the developed world, as well as Latin America, China, and India, exhibits low death rates.

Because both fertility and mortality rates are high in traditional, impoverished countries, the *difference* between them—natural population growth—is relatively low, often fluctuating around zero. At times, death rates may exceed birth rates, leading to negative growth. In such societies, families are large, and there are many children, but growth rates are kept low by malnutrition, diseases, and infant mortality. This kind of social structure typified the vast bulk of human existence: for tens of thousands of years, human growth rates worldwide were very low (Figure 7.1). Indeed, before the rise of industrialized, urban societies during the Industrial Revolution, high population growth was encouraged as a way to increase the **labor force**, enhance the division of labor, and raise standards of living. While relatively few societies in the world live in Stage I today, it still describes parts of central Africa, Brazil, or Papua New Guinea.

7.4b Stage 2: Early Industrial Society

Stage 2 of the demographic transition reflects societies in the earliest phases of industrialization, when manufacturing jobs begin to appear in urban areas. This was characteristic of Britain and the United States in the early nineteenth century, and of some countries in the developing world today, such as Mexico. As industrialization commences, changes in the labor market, incomes, consumption, and public health measures all have important demographic consequences.

Early industrial societies retain the high fertility rates found in preindustrial ones. The major difference is the decline in death rates, which produces gradually longer life expectancies. A common misconception about why mortality rates decline as societies industrialize is that it is due to better medical care, particularly hospital care and vaccinations. However, the evidence does not sustain this view; early hospitals were often filthy, and patients may have been more likely to die in them than if they stayed home. The introduction of vaccinations for many diseases typically came *after*, not before, declines in deaths occurred. In the United States, the vaccines for measles, scarlet fever, typhoid, diphtheria, tuberculosis, and pneumonia, which were invented from the 1920s to the 1940s, all occurred after the bulk of the decline in the death rate from each disease. Rather than doctors and vaccines, it was public health measures, particularly clean drinking water, that played the major role in lowering diseases. Moreover, the industrialization of agriculture, which led to cheaper food and better diets, was critical in lowering infant mortality rates.

As noted, infant mortality rates are a key part of death rates. These tend to drop earliest and most quickly during early industrialization. Frequently, premodern societies suffered infant mortality rates in excess of 200 infant deaths per 1,000 live births. Today, the highest infant mortality rates are found in sub-Saharan Africa and South Asia (Figure 7.10), where poverty, lack of adequate diets, diseases, inadequate drinking water, and insufficient health care services kill 10 percent of all babies in their first year.

In early industrial societies, because the death rate has dropped but the birth rate has not, the natural growth rate grows explosively (Figure 7.9). This situation is evident in a wide number of countries in the developing world today. In short, poor countries have rapid increases in population, not because they have more children than before, but because fewer people die young.

7.4c Stage 3: Late Industrial Society

Societies undergoing rapid industrialization, in which a large proportion or the majority of the population lives in urban areas, exhibit markedly lower birth, death, and growth rates. Death rates remain relatively low due to better diets, cleaner water, public sanitation, and health care. However, in the third stage, fertility rates also decline. Significantly, declines in birth rates always occur *after* declines in death rates; societies are much more welcoming of death control than birth control.

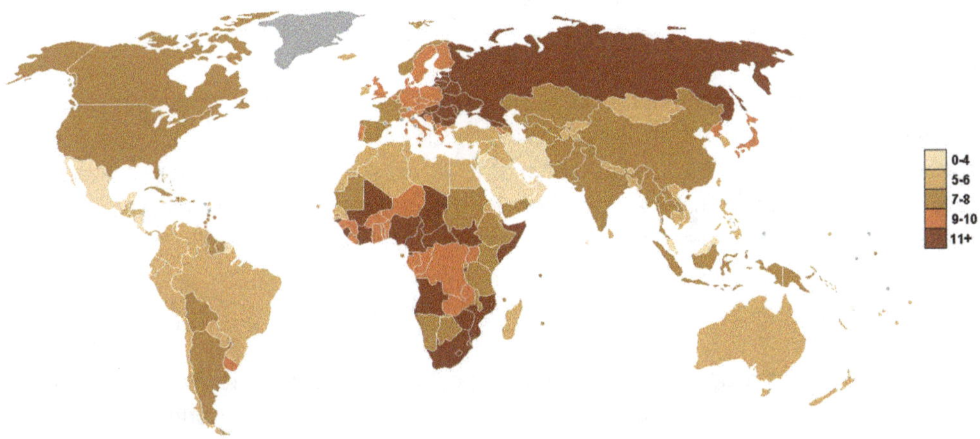

Figure 7.9 Map of world mortality rates.

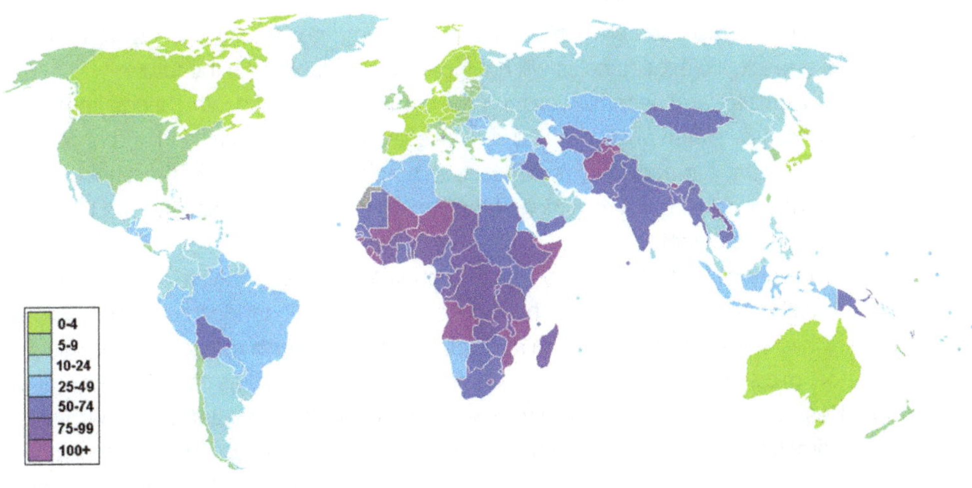

Figure 7.10 Map of world infant mortality rates.

The reasons why birth rates fall and families get smaller as societies become wealthier are important. They lie in the changing incentives that people face as their societies change from primarily rural to primarily urban; labor markets also change as labor becomes commodified. Urbanization and industrialization inevitably produce smaller families because the cost/benefit ratio of children changes over time. In societies in which a large number of women enter the paid labor force rather than work as unpaid workers inside it, the incentives to have many children decline. For women, who typically have primary responsibility for children, working outside of the home and taking care of young children in urbanized societies pose difficult challenges. Many women leave the labor market, if only temporarily, to take care of their children. As a result, they do not earn an income while staying at home, relying on their husbands for support. Economically, this process creates what economists call an opportunity cost to having children: the more children a family has, the longer a mother usually avoids working outside of the home. As women's incomes rise, the opportunity cost of children rises proportionately, leading to lower fertility rates.

In contrast to neo-Malthusian family planning, which ignores the social circumstances that shape birth rates, in the demographic transition, there is a clear link between labor markets and fertility. Getting women to work outside of the home in commodified labor markets is the most reliable form of birth control. As birth rates decline, the natural growth rate does as well. In sum, prosperous societies tend to have smaller families, leading to a shift from extended to nuclear families.

The higher levels of wealth, better diets, cleaner water, lower infant mortality, and improved medical care are reflected in increasing life expectancies. Between 1980 and 2020, the world's average life expectancy at birth increased from 61 to 68 years. In developed countries, average life expectancy is 74 years for males and 81 years for females; in developing countries, 65 years for males and 68 years for females; in least developed countries, 53 and 56, respectively.

7.4d Stage 4: Postindustrial Society

The fourth stage of the demographic transition, postindustrial society, reflects wealthy, highly urbanized countries. This set of conditions depicts Europe, Japan, and North America today. Death rates are very low, and life expectancies are high. Death rates obviously can never reach zero, which would mean that life expectancies become infinite. Future declines in the death rates will not exhibit much progress because they face diminishing marginal returns: the causes of death that are easy to control have been eliminated, and overcoming the remaining ones will be much more difficult. As a result, life expectancies will not increase substantially in the future.

As societies industrialize and become wealthier, the primary causes of death change from infectious diseases to lifestyle-related ones, particularly those associated with smoking and obesity, as well as, to a lesser extent, car accidents, suicides, and homicides. Worldwide, the leading cause of death is cardiovascular disease, a reflection of the world's rising levels of obesity and the abundance of food around the planet.

Birth rates also continue to decline in very wealthy countries, meaning that families grow smaller and many couples choose to go childless or have only one child. National incomes and population growth rates are inversely related. When birth rates drop to that of **death rates**, a society reaches **zero population growth** (ZPG). If birth rates drop below death rates, as they have in virtually all of Europe, Russia, and Japan, the society experiences **negative population growth**. Such countries are characterized by large numbers of elderly people, a high median age, and a relatively small number of children. Often, governments in such countries have population policies designed to increase the birth rate (e.g., subsidized childcare and long paid parental leaves).

Globally, uneven spatial development, the geographic signature of capitalism, reflects the legacy of colonialism, different national policies, role in the global economic system, and national cultures. Uneven development generates spatially varying patterns of natural population growth, the differences between the geographies of birth and death rates (Figure 7.11). The most rapid rates of natural increase today are found throughout the poorest parts of the developing world (Africa and the Arab and Muslim worlds) shown in increasingly lighter colors on the map to indicate higher growth. The economically developed nations, including North America, Japan, Europe, Australia, and New Zealand, in contrast, have low rates of population growth, often around ZPG or even negative.

The geography of the world's population growth has significant implications for the future. The average global natural growth rate has been slowly declining, but this still means that about 120 million people are added each year. Projecting into the next several decades, declining fertility levels will lead to slower rates of population growth. However, because there are large numbers of people of childbearing age, particularly in the developing world, the total world population is projected to rise to roughly ten billion by 2100 (Figure 7.12). The vast bulk of these additions will be in the developing world. However, these projections are based on assumptions about the future of fertility rates: if birth rates decline more rapidly than expected, the future increase in the world's population may not be as large as earlier projections.

The population explosion in the developing world will have huge impacts on the world's economy and environment. It is simplistic to reduce the world's problems to population growth, but that does not mean many problems are independent of it either. Population growth will accelerate agricultural cultivation and soil erosion, overfishing, poaching, deforestation, desertification, depletion of water, loss of biodiversity, and rural-to-urban migration. Further, by generating a large pool of poor people, population growth keeps wages low, not only in the developing world, but worldwide, as globalization pits the labor forces of the developed and less developed worlds against one another.

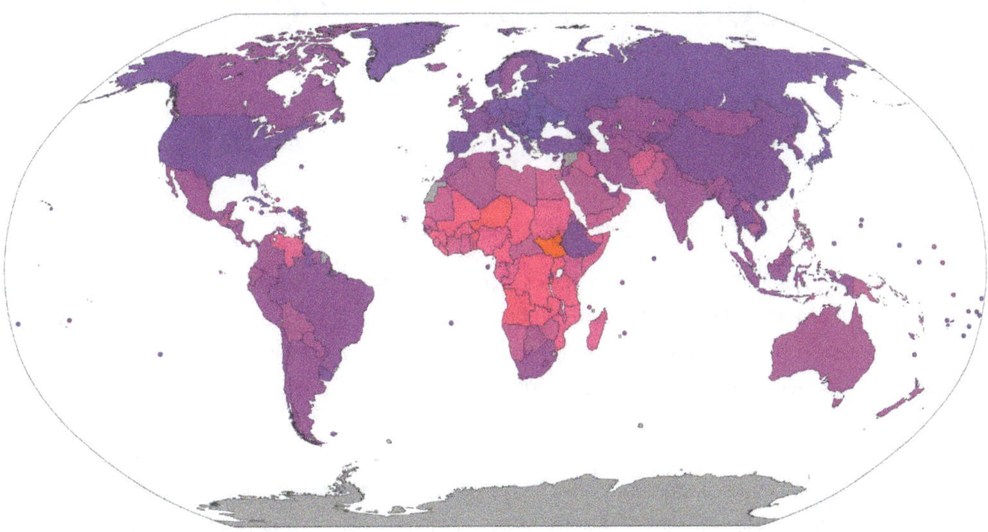

Figure 7.11 Map of world natural growth rates, 2021.

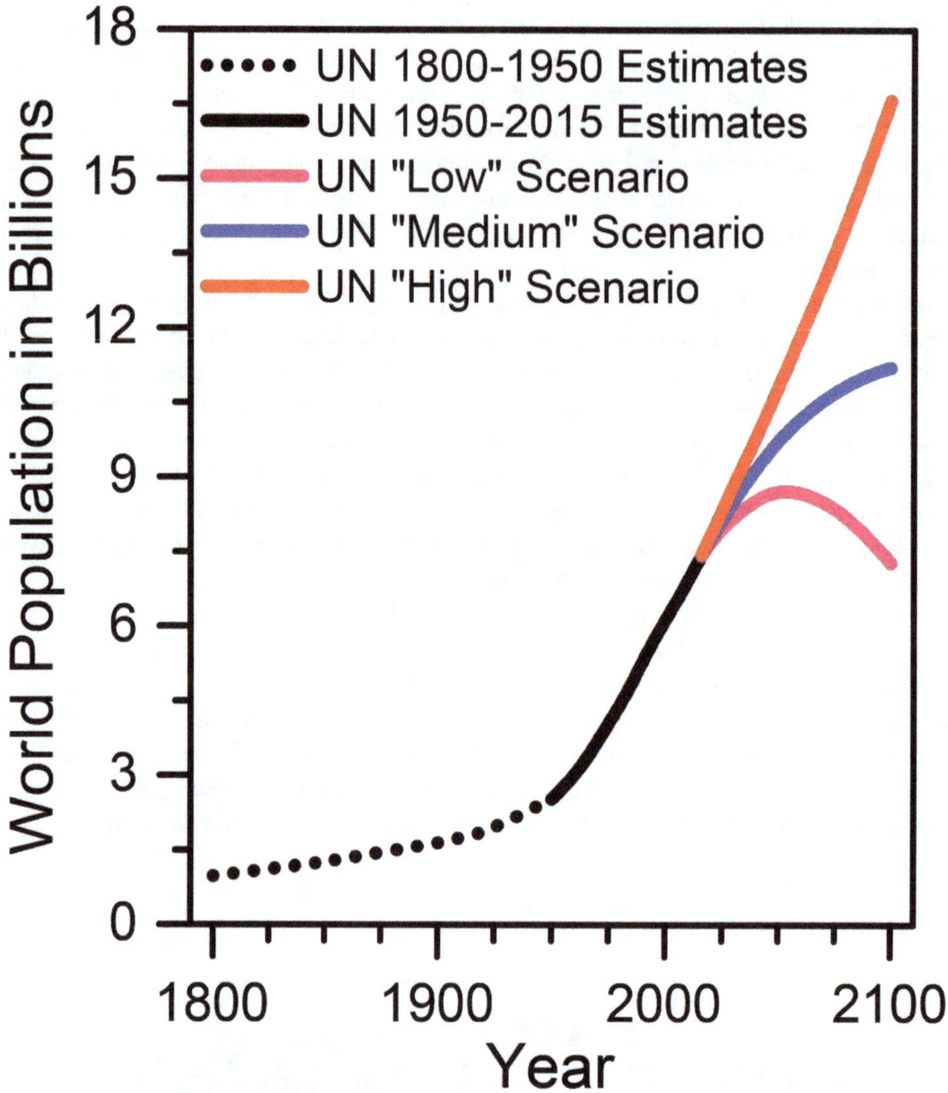

Figure 7.12 Projected world population levels.

7.4e Contrasting the Demographic Transition and Malthusianism

Clearly, the assumptions, analyses, and conclusions offered by the **demographic transition** and Malthusian theory are markedly different. Whereas Malthusian thought takes fertility for granted, assuming that people always reproduce uncontrollably, the demographic transition emphasizes that fertility is socially constructed, i.e., families do or do not have children depending on their social context and the economic costs and benefits of children. Whereas Malthusianism depicts the population as growing uncontrollably so that it inevitably exhausts available resources, the demographic transition predicts steadily declining levels of population growth as birth rates decline over time. The evidence supports this latter view. After rising for two centuries, the rate of world population growth is slowly declining (Figure 7.13). In the 1960s, it reached a peak of 2.3 percent,

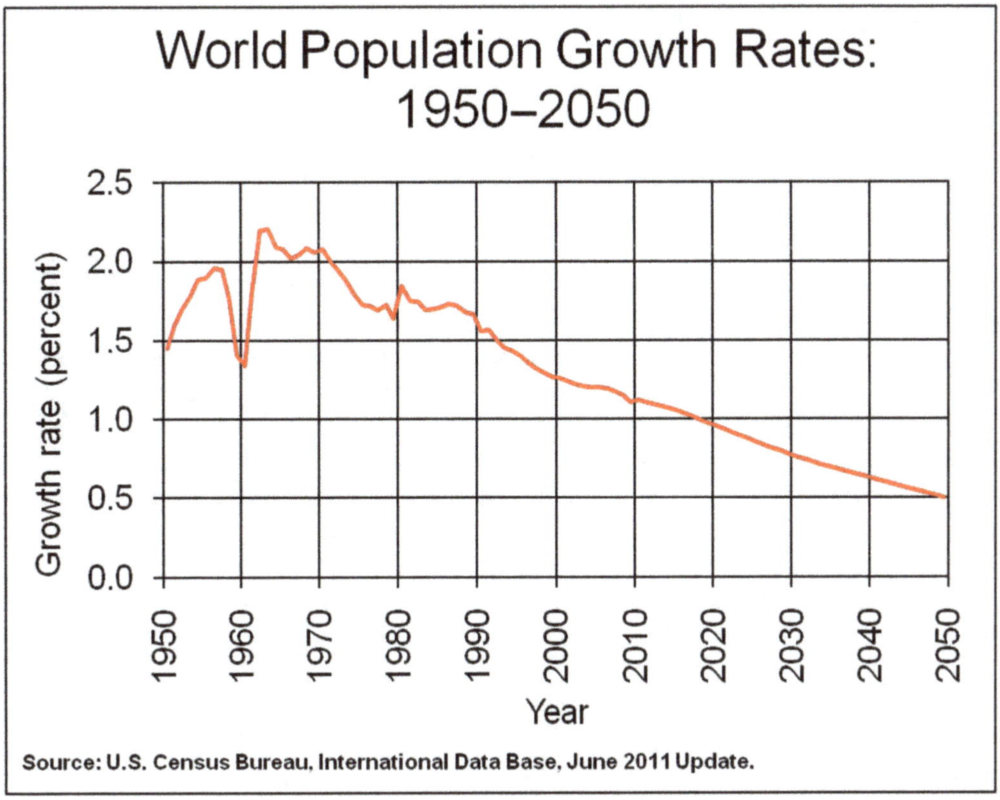

Figure 7.13 World population growth rates.

declining to 1.4 percent a year in the 1990s, and dropping further to 1.0 percent in 2020. However, the absolute size of the world's population will continue to increase because the population that generates fertility, i.e., young adults, is so large.

7.4f Criticisms of Demographic Transition Theory

Although the demographic transition is appealing because it links fertility and mortality behavior to changing socioeconomic circumstances, it has also been criticized. Some point out that it is based on the experience of the West and then applied to many non-Western societies, as if they must repeat the same sequence of stages that occurred in Europe, Japan, and North America. There is no inevitable reason that the developing world must follow in the trajectory of the West. The developing world is in many ways qualitatively different from the West, partly because of the long history of colonialism. In addition, demographic changes in the developing world have been much more rapid than those that occurred in the West in the past. Whereas it took decades, or centuries, for death rates in Europe to decline to their current levels, in some developing countries, the mortality rate has dropped radically in only one or two generations, in part because global food supplies have improved and because of the widespread availability of antibiotics. Because death rates do not vary geographically as much as birth rates, the geographic variation in natural growth around the world is primarily due to differences in fertility. These warnings are useful in reminding us to examine the historical context in which theories emerge and to be wary of simplistically applying models developed in one social and historical context into very different ones.

7.5 Population Structure

An important demographic feature of any population is its age-sex structure. The distribution of people by age and sex affects the needs of a population, its growth rates, and the supply of labor; therefore, it has significant policy implications. High fertility and low mortality rates mean that a rapidly growing population has a high proportion of young people under the working age. A young population also puts a burden on the education system. When young cohorts enter their working ages, rapid increases in jobs are needed to accommodate them. In contrast, countries with a high proportion of the elderly must develop retirement and health care systems to serve them. Therefore, as a population ages, its needs change from schools to jobs to medical care.

The age-sex structure of a country is often summarized through the **population pyramids**, which show the distribution of males and females of different age groups as percentages of the total population. They are usually divided into five-year age groups, in which the base represents the youngest group, the top represents the oldest. The shape of a pyramid reflects fertility and mortality rates and the effects of **baby booms**, migration, wars, and epidemics. It also reflects the potential for future population change.

Two basic types of pyramids may be distinguished. One is typical of developing countries with high birth rates and relatively few elderly, such as Egypt (Figure 7.14). Because fertility rates are high, there is a broad base of people under 5 years old. Natural growth rates in such societies tend to be high.

In contrast, the pyramid for economically developed countries, such as the United States, portrays an aging, slowly growing population (Figure 7.15). Its shape is the result of a history of slowly declining fertility and mortality rates, as well as substantial immigration. With low birth rates, few people have entered the base of the pyramid as babies; with low death rates, more people

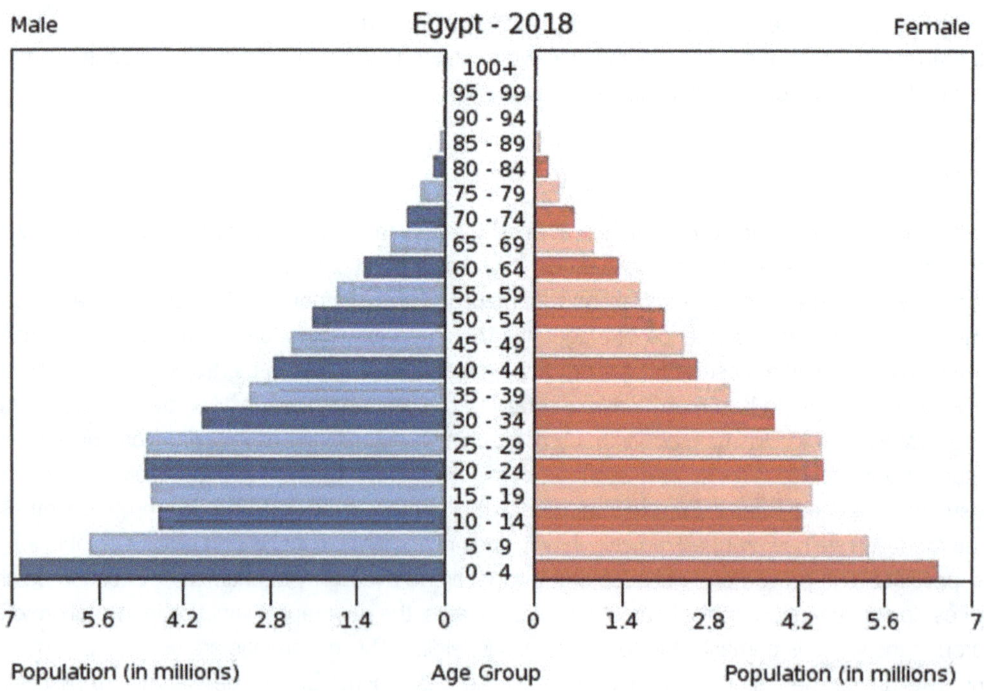

Figure 7.14 Population pyramid for Egypt.

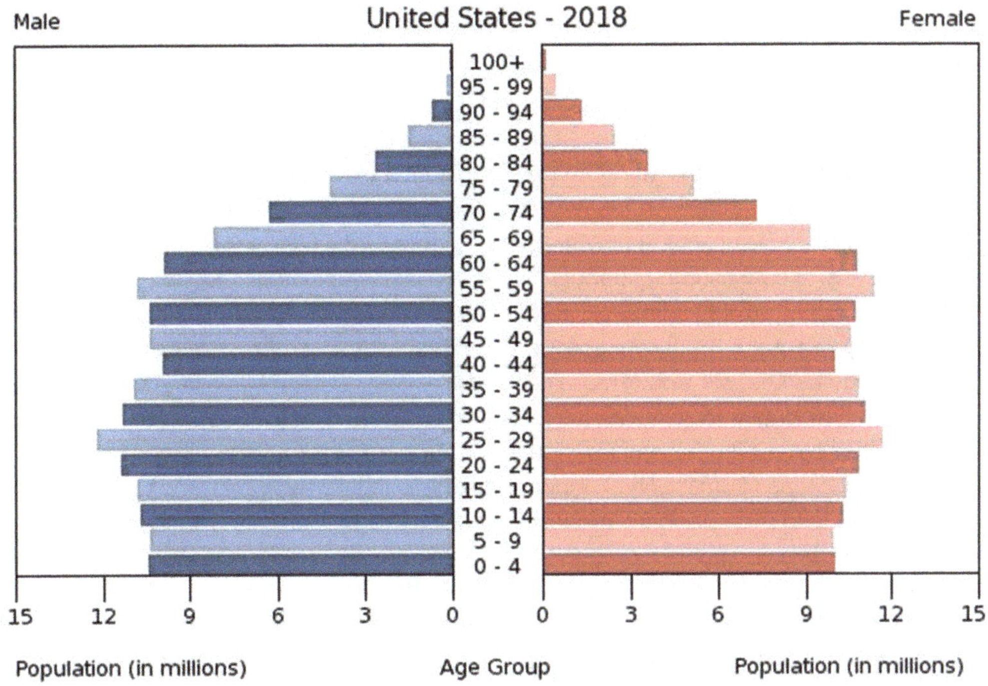

Figure 7.15 Population pyramid for United States, 2018.

survived until old age. Thus, the structure of a population pyramid reflects the stage of the demographic transition in which a country finds itself.

Some countries experience zero population growth (ZPG) or negative population growth. Population decline is an economic and social concern for many European countries, Russia, and Japan. Even in China, the birth rate has fallen sufficiently to cause concern for the government, which worries about who will fill the future labor force. Some states, such as Japan, have resisted immigration from developing countries to supplement their declining labor force. Therefore, demographic changes have a profound effect on immigration policies and the size and nature of the labor force, which in turn affects incomes and purchasing power.

7.6 Migration

Migration is a movement involving a permanent change of residence. It is a complex phenomenon that involves issues such as the reasons why people move and the effects on both origin and destination countries. Immigration refers to movements across international borders into a country (in-migration if within a country); emigration refers to international movements out of a country (out-migration if from one region to another in the same country). International migrations are greatly exceeded by internal movements, especially to cities.

7.6a Causes of Migration

The vast majority of migrants move for economic reasons, that is, to take better-paying jobs or to search for jobs in new areas. People also move to escape poverty and violence, to find a better life for their children. In short, migration is a response to uneven spatial development.

The causes of migration are often divided into **push-and-pull factors**. Push factors include high rates of unemployment, low wages, poverty, shortages of farmland, famine, or war. These factors are largely what drive migration from developing countries to developed ones. Pull factors include job and educational opportunities, relatively high wages, and the magnetic appeal of large cities and the wealth of social and entertainment possibilities there. Geographic differences in economic opportunities explain why so many young people leave rural areas, the vast movement of Europeans to North America in the nineteenth century, the influx of Latin Americans into the United States, the movement of South Asians and Filipinos to the Middle East, or the immigration of Arabs, Asians, and Africans into Europe.

Migrations can be voluntary or involuntary in nature, although at times the line between these is blurry. They reflect the historically specific cultural, economic, and political circumstances in which migrants live. The vast bulk of movements are voluntary, driven by the search for economic opportunities. Involuntary movements are forced or impelled, in which people have no choice; their transfer is compulsory. Examples include the African slave trade, in which twelve million people were kidnapped from their homelands and shipped to the New World, and the deportation of British convicts to Australia in the nineteenth century. In the nineteenth century, many Jewish victims of the Russian pogroms moved to the United States and the UK; others fled the Nazis in the 1930s. Civil wars in Central America in the 1980s led hundreds of thousands of Salvadorans and Nicaraguans to emigrate to the United States. In Africa, multiple civil wars have displaced millions of refugees into neighboring countries.

7.6b The Economics of Migration

Migration is an expression of a free market for labor, although markets are never as free for labor as they are for capital. Unrestricted migration is generally only possible within the same nation-state. Among countries, restricting the flow of immigrants is one of many ways states as well as markets shape economic landscapes (Chapter 3).

Migration has significant impacts on employment and wages, which is why it is such a contentious political issue. While the costs of labor in different regions and countries will never be equalized, the long-run price of labor in two places experiencing significant migration should become similar to one another. However, despite significant movements of people, wage rates do not seem to be converging among regions. Major variations occur in the wages paid among various regions of the United States and among countries of Europe, as well as in Latin America and India. One reason that wage differential rates exist is because of the uncertainty about opportunities; many workers in less developed countries do not know that jobs they may be qualified for in developed countries even exist. Moreover, the opportunity to move varies by sex and age; elderly people have less incentive to move, and men have more opportunities than do women. Consequently, the largest numbers of international labor migrants are young males, although in some cases, such as migration to the Persian Gulf states, women are a significant share. Finally, political and cultural differences are also important. Immigrants may not have the resources to afford living in the destination, or may not speak the language there. Many immigrants, including those to the United States, have faced xenophobic racism and ostracism.

Barriers to migration, including immigration restrictions, imperfect information, lack of skills, inability to afford transportation, and the powerful social ties that hold people in place, all prevent a free flow of labor among countries. Only a small fraction of the population in low-wage countries has the ability to gain access to higher incomes in developed nations.

All countries regulate immigration. The United States limits legal immigration to approximately 600,000 people annually, although a total of roughly 1.1 million enter the United States legally or illegally every year. About forty-five million immigrants lived in the United States in 2020, comprising 15 percent of the population. Of this group, an estimated total of 11,000,000 people live in the country illegally, often under constant threat of being caught and deported. The federal government spends billions of dollars annually to police the borders of the United States. Undocumented immigrants are a significant and contentious political issue in the United States: some argue that by paying income and sales taxes, immigrants generate more wealth than they consume, while others argue that they compete with unskilled American residents, many of whom are ethnic minorities, and drive down wages in that part of the labor market. Some immigrants are highly entrepreneurial: in the United States, companies such as Kraft, Intel, Google, Tesla, eBay, Pfizer, and Yahoo were all started by immigrants.

7.6c Consequences of Migration

Immigration has numerous demographic, social, and economic effects, especially because immigrants tend to be young adults and are often ambitious. Because most migrants are young adults, their departure increases the average age in the origin country, raises the death rate, and lowers the birth rate. For the destination, the opposite is true. In some cases, if migrants are retirees, their effect is to increase the average age, raise the death rate, and lower the birth rate, such as in Arizona and Florida.

Social conflict is a fairly frequent social consequence of migration. It often follows the mass movement of people from poor countries to rich ones. Serious tensions arose in Boston and New York after the Irish arrived in the 1840s; fleeing the potato famine, they were the first Catholics to arrive in the United States in large numbers. Similar issues have arisen with recent migrations of Cubans to Miami and Mexicans to the southwestern United States. Many immigrants are the subjects of racist xenophobia and become scapegoats for all the problems in their new country, especially during recessions. In Europe, for example, French, British, and German nationalists blame Turks, Arabs, Pakistanis, and other immigrants for unemployment. Migrants are often blamed for crime and diseases.

With few exceptions, immigrants contribute enormously to the economic well-being of places to which they come. For example, guest workers from Turkey and the former Yugoslavia were vital to the economy of West Germany in the 1960s and 1970s. Without Mexican migrants, fruits and vegetables in Texas and California would go unharvested, and service in restaurants and hotels would be much more expensive. Migrants to the United States also pay income and sales taxes, but illegal ones do not reap the benefits of programs such as Social Security.

Emigration can relieve problems of poverty by reducing the supply of labor. External migration reduced poverty somewhat in Jamaica and Puerto Rico in the 1950s and 1960s. However, emigration can also be costly. Some of the most skilled and educated members of the population of developing countries migrate to developed countries, a "brain drain" that can lead to shortages of skilled labor in the origin country. Finally, immigrants often send part of their earnings back home in the form of remittances to family members who stayed behind. Remittances are often a major source of income for impoverished villages in the developing world and can form a significant part of their GDP (Figure 7.16).

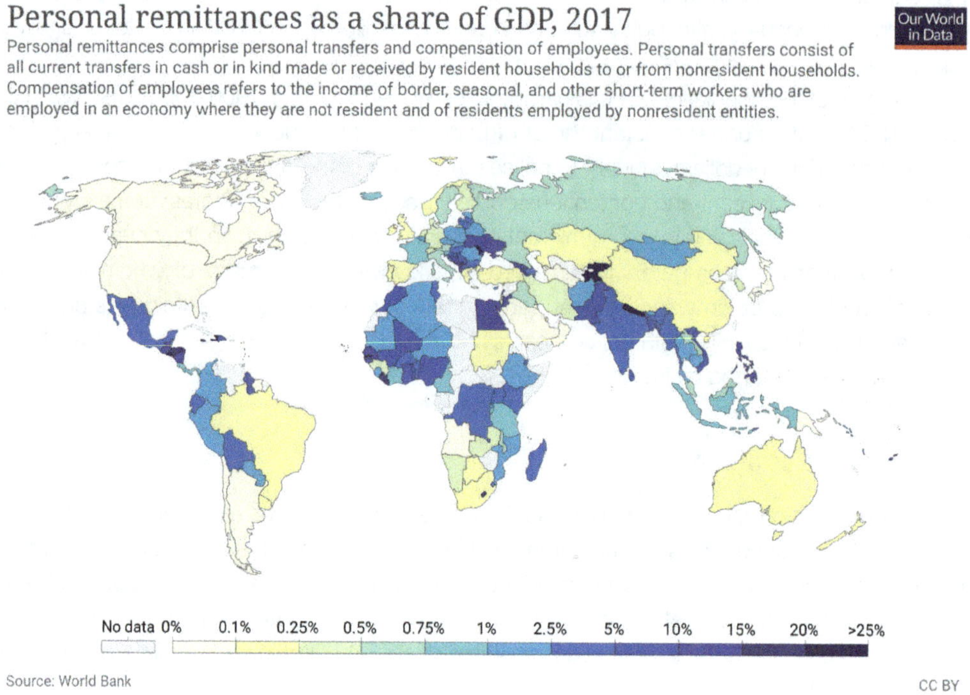

Figure 7.16 Remittances as a percent of GDP, 2017.

7.6d Patterns of Migration

Several great waves of migration have reshaped the world's economic geographies, particularly starting with the European colonial conquests that began in the sixteenth century. Two major examples are the transoceanic exodus of Europeans and the Atlantic slave trade to the New World (Chapter 2). These flows contributed greatly to a redistribution of the world's population. Roughly twelve million slaves from Africa were hauled by Europeans into the sparsely inhabited Americas through involuntary migration, which had been largely depopulated by the smallpox epidemic. The "triangular trade" network that connected Europe, Africa, and the Americas was vital to British economic development. Kidnapped Africans were purchased with British-manufactured goods such as guns, textiles, and whiskey; they were shipped across the Atlantic to work on plantations, where they produced sugar, cotton, indigo, molasses, and other products. Plantation owners and slaves provided a new market for British manufacturers, whose profits helped to finance Britain's Industrial Revolution.

The slave trade, however, was dwarfed by the intercontinental migration of Europeans to North America, which was voluntary. Mass emigration began in the 1820s and peaked around the First World War, when the annual flow reached 1.5 million people (Figure 7.17). The first migrants came from densely populated northwestern Europe, that is, Britain and Germany. Later, they came from poorer parts of Southern and Eastern Europe. Between 1840 and 1920, at least fifty million Europeans emigrated, mostly to North America, but also to Australia and New Zealand, Latin America (especially Argentina), and South Africa. Another large-scale intercontinental migration was the diaspora of overseas Chinese during the nineteenth and early twentieth centuries into Southeast Asia.

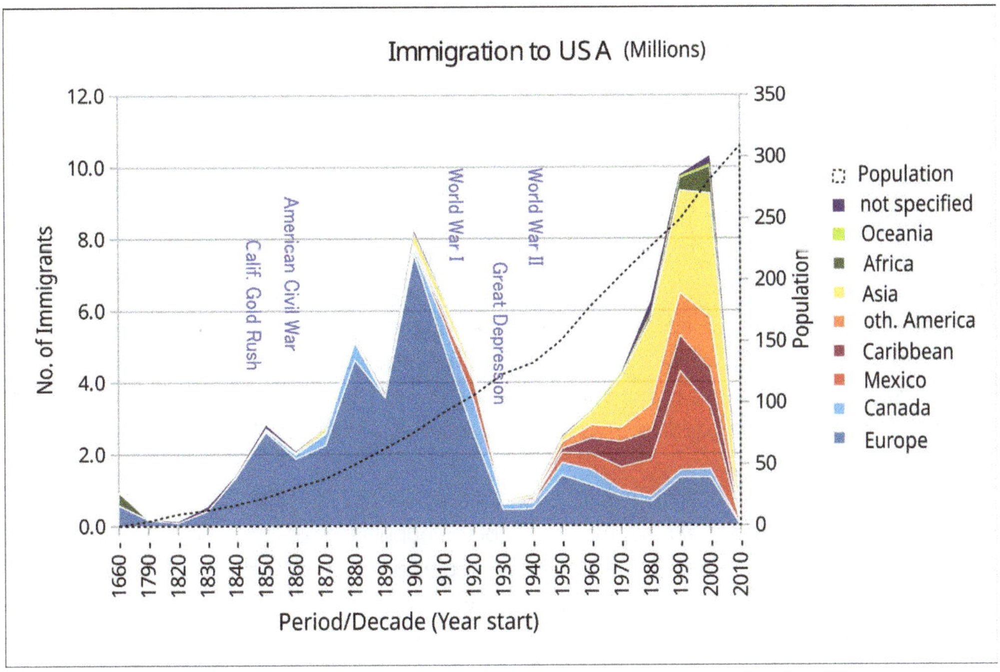

Figure 7.17 US immigration, 1660–2009.

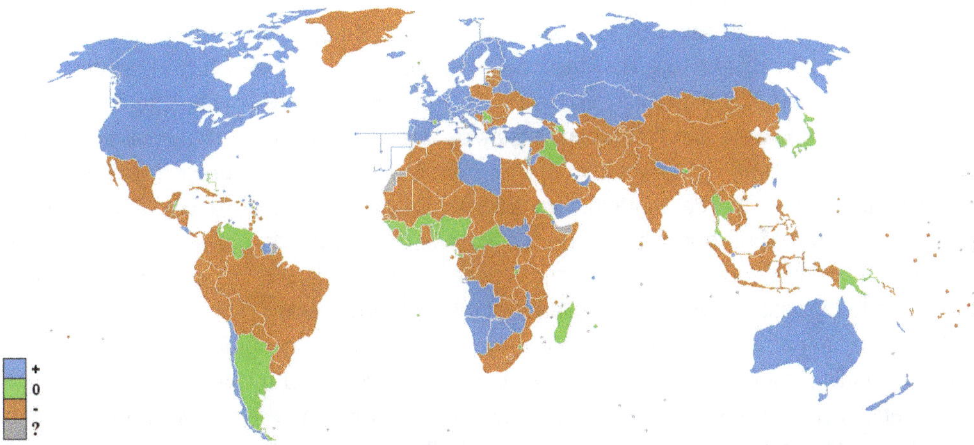

Figure 7.18 Global net migration rates, reflecting major source and destination countries.

Since the Second World War, the geography of intercontinental migration has changed. Instead of heavy flows from Europe to the New World, immigrants are overwhelmingly from developing to developed countries (Figure 7.18). Migration into Europe and to North America has been spurred partly by widening global economic inequalities and by rapid rates of population increase in the developing world. Immigrants form significant populations in many countries around the world, particularly in the United States, Western Europe, Australia, and South Asia.

The most important type of internal migration is rural-urban migration, which is usually driven by economic motives. The movement of farm workers to industrial urban centers was prevalent in developed countries during the nineteenth century, and is still the dominant form of migration in the developing world. Burgeoning cities have functioned as magnets attracting migrants in search of employment, often resulting in shantytowns and slums (Chapter 6).

7.7 Summary

People are the most important element in understanding the world economy. This chapter began by examining the uneven distribution of the planet's population. The vast majority of the eight billion people alive today live in the developing world, particularly in Asia, where more than half reside. The distribution of people is a reflection of centuries of uneven spatial development, particularly following the Industrial Revolution and the formation of global colonial empires from the sixteenth to the twentieth centuries, which dramatically concentrated populations in cities and along the coasts. Population density does not explain economic well-being: some of the world's poorest countries are sparsely inhabited, and some of the wealthiest, such as the Netherlands, are very densely populated.

The two major components of population change are migration and natural increase. Malthusian views of population change held that the growth of the population will inevitably outstrip the available resource base of the planet. While uncontained population growth cannot go on unchecked indefinitely, this view is simplistic in how it portrays why people have children. Malthusianism was derailed by the productivity growth of the Industrial Revolution, during which resources such as food grew more quickly than did population. By focusing on population growth as the only source of the world's problems, Malthusianism blames the poor for being poor and ignores other, more important political and economic forces, such as colonialism.

The demographic transition offers a superior analysis of population growth because it links fertility and mortality rates to the dynamics of industrialization, urbanization, and capitalism. This perspective explains why birth rates are high in poor countries and why, in the premodern world, natural growth rates were low. It also explains the declines in death and birth rates that occur with economic development, including why couples have fewer children as their incomes rise. Thus, unlike Malthusianism, the demographic transition embeds fertility in its historical and economic context. Its conclusions about the future of the world's population growth are also different. Although the world's population growth rate is declining, it is projected to increase for decades to come due to the vast and youthful population of the developing world.

In addition to natural population growth, migration shapes the geography of the world population. Excluding involuntary migration such as slavery or impelled migration (typically refugees from wars), spatial discrepancies in economic opportunities—uneven development—are the major forces driving migration. The causes include both push-and-pull factors, but typically involve unemployment rates and average incomes. The great transcontinental migration streams of the nineteenth century have given way to flows from the developing to the developed world today. Young people, particularly males, constitute the largest group of immigrants. Migration has important impacts on local labor markets, affecting the supply of labor, and thus wage rates.

Glossary

Baby boom: the generation born between 1945 and 1964 that had a huge impact on American society
Birth rate: the number of babies born per 1,000 people per year
Carrying capacity: a notion that originated in ecology concerning the maximum number of inhabitants that a given area or ecosystem can sustain that was imported into the social sciences

Club of Rome: an international organization composed of politicians, business leaders, and academics that was central to the emergence of neo-Malthusianism in the 1960s
Death rate: the number of deaths per 1,000 people per year
Demographic transition: the model of how birth, death, and growth rates change over time as societies industrialize and urbanize
Diminishing returns: the declines in output per extra unit of input (e.g., land, labor) that typically occur when production expands
Doubling time: the number of years it takes a population to double, roughly 70 divided by the growth rate
Fertility rate: the average number of babies born per woman in her lifetime
Human capital: the skills and educational level of a population
Infant mortality rate: the number of deaths of infants younger than one year per 1,000 babies
Labor force: the human beings available for a society to use as workers
Malthusianism: the view proposed by Thomas Malthus that populations will always grow more quickly than their food and resource base, resulting in catastrophe
Migration: the movement of people from one place to another; international movements are considered immigration
Mortality rate: the number of deaths per 1,000 people per year
Natural growth rate: the birth rate minus the death rate
Negative population growth: the rate of decline when deaths exceed births
Neo-Malthusianism: the updated version of Malthusian population theory that emphasized birth control and the ecological limits to growth
Net migration rate: the difference between in-migration and out-migration rates
Population pyramid: a diagram depicting the age and sex distribution of a population
Push-and-pull factors: the incentives for people to leave some areas and move to other ones
Zero population growth (ZPG): a society where the birth rate equals the death rate

Study Questions

1. Summarize the spatial distribution of the world's population.
2. Define crude fertility rates and mortality rates.
3. What are the mathematics of the four major components of population growth?
4. Why did Malthus have such a gloomy view of the future?
5. How do neo-Malthusians resemble and differ from Malthusians?
6. What are the four stages of the Demographic Transition?
7. What explains the world map of fertility rates?
8. Did doctors alone reduce mortality rates?
9. What are the major killers of Americans?
10. Why are the poorest countries the most rapidly growing?
11. What is a population pyramid, and how does it vary between developed and developing countries?
12. Summarize the major causes of international migration.
13. What are some consequences of international migration?

14 Where will the bulk of the world's population growth occur in the twenty-first century? Why?

15 What were the two largest periods of immigration in the United States?

Suggested Readings

Brettell, C. and Hollifield, J. (eds.) 2022. *Migration Theory: Talking across Disciplines*. Routledge.

Caldwell, J. 2007. *Demographic Transition Theory*. Springer Science.

Campisi, N., H. Kulu, J. Mikolai, S. Klüsener, and M. Myrskylä. 2020. Spatial variation in fertility across Europe: Patterns and determinants. *Population, Space and Place* 26(4):p.e2308.

Coddington, K. 2021. For political geographies of fertilities. *Environment and Planning C: Politics and Space* 39(8):1675–91.

de Haas, H., S. Castles, and M. Miller. 2020. *The Age of Migration: International Population Movements in the Modern World*, 6th ed. Guilford.

Fox, N. 2025. Climate change, global population and the capitalist axiomatic: Making sense of Malthus. *Globalizations* 22(5):857–74.

Hassan, M. 2020. *Population Geography: A Systematic Exposition*. Routledge India.

Hazen, H., H. Alberts, and K. Zaniewski. 2023. *Population Geography: Social Justice for a Sustainable World*. Routledge.

Kirk, D. 1996. Demographic transition theory. *Population Studies* 50(3):361–88.

Lewis, G. 2021. *Human Migration: A Geographical Perspective*. Routledge.

Lubbock, R. 2025. The geopolitics of passive revolution and the ghost of Malthus in the American Century. *European Journal of International Relations* 31(2):363–86.

Monnat, S., and I. Elo. 2023. Geographic inequalities in health and mortality: Factors contributing to trends and differentials. *Frontiers in Public Health* 11:1217803.

Montano, B., and M. García-López. 2020. Malthusianism of the 21st century. *Environmental and Sustainability Indicators* 6:100032.

Newbold, K. 2021. *Population Geography: Tools and Issues*, 4th ed. Rowman and Littlefield.

Ojeda, D., J. Sasser, and E. Lunstrum. 2020. Malthus's specter and the Anthropocene. *Gender, Place & Culture* 27(3):316–32.

Schapendonk, J., I. Van Liempt, I. Schwarz, and G. Steel. 2020. Re-routing migration geographies: Migrants, trajectories and mobility regimes. *Geoforum* 116:211–16.

8 Agriculture

Agriculture has long been a major component of the world's economic landscapes and essential to the survival of the human species. Throughout the vast bulk of human existence (95 percent or more), people survived as hunters and gatherers, not agriculturalists; they were food collectors, not producers. The Neolithic Revolution, roughly 10,000 years ago, saw the invention or discovery of agriculture, and paved the way for a social surplus and the rise of cities, denser societies, more refined divisions of labor, and technologies such as writing, pottery, and metalworking.

Of all human activities, agriculture has had the most profound impacts on the natural environment. Almost everywhere, the biophysical environment has been modified extensively by human beings, making it difficult to speak of nature as a phenomenon separate from human impacts. Vegetation has been changed most noticeably. Virtually all vegetation cover has undergone extensive clearing, burning, and the browsing of domestic animals, often resulting in soil erosion and land degradation. As agriculture intensifies, environmental destruction increases. The expansion of farmers into the rainforests of the Amazon, for example, is one major reason for the destruction of that ecosystem.

Medieval farming methods prevailed in Europe until capitalism industrialized farming. Market forces transformed food into a commodity to be bought and sold for a profit. As a result, farmers changed to crops that generated the highest rate of profit. Capitalism thus replaced **subsistence agriculture** with market-oriented agriculture. Crop rotation replaced the medieval practice of fallowing fields. New agricultural areas opened up in the Americas. Farm machines replaced or supplemented human or animal power. This transformation was part of a vast increase in the size of trading cities that depended on the countryside for food and raw materials.

Until the Industrial Revolution, agriculture produced little food per worker; as a result, most of the population worked on the land as farmers, peasants, and serfs. The industrialization of agriculture that started in Europe during the nineteenth century freed large numbers of people to move to cities and work in manufacturing and services. The shift of employment from agriculture to other sectors is one of the most remarkable changes in the world economy. In most industrialized countries, less than 2 percent of the labor force works directly in agriculture. In contrast, large shares of the labor force in many African and Asian countries are still engaged in farming.

8.1 Pre-industrial Agriculture

Agriculture originated in the domestication of plants and animals during the Neolithic Revolution, which began roughly around 10,000 years ago. Domesticated plants and animals depend on humans for their survival and reproduction. Several distinct centers of domestication exist across the world, and domesticated plants and animals have spread widely across trade routes over time.

For example, many Asian crops found their way into Europe (e.g., tea, peaches, plums, apricots), and Muslims were critical in introducing sugar, bananas, rice, and coffee to Spain and Portugal.

Since the Neolithic Revolution, societies across the globe fed themselves through several preindustrial agricultural systems, many of which are still used today. These forms of farming are labor-intensive, involving endless hours of backbreaking toil. In such societies, the bulk of the population lives in rural areas. Peasants are the unsung heroes of world history: their labor supported the cities and elites that dominated societies. Thus, while **preindustrial agriculture** was essential to the rise of civilizations, it continues to be a part of the world's agricultural landscapes today.

Preindustrial agricultural systems differ from industrialized ones in several respects. Most importantly, preindustrial agriculture relies on animate sources of energy derived from human or animal muscle power; because it does not use fossil fuels, it is considerably less energy-intensive in nature. Thus, preindustrial types of farming are much more labor-intensive. In societies that rely on preindustrial agriculture, the vast bulk of people are engaged as farmers or peasants. Because preindustrial societies are often not capitalist ones, food has not been fully commodified; this means that their agricultural systems are organized around production for subsistence rather than profit. Food is grown for local consumption rather than for sale on a market.

Preindustrial agricultural systems played an enormous role in the slave-based and feudal social formations that lasted for millennia. For example, Roman *latifundia*—large estates worked by slaves—formed the backbone of agricultural production in the empire. Later, the manorial system that formed the economic basis of feudal Europe involved peasants and serfs who rented land from large landowners, paying rent with a portion of their output. Several types of peasant-based production continue to be important in many countries.

Today, several forms of preindustrial agricultural systems are found throughout the world (Figure 8.1), with large variations in the crops grown, the methods used to plant and harvest them, their productivity, labor relations, and their relative vulnerabilities to drought or other hazards.

Nomadic herding is not technically a form of agriculture because it involves only the domestication of animals, not crops (Figure 8.2). Often, nomadic herders measure their wealth in terms of livestock (cattle, goats, or reindeer), and follow their herds in annual migratory cycles known as

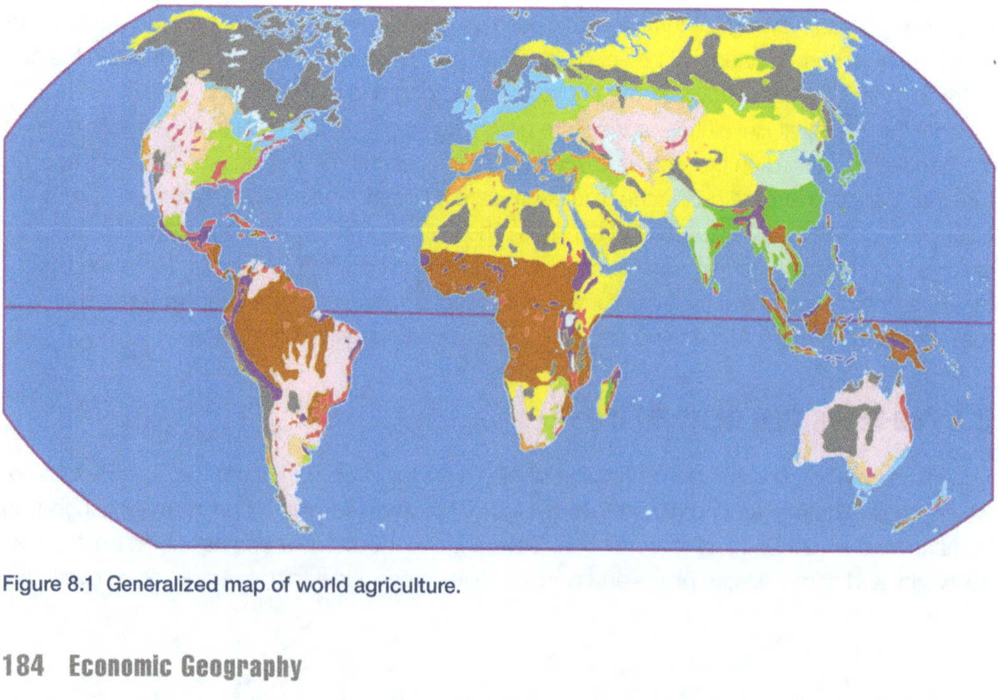

Figure 8.1 Generalized map of world agriculture.

Figure 8.2 Nomadic herders in Mongolia.

transhumance, the movement between summer pastures in higher elevations and winter pastures in lower ones. Nomadic herding has been gradually vanishing throughout the world over the last several centuries, but contemporary examples include the Masai of East Africa, Mongols in Mongolia and northern China, the Tuareg of northern Africa, and the Sami of northern Finland.

The most widely known and practiced form of preindustrial agriculture is slash-and-burn, also known as swidden or **shifting cultivation** (Figure 8.3). Swidden is found in tropical areas such as parts of Central America, the Amazon rainforest, west and central Africa, and parts of Southeast Asia. Roughly fifty million people continue to be fed this way in these regions. Due to heavy rainfall in tropical areas and the leaching of nutrients, tropical soils are generally poor, and most nutrients are stored in the biomass, i.e., the living tissues of vegetation. Slash-and-burn agriculture, therefore, begins by cutting down trees and bushes in a given plot of land and burning them, releasing nutrients into the soil through the ash. Crops are then planted for several years. However, because the rate of nutrient extraction by farming exceeds the rate of replenishment by the environment, the site can only be used for two to six years before the farmers must move on to a new site. Abandoned sites gradually recover if left unplanted for long periods. Should rapid population growth pressure farmers to reduce periods between plantings, the soil fertility may permanently decline. This form of farming was widely practiced in the Mayan kingdoms prior to the Spanish conquest in the sixteenth century, and declining soil fertility may have played a role in the collapse of the Mayan states.

A third form of preindustrial agriculture, Asian rice paddy cultivation, is widely practiced throughout a region stretching from East Asia, southern China, Southeast Asia, and eastern India (Figure 8.4). Rice is the staple crop for billions of Asians, and its cultivation goes back millennia. Young rice shoots require standing pools of water, and to create spaces in which this occurs, Asian societies carved terraces out of hillsides to control the flow of water with networks of dikes

Figure 8.3 Shifting cultivation in India.

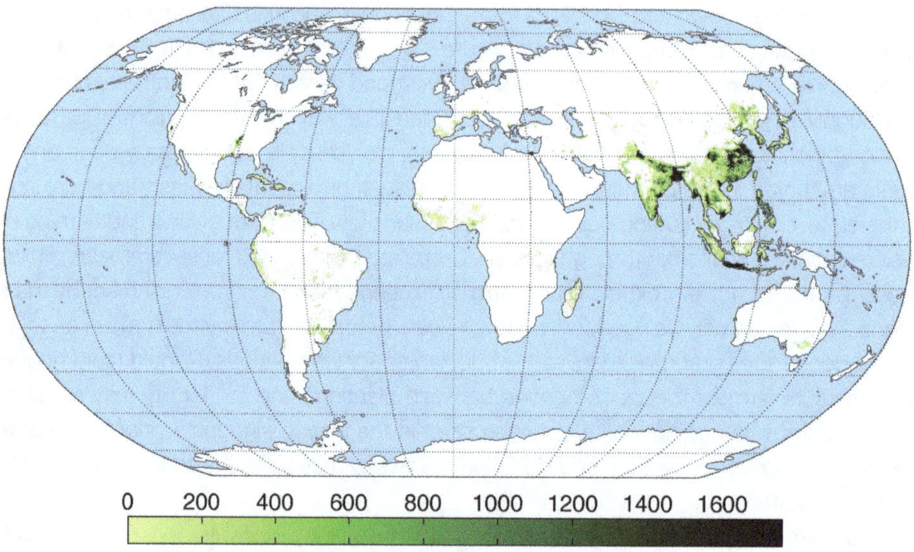

Figure 8.4 World rice yields.

Figure 8.5 Rice paddy cultivation in Nepal.

and levees (Figure 8.5). Often, small fish are grown in these pools of water as a source of protein. The planting and harvesting of rice are very labor-intensive, and there are significant seasonal variations.

Peasant agriculture occurs entirely in many developing countries where market relations have not fully penetrated agriculture. Peasant farmers are small-scale producers who invest little in equipment, irrigation, or fertilizers. It is largely subsistence in nature: farm families consume most of what they produce. Intensive peasant agriculture occurs in the extensive rice fields of South, East, and Southeast Asia, as well as parts of Africa and Latin America. Peasants earn a living by combining several activities, growing crops and livestock, maintaining beehives, hunting and fishing, cutting firewood, making charcoal, delivering water, and carrying sand for use in construction.

Preindustrial agricultural systems have functioned effectively for thousands of years, and continue to do so in many parts of the developing world. Despite their low productivity, preindustrial systems enjoy a diversity of crops (in contrast to industrialized monocultures) and freedom from dependence on pesticides and petroleum. Therefore, we should view these not as remnant forms of an earlier way of life but as historical adaptations to particular social and environmental contexts.

8.2 The Formation of a Global Agricultural System

By AD 1500, on the eve of European colonialism, agriculture had spread widely throughout the Old World and much of the New World. In Europe, the Middle East, Africa, central Asia, China, India, and Indonesia, farming and horticulture were common features of the rural economy. By the time of the first European voyages across the Atlantic, the cultivation of maize, beans, and squash in the New World had spread throughout Central America and eastern North America. These patterns of agriculture persisted until the era of European colonial conquests.

Gradually, European settlements assumed two forms: (1) farm-family colonies in the mid-latitude grasslands of North America, Australia, New Zealand, and South Africa; and (2) plantation colonies in the tropical regions of Africa, Asia, and Latin America. Farm colonization in North America relied on a large influx of European settlers, who introduced farm techniques, field patterns, and types of housing. This often involved the appropriation of land from Indigenous peoples and violence. In tropical areas, Europeans imposed a plantation system that did not require substantial settlement by settlers. Plantations are large-scale agricultural enterprises devoted to the specialized production of one tropical product raised for the market (Chapter 2). The plantation system relied on slaves or poorly paid labor. They were first developed in the 1400s by the Portuguese on islands near the tropical West African coast. Plantations produced tea, cocoa, coffee, tobacco, sugarcane, cotton, jute, and hemp. These crops were sold on the world market, and they were grown near the coasts to facilitate shipment to Europe. Thus, plantations constituted the first wave in the global commodification of agriculture.

Centuries of European colonialism reorganized agricultural land uses worldwide and made **commercial agriculture** a feature of much of the world. Gradually, commodified agriculture has displaced preindustrial farming systems around the world. Hunting and gathering have virtually disappeared. Nomadic pastoralists have steadily declined in numbers, often victims of national borders. Subsistence farming still exists in many developing countries, but only in areas where impoverished farmers eke out a living from tiny plots of land in quasi-feudal contexts.

8.3 The Industrialization of Agriculture

In the nineteenth and twentieth centuries, another agricultural revolution took place when industrialized agriculture became the dominant form in developed countries. Key aspects of industrialized agriculture are its extreme capital intensity, high energy use, and concentration of economic power. Although **industrial agriculture** dramatically increased productivity, it has also depleted water and soil resources, polluted the environment, and destroyed a way of life for millions of traditional farmers.

Because it relies on inanimate energy, industrialization made agriculture extremely capital-intensive and energy-intensive. Farmers rely on large amounts of chemicals (for pesticides and fertilizers), tractors, harvesters, airplanes, and other equipment, much of it very sophisticated, computer-controlled, and expensive, to keep labor inputs low and productivity levels high (Figure 8.6). Only 2 percent of the US labor force is employed in agriculture, and it not only feeds the other 98 percent of the population but also exports vast amounts of food internationally. The very productivity has resulted in rural depopulation. Industrialized agriculture is also becoming more common in parts of the developing world.

The industrialization of agriculture radically reduced the number of farmers in North America. In the United States, the number of farmers declined from 7 million in 1935 to 1.3 million in 2020. In Canada, the number of farmers dropped from 600,000 in 1951 to less than one-third that number in 2020. Europe witnessed similar trends. This reduction in farmers as a percentage of the population is a result of several factors, including waves of corporate mergers, the high cost of equipment, low crop prices, and high interest rates. Farming is difficult, often dangerous work. Meanwhile, the opportunity for higher-paying jobs in the cities has long lured young rural people from the land. Another serious problem is the encroachment of metropolitan areas onto neighboring farmland through the expansion of housing subdivisions and shopping centers. Suburban sprawl, encouraged by interstate highways that reduce commuting times to cities, has

Figure 8.6 Agricultural combine and tractor.

destroyed viable topsoil and farmland around many metropolitan areas in the United States as well as around many European cities.

Today, the percentage of a labor force engaged in primary economic activities (agriculture, logging, fishing, mining) is a common measure of economic development around the world (Figure 8.7): Economically advanced countries have a small share of their workers in agriculture, whereas in much of the developing world, significant proportions of workers are engaged in farming. In most of Africa and South Asia, for example, more than 60 percent of the employed population works as peasant farmers.

8.4 Commercial Agriculture

Commercial agricultural landscapes reflect the dominance of capitalist social relations. These are spread throughout the United States and Canada, Argentina, and portions of Brazil, Chile, Europe, Russia and Central Asia, South Africa, Australia, New Zealand, and portions of China.

Agriculture in the United States is an excellent example of the contemporary capitalist system of food production. The American agricultural belt developed in the nineteenth century as part of the spread of the European "frontier" across North America, which involved the annihilation of Indigenous peoples and the expropriation of their lands. In the late nineteenth century, railroads and steamships dramatically lowered transport costs to the markets along the East Coast and Europe. As a result, agricultural trading centers and ports, such as Chicago, St. Louis, Kansas City, Memphis, and New Orleans, grew rapidly. By the start of the twentieth century, the

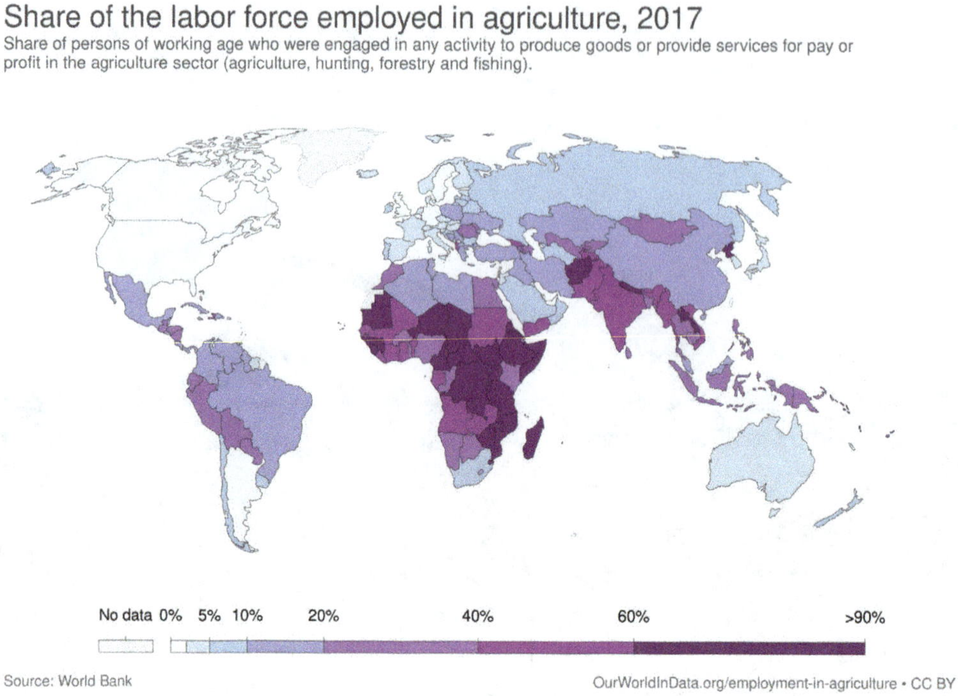

Figure 8.7 Percentage of labor force employed in agriculture.

United States had become the world's largest supplier of wheat and other foodstuffs to Europe. The only other significant producer, Russia, withdrew from world markets following the revolution there in 1917. Consequently, the North American agricultural belt flourished across the Ohio and Mississippi River basins into the Great Plains, extending into central Canada.

US **agribusiness** today is a huge and highly productive industry dominated by a handful of large food corporations such as ConAgra, Bunge, Cargill, Dole, Nabisco, General Mills, General Foods, Archer Daniels Midland, and United Brands. These firms typically control the entire food chain from "seedling to supermarket." Whereas many people hold to outdated popular images of small family farmers, in reality, American agriculture is organized around a small handful of large firms. These usually do not own the farmland but control the food production and processing (e.g., canning), distribution, price and cost information, and marketing.

Agriculture's major backward linkages—its purchases from other sectors—include petroleum and machinery. Notably, labor is only a small part of the production costs, given how capital-intensive this sector is. The forward linkages of this sector—sales to other industries—include the food processing sector and meat production; a large share of cereals, especially corn, is grown for animal feed.

Commercial agriculture includes a variety of types of farming that produce diverse outputs: mixed crop and livestock farming, dairy farming, grain farming, cattle ranching, Mediterranean cropping, and horticulture and fruit farming.

Mixed crop and livestock farming, the most common type of commercial agriculture, is found in Europe, Russia, Ukraine, North America, South Africa, Argentina, Australia, and New Zealand. The primary characteristic of this system is that farmers' largest source of revenue is beef cattle and hogs, although they also earn income from milk, eggs, veal, and poultry. The majority of farmland is devoted to the production of crops such as corn, which is mostly fed to cattle. American corn is highly subsidized by the federal government due to the political power of agribusiness. It is therefore produced in enormous quantities. While some corn is consumed by the population, including corn oil and margarine, most is fed to cattle or hogs. Corn is widely used in sweeteners in many processed foods, including soft drinks and ketchup. Another important crop in mixed crop and livestock farming regions of North America is the soybean, which is used mainly for animal feed.

Dairy farming accounts for much farm acreage in the northeastern United States and northwestern Europe, and accounts for 20 percent of the total output of commercial agriculture. Ninety percent of the world's milk supply is produced in these few areas of the world. Some dairy farms produce butter and cheese as well as milk. New Zealand devotes three-fourths of its dairy farms to cheese and butter production, whereas three-fourths of the farms in Britain produce milk. Dairy farms are relatively labor-intensive because cows must be milked twice a day. This milking is done with automatic milking machines.

Only a few places in the world support large grain-farming operations, including China, the United States, Canada, Russia, Ukraine, Argentina, and Australia. Wheat is the primary crop used to make the world's flour and bread, and is the most important for world food production (Figure 8.8). Wheat is the leading international agricultural commodity transported among nations. The United States and Canada are the leading exporters of grains and together account for 50 percent of wheat exports worldwide. The North American wheat-producing region has labeled the world's breadbasket because they still provide the major source of food to many deficit areas, including several countries in Africa. Other important grains include barley, oats, rye, and sorghum. Grains are not particularly perishable and can be shipped long distances, suturing grain-producing areas into the world economy.

The United States stands as the world's only agricultural superpower and plays a unique role in the global food production system. Agricultural exports generate more than $170 billion in export

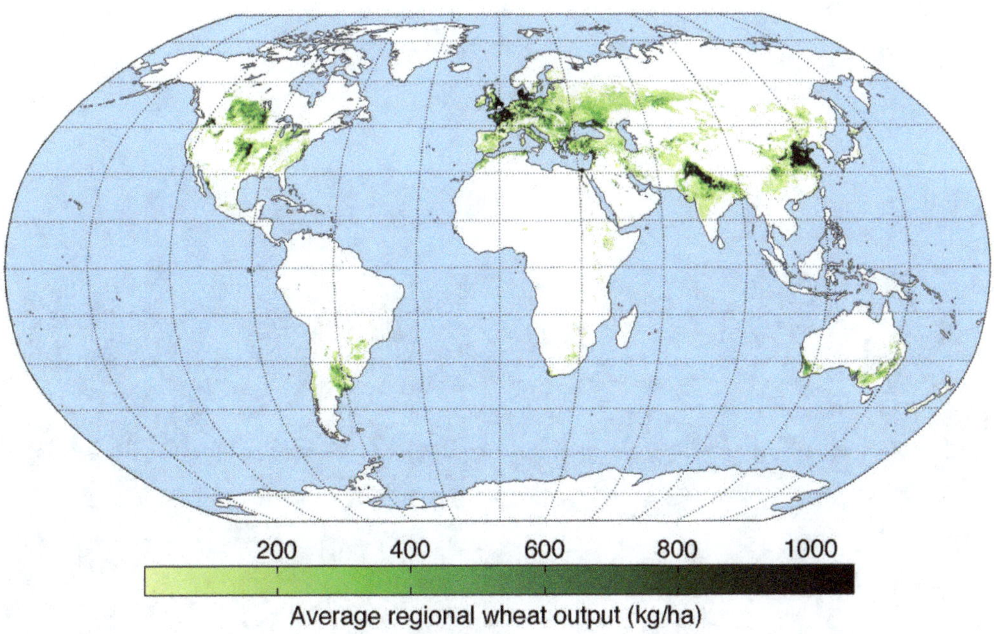

Figure 8.8 World wheat output.

revenues annually, and the United States exports 20 percent of all food traded internationally. As with other economic sectors, agriculture is thoroughly globalized.

In North America, the Spring Wheat Belt lies west of the mixed crop and livestock farming area of the Midwest, and includes Minnesota, North Dakota, South Dakota, and Saskatchewan. Winters are severe in the Spring Wheat Belt; seeds planted in the fall would freeze in the ground, so instead, spring wheat is planted in the spring and harvested in the fall. Major cities in the Great Plains, such as Minneapolis and St. Paul, were established as centers of flour milling and distribution. Just south of the Wheat Belt is the Winter Wheat Belt centered in Kansas, Colorado, and Oklahoma. Winter wheat is planted in the fall, sprouts in the spring, and is harvested in early summer.

Cattle ranching is practiced in parts of the world where crop farming is not feasible because of a lack of rainfall. Cattle ranching is an extensive agricultural undertaking because so much land is needed to raise cattle. When they mature, cattle are penned near cities in fattening pens called feedlots (Figure 8.9).

Many cities arose across the western United States because they contained slaughterhouses and stockyards, such as Denver, Dallas, Chicago, Kansas City, and St. Louis. Ironically, given the frequent hostility of cattle ranchers to the government, today 60 percent of cattle grazing occurs on land leased from the federal government, with ranchers paying fees well below market rates, a trend that defies the myth of the "free market." Some extremely large ranches are owned by meat-packing companies that can fatten the cattle, slaughter them, and package the meat on the same ranch.

Several Latin American countries have significant cattle-ranching industries, such as Brazil, Mexico, Argentina, and Uruguay. In some areas, rainforest is cut down to make room for ranches;

Figure 8.9 Cattle feedlot.

meat consumption is a primary driver of tropical deforestation. Today, ranching worldwide has become part of a vertically integrated agribusiness meat-processing industry.

Finally, commercial agriculture includes horticulture, or the growth of fruits and vegetables. Many truck farms are located as close as possible to large urban areas to minimize transport costs to market. The products are shipped daily to the cities for direct consumption or for packing and freezing. This sector of agriculture is labor-intensive and relies heavily on immigrant labor to pick fruits and vegetables. Many workers are undocumented and live and work under brutally exploitative conditions for very low pay (farm workers are not covered by minimum wage laws). Politically powerless, they suffer at the hands of unscrupulous employers, who may cheat them out of their pay and expose them to pesticides. Cheap fruits and vegetables come at the high price of human misery (Figure 8.10).

8.5 US Agricultural Policy

The state of agriculture is far removed from the mythical "free market" because it has long involved active government intervention of various sorts. The early history of American farming was typified by small family-owned farms that served local markets. In that competitive market, farm prices were relatively stable, although persistent overproduction in the late nineteenth century depressed prices and bankrupted many farmers. By the early twentieth century, waves of technological and organizational change transformed farming. Farms became much larger and highly mechanized. Improved transportation to the markets meant that US farmers now served a much wider market area, including other countries. However, many farmers went bankrupt during the 1920s and 1930s with the twin economic and environmental catastrophes of the Great Depression. After the Second World War, farmers' fortunes declined in the 1950s and 1960s, until the US government agreed to major grain trade agreements with the Soviet Union in the 1970s, which increased American exports. Since then, world markets for US grain have shrunk as many countries have become able to produce more of their own food. As a result of the Green Revolution, India, formerly a net food importer, is now a net food exporter. When interest rates are high, the costs

Figure 8.10 Farm workers picking strawberries in Georgia.

of borrowing fertilizer, seeds, and equipment become prohibitive. Meanwhile, the yield from agriculture has increased enormously during the last century. Technological improvements and hybrid seeds have increased yields so much that US agricultural productivity is the highest in the world. As a result, the quantity of farm products has increased much more rapidly than demand, which has pushed down prices. Industrialized agriculture thus exemplifies the capitalist tendency toward overproduction.

The result has been relatively low returns to farm families, which often spells disaster for farmers. With lower prices and increased quantities, fewer farmers can afford the rising costs for machinery, fertilizer, transportation, and labor. World crop prices fluctuate over time, but have not increased significantly. The result for US farmers has been a continuing decline in the return on their investment. As a result of persistent overproduction and low prices, there has been a large movement of farm families and farm labor away from the farm, which is a major reason for rural-to-urban migration and rural depopulation. In 1910, 35 percent of the US population lived and worked on farms; by 2020, this dropped to less than 1 percent.

In a competitive market, resources would have shifted away from agriculture into other activities. However, because of the US government price supports for farms, this has not been the case. In 1933, with farming in crisis during the Great Depression, Congress passed the Agricultural Adjustment Act to assist American farmers. The act raised farm prices so that farmers could enjoy a "fair price," or *parity price*, for their products. A parity price was defined as the amount that would ensure a reasonable difference between the prices at which farmers could sell their products and the costs they would incur to run a farm. Admitting that markets had created instability in agriculture, the federal government stepped in to establish a subsidy program, or a price floor, for key agricultural commodities, a guaranteed price above the market price. For example, the government bought substantial amounts of corn and wheat from farmers and sold them at what the market would bear. It stores many of these commodities in publicly owned storage facilities; thus, public funds are used to encourage farmers to grow more than the market can consume and also to store the surplus. By far the most heavily subsidized crop is corn, most of which is used for cattle feed.

Price supports have enormous impacts on agriculture. They mean that the market cannot arrive at an equilibrium price through supply and demand. Because the government price exceeds the market price, farmers produce a larger amount of crops than consumers are willing to buy, resulting in a surplus. Considerable quantities of agricultural land are under cultivation because of price supports. Conversely, buyers pay more and buy less than they would if market conditions prevailed. Moreover, farmers' incomes are artificially raised by government subsidies, and consumers' incomes are lowered. But price supports are very popular among farmers, who lobby strenuously to keep them even as they object to other types of government spending.

For years, the government price-support programs have kept inefficient farmers in business. Government subsidies, however, which amount to more than $100 billion annually, favor large corporate farmers over small, family-owned ones. Because subsidies are proportionate to a farm's output, the subsidy program benefits the largest farmers, the corporate agribusinesses, not the small family farms. The US corn industry is the largest recipient of government payments. This process has, in effect, continued to force small farmers off the land. Due to the political power of agribusiness, the federal government has found it politically impossible to reduce **agricultural subsidies**, although doing so would lower prices, eliminate overproduction, and reduce the hostility toward the United States that subsidized agricultural exports generate.

The United States is not the only country to subsidize farmers. Half of the budget of the European Union, for example, is dedicated to its Common Agricultural Policy, which subsidizes farmers in

France, Germany, Spain, and other countries. Japanese farmers are heavily subsidized to grow rice in a country with little arable land, and Japanese consumers pay prices much higher than the world average as a result. In 2020, the world's developed countries spent more than $800 billion in agricultural subsidies, flooding the world with cheap food and bankrupting millions of farmers in the developing world. Agriculture exemplifies the powerful role of the state in market-dominated societies; there is certainly no "free market" in farming.

8.6 Sustainable Agriculture

Industrialized agriculture has had enormous impacts on the natural environment over the last century. Many of these impacts result from the monocultures that rely heavily upon mechanical and chemical inputs (e.g., tractors, hybrid seeds, herbicides, and pesticides). Industrial agriculture requires enormous amounts of energy in the form of fossil fuels for tractors, petrochemical-derived fertilizers to increase yields, and the petroleum required to transport food. Agribusiness consumes vast amounts of water, which has depleted aquifers in places like the US Midwest. The use of pesticides has grown rapidly since the Second World War, causing thousands of human deaths annually and risks due to cancer, poisonings, and neurological damage. Pesticides also harm wildlife, including microorganisms, fish, birds, and mammals. The health of soils is damaged as well. With inorganic fertilizer use, the organic matter of the soil is not replenished, and despite chemical substitutes, many areas have seen a gradual decline in soil productivity. The use of pesticides and herbicides has led some insects, weeds, and fungi to evolve resistance. Livestock operations generate huge amounts of waste that often pollutes local water supplies. Meat production is a leading cause of anthropogenic climate change. Overuse of antibiotics in industrial livestock production has led pathogens to become antibiotic-resistant, which reduces the efficacy of antibacterial drugs. These problems are compounded by the overall loss of genetic diversity in agriculture.

The long-term ecological sustainability of industrialized agribusiness is highly questionable, given its contributions to environmental destruction and climate change. In response, increasing attention has focused on **sustainable agriculture**, which is centered on social, ecological, and economic health for food-producing communities. There are numerous locally based, low-input farming systems that follow the principles of sustainable agriculture. While sustainable agricultural systems are frequently labeled as "alternative," in reality, these approaches are actually quite normal and are practiced by millions of farmers in the developing world.

The essential approaches to sustainable agriculture include organic farming, agroecology, urban gardening, and community-supported agriculture. There is an emphasis on recycling, composting, and minimizing the use of fossil fuels and ecological impacts. The local knowledge that farmers possess is a rich and reliable source of information regarding agroecosystems, and reduces the costly external inputs found in industrialized agriculture. Farmers' local knowledge can improve nutrient cycling, crop productivity, and the conservation of soil, water, and energy. Sustainable agriculture seeks to reduce or eliminate external, nonlocal, and nonrenewable inputs (e.g., petroleum, pesticides), which have high financial and environmental costs. Sustainable agricultural systems are often relatively labor-intensive.

8.7 Summary

The invention or discovery of agriculture around 10000 BCE, in multiple areas of the world, marked a momentous change in human existence, enabling settled communities, class-based societies, the state, writing, metal working, and a more specialized division of labor. Agriculture necessarily modifies the land and thus involves the large-scale transformation of nature.

Agriculture in developing countries is primarily of the subsistence variety, using peasant labor. Subsistence agriculture includes shifting cultivation, pastoral nomadism, and intensive rice farming in paddies. Developing countries often have more than 50 percent of their labor force engaged in agriculture and use relatively little machinery.

In contrast, since the Industrial Revolution, agriculture in developed countries has employed a tiny percentage of the labor force because it uses large amounts of inanimate energy, mechanized equipment, and high quantities of fertilizers. As a result, outputs per worker are as much as fifty times greater than in preindustrial agriculture. Industrialization of farming, a process still underway in much of the world today, freed millions from the boring and backbreaking work on the farms. Farmers raising crops and livestock are part of a vast food-producing system that includes machinery manufacturing firms; fertilizer, pesticide, and energy suppliers; grain mills; slaughterhouses; food processing facilities; and wholesale and retail distributors. Commercial agriculture includes mixed crop farms, dairy farming, commercial grain farming, cattle ranching, and Mediterranean horticulture.

Like much of the capitalist economy, agriculture has been shaped by the state. Far from being a free market, agriculture is heavily subsidized by governments in programs that reflect the political power of agribusiness, including in the United States and the European Union. Price floors and subsidies lead to the chronic overproduction and low prices that farmers often face, and subsidies go disproportionately to large corporations that dominate the industry rather than to small family farms.

Because industrialized agriculture is so enormously energy-intensive and has such serious environmental consequences, the long-term ecological viability of commercial farming is questionable. In response, many farmers worldwide are experimenting with sustainable agricultural practices, which use fewer chemicals, less energy, more local inputs, more labor, and more local knowledge.

Glossary

Agricultural subsidies: government payments to farmers when the government-set price exceeds the market price

Agribusiness: farming conducted by large corporations

Capital-intensive production: production that uses lots of machinery and little labor

Commercial agriculture: farming conducted for a profit

Industrial agriculture: farming that utilizes inanimate energy and extensive machinery

Preindustrial agriculture: farming that relies on animate energy and peasant labor

Shifting cultivation, also called swidden and slash-and-burn agriculture: a preindustrial farming method that involves burning the vegetation cover to release the nutrients and then using the burnt area to grow crops for a limited time

Subsistence agriculture: farming in which crops are not sold on the market but are used almost entirely for domestic consumption

Sustainable agriculture: agriculture designed to minimize environmental impacts, including minimal water use and renewable energy

Study Questions

1 Describe the impacts of European colonialism on global patterns of agriculture.

2 What are plantations?

3 What is nomadic herding? Slash-and-burn agriculture? Intensive paddy agriculture?

4 How do preindustrial and industrialized agriculture differ?

5 What is the main characteristic of commercial agriculture?

6 How have the number and size of US farms changed over time?

7 What are the causes and impacts of agricultural subsidies?

8 How does sustainable agriculture differ from commercial, industrialized forms of farming?

Suggested Readings

Akinola, R., L. Pereira, T. Mabhaudhi, F. De Bruin, and L. Rusch. 2020. A review of indigenous food crops in Africa and the implications for more sustainable and healthy food systems. *Sustainability* 12(8):3493.

Anderson, C., J. Bruil, M. Chappell, C. Kiss, and M. Pimbert. 2021. *Agroecology Now! Transformations towards More Just and Sustainable Food Systems*. Springer Nature.

Blakeney, M. 2022. Agricultural innovation and sustainable development. *Sustainability* 14(5):2698.

Debonne, N., M. Bürgi, V. Diogo, J. Helfenstein, F. Herzog, C. Levers, F. Mohr, R. Swart, and P. Verburg. 2022. The geography of megatrends affecting European agriculture. *Global Environmental Change* 75:102551.

El Bilali, H., C. Strassner, and T. Ben Hassen. 2021. Sustainable agri-food systems: Environment, economy, society, and policy. *Sustainability* 13(11):6260.

Hammelman, C., K. Reynolds, and C. Levkoe. 2020. Toward a radical food geography praxis: Integrating theory, action, and geographic analysis in pursuit of more equitable and sustainable food systems. *Human Geography* 13(3):211–27.

Kirby, R. 2016. *American Farms, American Food: A Geography of Agriculture and Food Production in the United States*. Lexington Books.

Morgan, W., and R. Munton. 2023. *Agricultural Geography*. Routledge.

Norton, G., J. Alwang, and W. Masters. 2021. *Economics of Agricultural Development: World Food Systems and Resource Use*. Routledge.

Robinson, G. 2014. *Geographies of Agriculture: Globalisation, Restructuring and Sustainability*. Routledge.

Rosa, L., D. Chiarelli, M. Rulli, J. Dell'Angelo, and P. D'Odorico. 2020. Global agricultural economic water scarcity. *Science Advances* 6(18):eaaz6031.

Sumberg, J., and K. Giller. 2022. What is "conventional" agriculture? *Global Food Security* 32:100617.

Tornaghi, C. 2014. Critical geography of urban agriculture. *Progress in Human Geography* 38(4):551–67.

9 Manufacturing

Manufacturing involves the transformation of raw materials into finished goods. Manufacturing produces goods that make life easier, safer, and more comfortable. It provides employment and generates economic growth. It has played a central role in capitalist economies since the Industrial Revolution began in Britain in the late eighteenth century (Chapter 2), when manufacturing created the working classes of Europe, North America, and Japan.

Manufacturing firms decide what to produce, where, and how. They gather together the raw materials and semi-finished inputs at a plant, combining the inputs to produce a finished product, and distribute the finished product to wholesalers. The assembly phases require the transportation of raw materials. The production phase involves combining land, labor, capital, and management, factors that vary widely in cost and productivity among places.

Changing raw materials into goods increases their use or value. For example, flour milled from wheat is more valuable than raw grain; bread, in turn, is worth more than flour. This process is called **value added** by manufacturing. The cost and productivity of labor, and the availability of skills, play an important role in high-value-added manufacturing.

9.1 Major Concentrations of World Manufacturing

Manufacturing is very unevenly distributed around the world. Four major areas account for approximately 80 percent of the world's manufacturing output (Figure 9.1): North America, Europe, Russia, and East Asia (Figure 9.2).

9.1a North America

North American manufacturing is largely concentrated in the northeastern and midwestern United States and southeastern Canada, a region often called the **Manufacturing Belt**, which accounts for one-third of the North American population, two-thirds of North American manufacturing employment, and one-half of its industrial output. The belt extends from New England along the Great Lakes to Milwaukee, south to St. Louis, and eastward along the Ohio River valley to Philadelphia (Figure 9.3).

This area grew rapidly in the late nineteenth century, with ready access to raw materials from the continental interior as well as European markets. As capital flooded into the region, much of it foreign in origin, the Manufacturing Belt developed a large labor pool comprised largely of immigrants from Southern and Eastern Europe. The transportation system included the St. Lawrence River and the Great Lakes, which were connected to the East Coast and the Atlantic Ocean by the Mohawk and Hudson Rivers. This transportation system allowed the easy movement of bulky and

Figure 9.1 Manufacturing as percent of GDP, 2017.

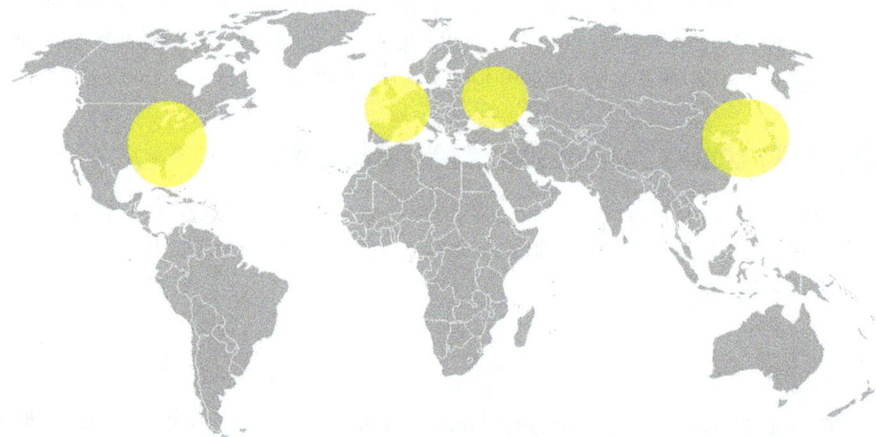

Figure 9.2 Four major world concentrations of manufacturing.

heavy materials. Canals and railroads supplemented the rivers and lakes. The first major factories in the Manufacturing Belt were the textile mills of the 1830s and 1840s that were clustered along the rivers of southern New England. When coal replaced running water as the major source of power between 1850 and 1870, and when railroads began to connect cities over land, factories were freed from the rivers. Between 1850 and 1870, many urban areas saw rapid growth in their industrial production, including New York, Boston, Philadelphia, Pittsburgh, Cleveland, and Chicago. Factories in the Manufacturing Belt could secure easy railroad and waterway access to major resource deposits such as Appalachian coal and the Lake Superior region's iron mines. Also,

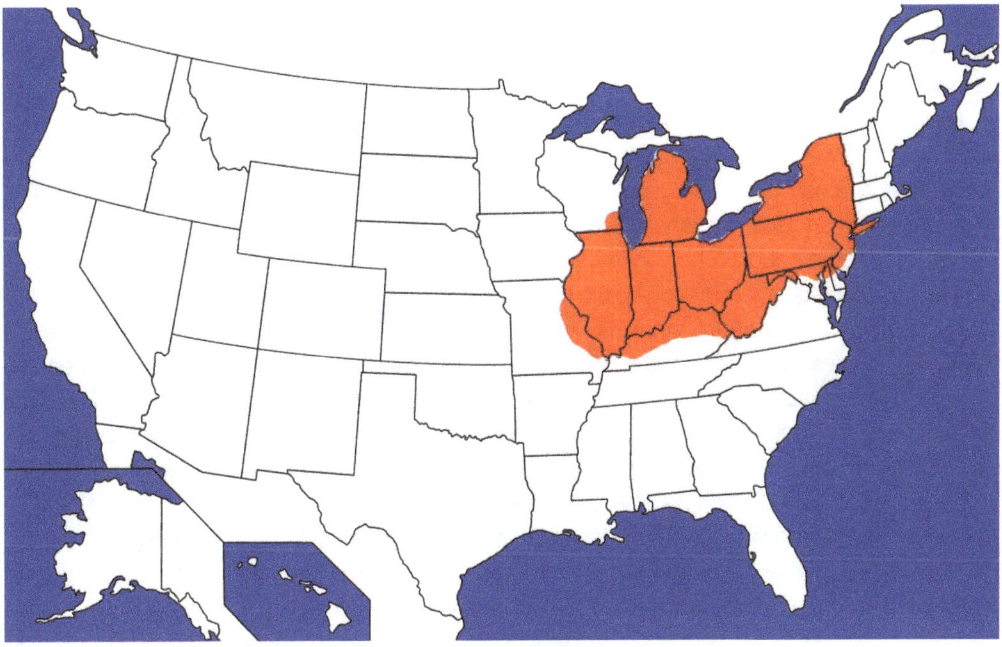

Figure 9.3 The Manufacturing Belt.

they could locate near suppliers of industrial machines, which lowered transport costs, and they could be close to major markets for their goods. In short, agglomeration economies (Chapter 3) accounted for the concentration of manufacturing activity in the Manufacturing Belt. This highly concentrated pattern of production served the country and much of the world well for about a hundred years between the 1870s and 1970s.

Within the North American Manufacturing Belt, several districts formed. The oldest is southern New England, which historically was the American textile manufacturing center of the nineteenth and early twentieth centuries. Cotton was brought from the southern states to be turned into textiles, which were then turned into garments in New York City. As the region developed, wages became higher, and the textile industry moved to the South in the 1930s in a classic example of relocation during a new Kondratief wave (Chapter 3). Today, New England's manufacturing centers on electronic products.

The Middle Atlantic district stretches from New York to Baltimore, Philadelphia, and Wilmington, Delaware. New York City is the largest market and has the largest labor pool in the region. Because of the agglomeration economies it offers (Chapter 3), many corporate headquarters are located there. It is also near financial, communications, and news and media industries, which are important for advertising and distribution. In addition, it is the headquarters of the North American publishing industry. The Mohawk River valley district, including cities such as Buffalo, Rochester, Syracuse, Utica, Schenectady, and Albany, produces electrical machinery and optical machinery. Abundant electrical power produced by Niagara Falls provides inexpensive electricity to this district and explains the location of the aluminum industry, which requires large quantities of electricity.

The Pittsburgh–Lake Erie district, centered in western Pennsylvania and northeastern Ohio, is the oldest steel-producing region in North America. Pittsburgh was the world's largest steel-producing center in the nineteenth century because of the coal supplies in nearby Appalachia and iron ore deposits in northern Minnesota that could be accessed by water over the Great Lakes.

This region has long produced iron and steel, electrical equipment, machinery, and machine tools. Cleveland was long a major producer of steel and automobiles, and Akron, Ohio, was the rubber-producing capital of the world.

The western Great Lakes industrial region is centered on Detroit and Chicago. Toledo, Ohio, and Milwaukee were long centers of production of transportation equipment, iron and steel, automobiles, fabricated metals, and machinery. Detroit, of course, long held the preeminent position of automobile manufacturing, and Chicago produced more railroad cars, farm tractors and implements, and food products. The convergence of railroad and highway transportation routes in this area makes this region readily accessible to the rest of the country.

Canada's most important industrial region stretches along southern Ontario near the St. Lawrence River Valley to the northern shore of Lakes Erie and Ontario. This area is located near the largest Canadian markets, has abundant skilled labor, and accesses inexpensive electricity from Niagara Falls. It produces iron and steel, automobiles, machinery, chemicals, processed foods, pulp and paper, and aluminum. For example, Toronto is the largest automobile-assembly location in Canada, whereas Hamilton is Canada's leading iron and steel producer.

The growth of manufacturing in Sunbelt states after the Second World War changed the geography of American manufacturing significantly. The 1960s marked the start of the steady gain of manufacturing employment in the South and Southwest. Labor costs in the South were lower because it was less skilled and less unionized, in part because southern states are "right-to-work" states, meaning that unions cannot force employees at an establishment to join (Chapter 3). The Southern manufacturing region stretches from Virginia through North Carolina, western South Carolina, northern Georgia, northeastern Alabama, and northern Tennessee. Birmingham was long the largest iron and steel center of the Southern United States, although it never rivaled the large northern centers such as Pittsburgh. Textiles are the main product, the industry having moved from the Northeast to the South in the 1920s to take advantage of less expensive, nonunion labor. However, the region's textile industry has been besieged for decades by foreign competition and job losses. The Gulf Coast manufacturing region stretches from southeastern Texas to the Florida panhandle, including cities such as Houston, Baton Rouge, Mobile, and Pensacola. Because of the large oil and gas fields along the Gulf Coast, petroleum refining and chemical production are found here.

California has high labor costs, yet it witnessed increases in manufacturing employment after the Second World War. Much of its growth was driven by military spending, which was central to the region's aerospace and electronics sectors. Since the 1960s, the district has lost most of its traditional, highly unionized industries, such as steel and rubber, but has attracted a cluster of high-tech industries and associated services tied strongly to enormous defense and military contracts from the US government. Today, aircraft, apparel manufacture, and petroleum refining are still found in Los Angeles, whereas San Diego specializes in pharmaceuticals, biotechnology, and military equipment industries. The San Francisco Bay Area is another important West Coast manufacturing center, particularly for computer software. Silicon Valley, the world's largest manufacturing area for software, is located just south of San Francisco in the Santa Clara Valley (Figure 9.4).

The Pacific Northwest manufacturing district includes the cities of Seattle and Portland. Boeing Aircraft is the single largest employer, followed closely by the paper, lumber, and food-processing industries. A more recent addition is Microsoft, the behemoth of the software industry.

Figure 9.4 Silicon Valley, the world's largest center of computer software production.

9.1b Europe and Russia

Europe has several of the world's industrial regions (Figure 9.5). They are mostly located in a belt starting in Scotland, extending through England, to the mouth of the Rhine River in the Netherlands, through the Ruhr region in Germany, northern France, and into northern Italy. Large supplies of iron ore and coal and highly skilled supplies of labor are vital to the success of these industrial regions.

Although the Industrial Revolution started in Britain in the mid-eighteenth century and made it the economic superpower of the nineteenth century, by the late twentieth century, Britain's outmoded factories, high labor costs, slow productivity growth, and deteriorating infrastructure reduced its overall global competitiveness, leading to widespread deindustrialization. In contrast,

Figure 9.5 Europe's major manufacturing regions.

Germany and Japan rebuilt after the Second World War, modernizing their factories and infrastructures at the same time.

The largest European manufacturing area today lies in the northern European countries of Belgium and the Netherlands, northwestern Germany, and northern France. This region's backbone has been the iron and steel industry because of its proximity to coal and iron ore fields. The Rhine River, the largest waterway of European commerce, empties into the North Sea in the Dutch city of Rotterdam, one of the world's largest ports. The Alsace-Lorraine region in southwestern Germany and eastern France is well situated for distribution to population centers throughout Western Europe. The main cities include Frankfurt, Stuttgart, and Mannheim. Frankfurt became the financial and commercial center of Germany. Stuttgart is a center for precision goods and Mercedes-Benz, Porsche, and Audi automobiles. Mannheim, located along the Rhine River, is noted for its chemicals, pharmaceuticals, and inland port facilities. The region also produces iron and steel.

The Po River valley in northern Italy includes Turin, Milan, and Genoa, and includes more than 70 percent of Italy's industries and 50 percent of its population. This region produces iron and steel, automobiles, textiles, and foods. The Emilia-Romagna region is Italy's high technology center, producing high-value-added textiles, industrial ceramics, processed foods, and electronic equipment.

Ukraine's industrial region relies on the rich coalfield deposits of the Donets Basin, with nearby Odessa serving as the principal Black Sea port. The area is collectively known as Donbass. Like the German Ruhr, Ukraine's industrial district is near iron ore and coal mines, a dense population, and a large agricultural region and is served by good transportation facilities.

Russia has several industrial districts (Figure 9.6). The Moscow industrial region has a large, skilled labor pool and is known for its output of textiles. East of Moscow is the Volga region, extending northward from Volgograd on the Volga River, a chief waterway of Russia, which is linked by canal to the Don River and the Black Sea. It is the principal location of substantial oil and gas production that are Russia's primary exports. Recently, oil and gas fields were discovered in Siberia.

East of the Volga region lie the Ural Mountains that separate the European and Siberian parts of Russia. The Urals have large deposits of industrial minerals, including iron, copper, potassium, magnesium, salt, tungsten, and bauxite, and provide Russia with iron and steel, chemicals, machinery, and fabricated metals.

9.1c East Asia

East Asia, the most rapidly growing part of the world economy since the Second World War, has several industrial regions (Figure 9.7).

In the period following the Second World War, Japan rebuilt itself to become a major economic power. Today, it has the world's third-largest GNP and is one of the world's leading producers of electronics, steel, ships, and automobiles. Japan is practically devoid of significant raw materials, dependent on imported raw materials to fuel its economy. Japan has abundant human resources;

Figure 9.6 Manufacturing centers of Russia.

Figure 9.7 East Asia manufacturing regions.

however, there are 124 million Japanese, one-quarter of whom are crammed into an urban-industrial core near Tokyo-Yokohama. When a strong work ethic is combined with a high level of collective commitment, a first-rate educational system, and generous government support, Japan retains a highly competitive position in the world economy. Over the last twenty years, however, the country has suffered from persistent economic stagnation.

Although permanent workers in large firms are well paid, especially when viewed in relation to part-time workers in small and medium-sized firms, the relative cost of labor is lower in Japan than in many other industrial countries. Savings are encouraged by the state, providing ample investment funds. The government engages in aggressive economic planning, the center of which is a unique collaborative partnership between the Ministry of International Trade and Industry (MITI) and private firms. MITI has subsidized research and encouraged troubled industries such as steel making and shipbuilding to relocate to the newly industrializing countries (NICs), where labor costs are lower.

China, which began to industrialize rapidly in the 1980s, is now the world's largest manufacturing country, with industrial centers on its Pacific coast. These include light manufacturing in Hong Kong and Guangdong province in southern China. China is the world's largest producer of toys, steel, cement, automobiles, apparel, electronics, solar panels, and many other goods. This growth was largely fueled by foreign direct investments by American, Japanese, Taiwanese, and European companies. China's economic machine uses enormous amounts of coal and is the world's largest producer of gases linked to global warming.

Other Asian countries have become important centers in their own right. South Korea is the eighth-largest economy in the world, a major producer of electronics, steel, autos, and ships. Taiwan is an important center for apparel and semiconductors. Indonesia, Thailand, and Malaysia have become newly industrialized countries producing apparel and electronics. Bangladesh has become an important player in the global textiles market, which comprises 80 percent of its exports. Even long-impoverished India has witnessed the gradual emergence of manufacturing centers, largely along its west coast, which produce textiles, steel, software, and movies.

9.2 Deindustrialization

In the late twentieth century, capitalism underwent yet another of its periodic restructurings that radically changed the industrial geographies of the globe. A new international division of labor arose, characterized by declining manufacturing in the developed world and the rise of the newly industrializing countries (NICs). Rates of world economic growth declined, and the long boom following the Second World War drew to a close. This round of change began with a recession in 1974–1975, which was followed by oil shocks in 1973 and 1979. As corporations restructured or went bankrupt amid rising international competition, tens of millions of workers were laid off. **Deindustrialization** affected all of the world's developed countries: Europe, North America, and Japan all lost millions of well-paying jobs, which marked a steady erosion of the middle class.

In the United States, the Manufacturing Belt experienced severe manufacturing job losses even as Sunbelt states saw manufacturing job gains. As their economies collapsed, once-vibrant Manufacturing Belt cities became economically stagnant regions, with devastating social impacts. Many areas saw closed factories, bankrupt businesses, depressed real estate markets, and struggling blue-collar communities. People laid off by plant closures often lose not only their incomes but also their accumulated assets. Many suffered depression, psychological problems, and marital problems. Unskilled workers are particularly likely to bear the costs of deindustrialization. Because African Americans were disproportionately concentrated in manufacturing, they were particularly hard hit. The deindustrialization of inner cities was largely responsible for the creation of the impoverished ghettoes there.

A few old industrial cities, such as Pittsburgh, successfully created new economic niches in services. Southern New England underwent a new round of industrial expansion based on electronics, one that took advantage of its pool of highly skilled workers, as the geography of the new Kondratief wave settled into the older geography of textiles (Chapter 3). Manufacturing job losses in the Manufacturing Belt were occasionally matched by employment increases in the Sunbelt, reflecting another round of uneven spatial development as some areas were abandoned by capital and others attracted it.

Several factors explain the decline in manufacturing in the developed world. First, wages in Europe and North America are much higher than in the developing world, creating an incentive for firms to relocate abroad. In addition, manufacturing firms faced high pension costs for their workers and significant costs associated with providing health care. (The average automobile company spends more on health insurance per car produced than it does on steel.) Second, technological changes in manufacturing made it more capital-intensive, reducing the number of workers needed per unit of output. For this reason, although the United States lost many manufacturing jobs, its total industrial output stayed constant. Indeed, most jobs lost to deindustrialization fall victim to technological changes. Third, many manufacturing firms failed to reinvest sufficiently in research and development to remain competitive on the world stage; rather, many firms spent their revenues on high stock dividends. Finally, inadequate public investment in education and infrastructure led to shortages of human capital in emerging occupations and high transportation costs. The US infrastructure is in serious disrepair and in need of enormous investments.

Conversely, manufacturing jobs and output increased rapidly in lower-wage industrializing countries. The most rapid growth occurred in East and Southeast Asia, the region with the fastest rate of economic growth since the Second World War. Japan, with high wages but low labor militancy, set the model for much of this region. By the 1970s, South Korea, Taiwan, Hong Kong, and Singapore followed suit, and Southeast Asian countries (Thailand, Indonesia, Malaysia) formed a third generation of NICs. As the NICs industrialized, they acquired a productive capacity first in unskilled, light forms of manufacturing such as textiles and garments, then in more skilled, higher-value-added industries such as ships, steel, automobiles, and electronics. The "sunrise" industries of the developing world thus corresponded with the "sunset" industries of the West. In short, deindustrialization in some places and **industrialization** in others are mirror images of each other; both reflect the dual faces of uneven spatial development. These changes are massive and have reshaped the lives of roughly two billion people, lifting many out of poverty and catapulting them from rural villages into industrial cities.

9.3 The Dynamics of Major Manufacturing Sectors

Because different forms of manufacturing have unique technological, labor, and locational requirements, and because governments across the world regulate different industries in different ways, the changing geography of global manufacturing must be understood on an industry-by-industry basis. Five sectors are examined here—textiles and garments, steel, automobiles, electronics, and biotechnology—which illustrate the similarities and differences that exist among manufacturing sectors as well as their geographies.

9.3a Textiles and Garments

Textiles involve the creation of cloth and fabric, whereas the garment industry uses textiles to produce apparel. Roughly one-half of the output of the textile industry is used in apparel; the rest is used in sectors such as carpets and industrial fabrics (e.g., upholstery). Textiles and garments are the classic low-tech industry: unskilled, labor-intensive, relatively unsophisticated technologically, with small firms and few **economies of scale** (Chapter 3). The industry has low start-up costs and is thus easy to enter and exit, and it is highly competitive, leading producers to pay very low wages. Thus, textiles have long been among the most exploitative industries.

Textiles were *the* leading sector of the Industrial Revolution (Chapter 3), fueling the rise of a manufacturing base in cities. Then, as today, the industry relied on young women workers. By the early nineteenth century, it began to flourish in other parts of Europe, including Lyon in France and northern Italy. In the United States, textiles were the economic basis of manufacturing cities of southern New England; in the 1920s and 1930s, it moved to the Piedmont states in the South, which remains the core of American textile production. In the late nineteenth century, textile production also sprouted up in Japan, the first non-Western country to industrialize. Concentrations of garment producers are found in New York, a remnant of its heyday in the early twentieth century, as well as Los Angeles, where sweatshops use immigrant labor.

Textile and garment employment in the economically developed world declined steadily from the 1970s onward as the industry moved overseas. Lured by low wages in the developing world, the industry moved to East Asia. Today, major textile producers include China, where the garment industry has made it the world's largest producer, and India and Bangladesh; in all of these, working conditions resemble those of Britain two hundred years ago. A small industry has arisen in Central America, where sweatshops are found in Honduras, Guatemala, and El Salvador. Everywhere, the industry is notorious for its exploitation of young, female workers, who work long hours for very low wages. By constantly looking for cheap, easily exploitable labor, this industry has a very fluid geography over time.

9.3b Iron and Steel

The steel industry plays a hugely influential role in the economies of industrialized countries. Iron and steel production generate outputs essential to many other sectors, including automobiles, ships, aircraft, buildings, dams, industrial and agricultural machinery, pipes, tubing, wire, tools, furniture, and many others. The inputs into steel making are relatively simple and include iron ore purified into pig iron, large amounts of energy, and limestone, which is used in the ore purification process. Steel production is extremely capital-intensive, requiring very large investments. Because the industry's barriers to entry are high, it has been very oligopolistic: a few large firms dominate the sector. Unlike textiles, it relies on a highly skilled, entirely male labor force. Transport costs are high in this sector and have shaped the location of firms.

The historical geography of steel began in the British Midlands cities such as Sheffield, Manchester, and Birmingham during the early stages of the Industrial Revolution. A similar cluster of steel firms arose in the Ruhr region in Germany on the banks of the Rhine River. In the eastern United States, the earliest iron and steel producers were small and localized, using wood and charcoal as fuel. In the Kondratief wave that began in the late nineteenth century, however, numerous changes unfolded. The Bessemer open hearth furnace made the production of steel cheap, but forced firms to use huge amounts of energy in plants open twenty-four hours per day. The American steel industry came to center on the Manufacturing Belt, above all, Pittsburgh, which became the largest steel-producing city in the world. Other steel cities included Hamilton, Ontario; Buffalo, New York; Youngstown and Cleveland, Ohio; Gary, Indiana; and Chicago. Locating near the southern shores of the Great Lakes offered easy access to coal from Appalachia, iron ore from Minnesota, and cheap transportation by water and railroads. By clustering there, steel firms created agglomeration economies (Chapter 3), such as ready access to specialized labor and suppliers of parts and services. After waves of bankruptcies and mergers, the industry became increasingly oligopolized, as exemplified by the rise of the U.S. Steel Company under Andrew Carnegie, when 30 percent of American steel output was generated by one firm in 1900.

The United States dominated the world's steel industry in the early and mid-twentieth century, producing as much as two-thirds of the global output after the Second World War. By the 1960s, however, new competitors had recovered from the war, including Western Europe (Germany, France, and Spain) and Japan. By the 1990s, some developing countries became steel exporters (Brazil and South Korea). As a result, the US share of global steel declined to 8.0 percent of the world total in 2023. Conversely, the steel industry has grown rapidly in East Asia, particularly China, by far the world's largest producer of steel, with Japan in second place (Figure 9.8).

The decline of the US steel industry created severe problems for families living in the deindustrialized steel towns. Waves of plant closures led to high unemployment, depressed property values, and out-migration. Today, former centers of steel production in the Manufacturing Belt produce relatively little steel; Pittsburgh produces none at all. Many steel-producing firms went out

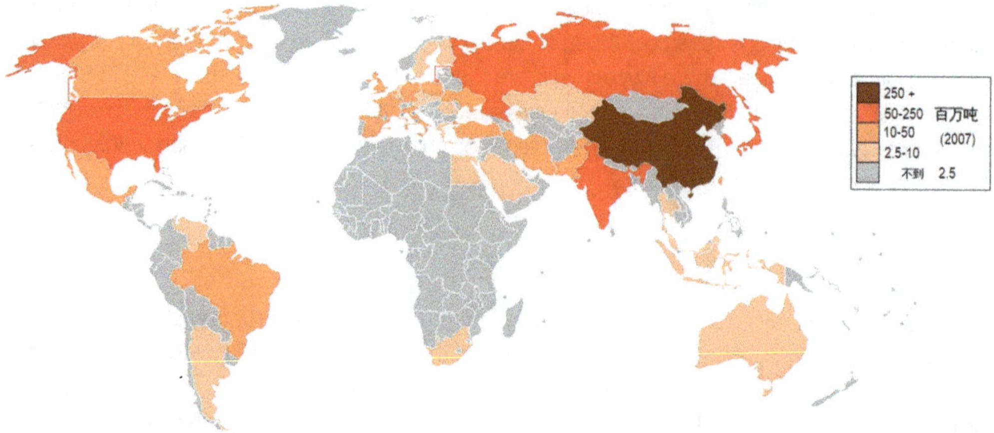

Figure 9.8 Global steel production.

of business as the global capacity exceeded global demand. In Europe, government subsidies led steel mills to stay open longer.

Other than closing steel plants, the industry responded to the crisis by calling for protectionism from imports. It also introduced new computerized technologies, such as highly automated minimills that use scrap metal as inputs, are not unionized, and produce specialized outputs for niche markets.

9.3c Automobiles

From its birth in the early twentieth century, the automobile industry has had enormous impacts on cities, economies, and everyday life across the globe. Henry Ford standardized the European invention of auto production, introduced the moving assembly line, and created a highly detailed division of labor that made automobiles affordable to the middle class. He paid high wages to reduce turnover rates for his workers. This model of production was so successful that it became known as "Fordism" and was widely imitated by other sectors. The American automobile industry centered upon the Detroit region, adding another round of investments to the Manufacturing Belt. Detroit enjoyed several other advantages, including the legacy of the earlier wagon and buggy industry, the presence of many parts-producing firms, the rubber producing capitol in nearby Akron, rail lines, and Great Lakes shipping routes that brought raw materials cheaply, and a large supply of skilled labor, including many immigrants.

Throughout the twentieth century, the market for automobiles expanded rapidly as average incomes rose and as cities increasingly developed in the form of low-density suburbia. Some auto companies purchased public mass transit systems and destroyed them, as in Los Angeles, to encourage people to buy cars. In the 1950s and 1960s, the federal government undertook construction of the Federal Interstate Highway System, which greatly increased the demand for cars. The auto industry also underwent a process of oligopolization, becoming steadily concentrated in a small group of capital-intensive corporations: the world's ten leading automobile firms produce over 70 percent of its automobiles. In the United States, General Motors, Chrysler, and Ford account for 95 percent of national automobile output. However, since the 1970s, the US auto

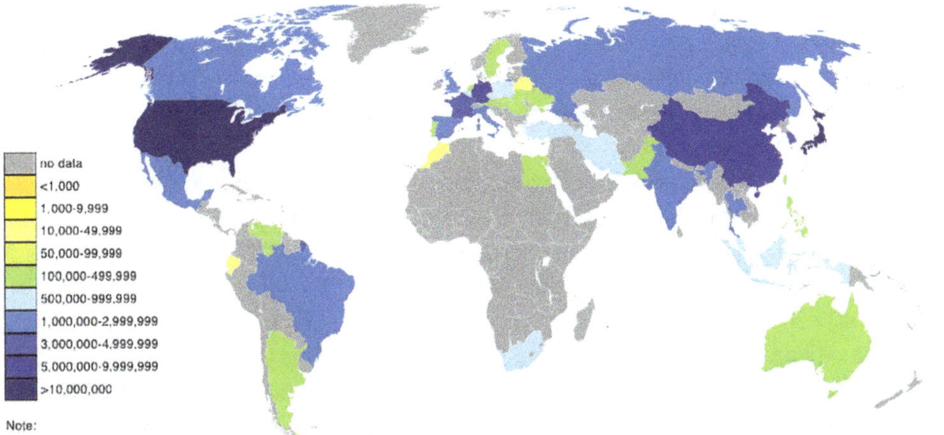

Note:

- I interpreted France in the data as including the overseas departments of Reunion, Guadeloupe, Martinique, and French Guiana as they are integral parts of France.

Figure 9.9 Global automobile production.

industry has been steadily besieged by foreign competition. The industry laid off tens of thousands of workers, mostly in the Midwest. Today, American automobile production remains distributed in a belt of cities stretching from Detroit through Ohio, Indiana, and into the South; most of these centers are **assembly plants**, making cars out of parts produced across the globe.

Today, automobiles are produced in many countries, and the industry has expanded in capacity as the growing middle class in many developing countries wants cars. Figure 9.9 shows the world distribution of automobile production and assembly. Three major clusters of automobile production exist—Japan, the United States, and Europe. In 2022, Europe accounted for 25 percent of the world's automobiles; Japan, 20 percent; and the United States, 19 percent. Japanese firms also invested in the United States, setting up factories in much of the Midwest that offered them access to the American market, low transport costs, and freedom from fluctuating exchange rates and protectionism. China has become a significant producer to serve its enormous domestic market. Smaller production centers exist in Brazil, Russia, South Korea, and India.

The globalization of automobile production is evident in the vast networks of subcontracts that stretch across many countries. Increasingly, it is difficult to say whether a car is "made" in a single country. One automobile might use engines produced in Canada or Mexico, electronic parts from Malaysia, software from the United States, upholstery from China, carburetors from Brazil, and so forth. Roughly eighteen countries are involved in the making of the Toyota Corolla, for example.

9.3d Electronics

With roots that go back to the nineteenth century, the electronics industry grew steadily. The radio was produced in 1901, but the modern electronics industry was not born until Bell Telephone Labs created the transistor in 1948. The transistor replaced the vacuum tube, which had been used in most radios, televisions, and other instruments. In the late twentieth century, it underwent a massive series of changes when the microelectronics revolution began. Microelectronics miniaturized the technology, making it cheap and greatly increasing its uses. The microelectronic transistor acted as a semiconductor for electric current. By 1960, the integrated circuit allowed transistors to be connected on a single small silicon chip. Thus was born the microprocessor, which could do the work of a roomful of vacuum tubes. A major part of these changes was the shift from analog to digital information, which is stored, communicated, and analyzed in binary code, which made it ideal for computers. The microprocessor made possible the microcomputer, which in turn revolutionized the analysis of information, with enormous effects on office work. As the industry generated an endless series of technological improvements, computing power and memory rose exponentially.

Simultaneously, new applications for semiconductors arose, including calculators, computers, telephones, industrial robots, aircraft-guidance systems, automobiles, and consumer products. The electronics industry, like textiles, steel, and automobiles before it, is now regarded as the hallmark of industrial prowess and technological success.

For several decades, from the 1960s through the 1980s, the United States dominated the field of semiconductor production. However, in the 1990s, Japan became the world's largest producer. Global production of electronics parts, which includes semiconductors, integrated circuits, and microprocessors, is dominated by East Asia; much of the television production that formerly occurred in the United States, Germany, and the UK now takes place in China, Korea, Taiwan, and Vietnam.

9.3e Biotechnology

Biotechnology is the application of cellular processes to solve health problems, develop products and services, or modify living organisms such as crops. After the discovery in 1973 of recombinant DNA, biotechnology became a rapidly growing industry worldwide. It has extensive linkages to agriculture, health care, energy, and environmental sciences. There is a wide range in the size of firms in this industry, including single proprietorships and large firms. Because it uses highly skilled labor, the industry's salaries are high.

Small biotech companies rely on venture capital to make their research commercially viable. Given the high costs and large amounts of research necessary to create their products, there is a long lag between R&D and commercial deployment (often twelve to fifteen years of preclinical development). Only a small handful of biotechnology innovations lead to a successful commercial product. Venture capitalists look for an experienced management team when deciding where to invest, and often serve on the boards of directors of young biotech firms.

The government has been actively involved in creating biotechnology complexes. Many countries' national science policies target it as a growth sector. The success of biotechnology firms is heavily affected by federal research funds through institutions such as the National Science Foundation (NSF), the National Institutes of Health (NIH), and the Food and Drug Administration. Federal policies regarding intellectual property rights, subsidies for medical research, and national health care programs are also important. State-level regulatory policies, educational systems, taxation, and subsidies also play a role.

Biotechnology firms tend to cluster in districts to reap the agglomeration economies there. Europe, for example, has the BioValley Network situated among France, Germany, and Switzerland. In Britain, Cambridge plays this role. Similarly, Denmark and Sweden formed Medicon Valley. The US biotechnology industry is dominated by a small group of cities such as Boston, San Diego, Los Angeles, San Francisco, New York, Philadelphia, Seattle, Raleigh-Durham, and Washington, DC. All of these places have excellent research universities with medical schools, skilled labor forces, good infrastructures (particularly fiber optics and airports), and diverse social and recreational environments.

Biotechnology firms agglomerate because in a research-intensive industry, the knowledge base is complex and rapidly changing, and innovation is created by networks of learning rather than individual companies. In a highly competitive environment where the key to success is new product formation, the success of biotechnology firms owes much to their strategic alliances with universities and pharmaceutical firms. Long-distance ties tend to be complementary, not substitutes for, locations in clusters where tacit knowledge is produced and circulated face-to-face.

Because they rely heavily on a scientifically talented workforce, a critical part of the locational needs of biotech firms is the presence of research universities and scientists. Most founders of biotech companies are research scientists with academic positions. Because specialized knowledge is generated and shared within close contacts, the creation of localized pools of knowledge depends on frequent interactions of colleagues in different and related firms. Successful biotechnology firms often recruit highly accomplished scientific "stars" with the necessary skills but also the vision and personality to understand the business dimensions.

9.4 Flexible Manufacturing

Part of the massive transformation of capitalism in the late twentieth and early twenty-first centuries, which included accelerated globalization, deindustrialization, the rise of the NICs, massive deindustrialization in the developed world, and the microelectronics revolution, was a widespread technological and organizational change in manufacturing. Several terms are often used to describe this change, including the shift from "Fordism" to "post-Fordism" as well as "flexibilism" or "the flexible economy" (Table 9.1).

9.4a Fordism

As the name implies, Fordism as a form of production is named after the American industrialist Henry Ford, who played a key role in the mass production of automobiles in the early twentieth century. Fordism involved standardized job tasks, interchangeable parts, and the moving

Table 9.1 **Contrasts Between Fordism and Post-Fordism**

Fordism	Post-Fordism
Vertically integrated	Vertically disintegrated
Long-run contracts	Short-run contracts, JIT
Large firms	Small firms
Economies of scale	Specialized output
Competitive producers	Cooperative networks
Product price	Product quality
Mass consumption	Segmented markets
Unionized	Nonunionized
Unskilled labor	Skilled labor
Routinized work	Varied tasks
National linkages	International linkages

assembly line. Ford's methods were so successful in producing cars cheaply that they were widely imitated by other industries. By the mid-twentieth century, Fordism was the predominant form of manufacturing throughout the developed world.

Some argue that Fordism began in the late nineteenth century, during the wave of technological change in the 1880s and 1890s, when mass production displaced the older, more labor-intensive mode of artisanal production. During this period, glass blowing, barrel making, and the production of rubber goods such as bicycle tires became mechanized, and the Bessemer open hearth furnace for making steel was invented. Fordism accelerated this process to include a highly refined division of labor within the factory, so that every worker performed highly repetitive tasks. Ford applied the insights of Frederick Taylor, founder of industrial psychology, and used time-and-motion studies to organize workers' jobs in the most efficient and cost-effective manner. In breaking down complicated jobs into many small, repetitive ones, Fordism redesigned work in ways that suited unskilled workers, including the waves of immigrants then arriving in the United States. Ford also paid high wages to his workers to reduce turnover.

Fordism was also a social contract between capital and labor. In this bargain, firms tolerated labor unions (such as the Congress of Industrial Organizations that arose in the 1930s). In return, strikes were kept to a minimum. Fordism became solidified in the context of the Great Depression and came to include a minimum wage. Others argue that Fordism was the backbone of the great economic boom following the Second World War. Whenever it began, Fordism was the specific form of capitalism that dominated the twentieth century.

Fordism centered on the mass production of goods, in which companies derived economies of scale to keep production costs low. Mass production was accompanied by mass consumption as the costs of goods declined. Thus, mass consumption, encouraged by advertising, may be seen as the demand side of Fordism. Firms operating in this way were large and vertically integrated, controlling the steps of production from raw material to final product. For example, Ford's automobile factories had raw materials like coal and iron ore enter one part of the factory, and cars came out the other side. Fordism was the form of industrial capitalism that gave rise to the great manufacturing complexes of the US Manufacturing Belt, the British Midlands, the German Ruhr, and the Inland Sea industrial area of Japan.

While Fordism worked quite successfully for almost a century, it ultimately reached its social and technical limits. Productivity growth in the 1970s began to slow dramatically. The petrocrises and growing competition from the NICs led to innumerable plant closures. Because wages and salaries were tied to the growth of productivity (but are not anymore), these changes led to the declining earning power of American workers and the steady erosion of the middle class. Rates of profit in manufacturing began to fall throughout the late twentieth century, and firms faced the choice of shutting down, moving overseas, or redesigning themselves with a new set of production techniques. As Fordism began to implode, it gave way to post-Fordist, **flexible production** techniques, which became widespread throughout the economy.

9.4b Post-Fordism/Flexible Production

Post-Fordism emerged as a markedly different way of producing goods that was better suited to the highly competitive, globalized world of late-twentieth-century capitalism. Also called flexible manufacturing, post-Fordism also produces goods cheaply, but uses computerized technologies to manufacture in small quantities designed for niche markets. A flexibly automated system can turn out a small batch of a product as efficiently as a mass-assembled commodity. This system

Figure 9.10 Industrial robots used in automobile production, a hallmark of post-Fordism.

appeared on the heels of the microelectronics revolution; indeed, the computerization of production represents capitalism's response to the crisis of profitability that occurred as Fordism began to decay. Post-Fordism also reflected the dire needs of American firms to increase their productivity in the face of intense international competition.

The most important dimension of this new system is the flexibility of the production process, including both the organization and management of labor within the factory and the flexible relationships among customers, supplier firms, and the assembly plant. In contrast to the large, vertically integrated firms that characterized Fordism, under flexible production, firms are relatively small and rely on computer algorithms to produce small quantities of goods. Rather than produce for mass consumption, their output is sold in specialized markets that reflect the division of consumption along different class, ethnic, and gender lines. In short, microelectronics obviated the need for economies of scale.

Two characteristic features of post-Fordism are the use of robots and just-in-time inventory systems. As they declined in cost and improved in sophistication, robots became increasingly feasible, particularly in automobile production (Figure 9.10). Japan invented just-in-time (JIT) manufacturing systems shortly after the Second World War when it adapted US practices to its national context. Pioneered by the Toyota Corporation, JIT negated the need for large warehouses to store parts, which saved land costs in a country in which the price of land is high. JIT is a method of organizing on-the-spot manufacturing and supply relationships among companies to reduce delivery times. Various stages of the production process are completed only when dictated by the market, not before and not later, and the parts are supplied with minimal warehousing time and cost. This system reduces idle machinery and raises productivity. Quality control (i.e., minimizing defect rates) became very important.

As automation displaced workers, many firms engaged in significant downsizing, shedding entire divisions of their companies and focusing on core competencies. Middle management suffered job losses, and corporate hierarchies became flatter. Simultaneously, the use of subcontracts grew rapidly. Firms inevitably face a "make or buy" decision, that is, a choice whether to purchase inputs from other firms or to produce them themselves. Under the Fordist system, many firms produced their own parts (i.e., decided to "make" rather than buy), which is a form of **vertical integration**. Their economies of scale justified such investments. Large firms, for example, would frequently have their own parts producers, trucks, or printing shops. With the rise of post-Fordism, however, and the uncertainty created by the rapid technological and economic changes of the late twentieth century, firms increasingly opted to buy rather than make (i.e., to purchase inputs from specialized companies). This meant that post-Fordism saw a large round of **vertical disintegration** in which companies specialized in some stages of the production process. The growth of subcontracts represented the wave of vertical disintegration that washed over the economy.

Flexible production also utilized information technologies extensively, which gave them more sophisticated control over the process. The new electronic computer-aided production techniques permit rapid switching from one process to another and allow production to be tailored to the requirements of small groups of customers. Traditional automation is geared to high-volume standardized production; the newer flexible manufacturing systems are not.

As firms downsized, adopted subcontracts, and engaged in vertical disintegration, inter-firm linkages proliferated rapidly. To survive, many firms entered into cooperative agreements with other companies, such as strategic alliances. Many firms became embedded in dense urban networks of interactions that involved face-to-face linkages; in such interactions, the roles played by noneconomic factors such as tacit knowledge, learning, reflexivity, conventions, expectations, trust, uncertainty, and reputation are important. Post-Fordism thus exposed the culturally-embedded nature of economic linkages.

Post-Fordist production, which varies among different regions, came to dominate the electronics industry, automobile manufacturing, and the minimills in the steel industry. Geographically, this regime became closely associated with the new industrial spaces such as Silicon Valley, Italy's Emilia-Romagna region, Germany's Baden-Württemberg, the Danish Jutland, and the British electronics region centered upon Cambridge. Increasingly, post-Fordism is giving rise to other industrial geographies, including in the developing world.

9.5 Summary

This chapter began with an overview of the geography of global manufacturing. Four major regions of the world account for 80 percent of the world's industrial production: North America, Europe, western Russia, and East Asia. As the post–Second World War boom ended, the textiles, automobile, and electronics industries shifted from the developed world to the developing world, particularly East Asia. Multinational corporations moved their production because of new technological innovations in transportation and communications, and their heightened ability to exploit low-wage labor abroad. With improved air freight, containerization, and telecommunications, firms can produce and ship products faster, cheaper, and with fewer losses.

Deindustrialization led to the loss of millions of manufacturing jobs in North America and Europe. The process played out unevenly and was most severe in the old Manufacturing Belt. Some firms moved to the American South and the West, although even there, industrial job losses continue. This process devastated local communities, which plunged into high unemployment, falling family incomes, and limited purchasing power. Deindustrialization marked the beginning of the decline of the middle class.

Lastly, this chapter described the massive set of changes that rocked the world economy in the late twentieth century, and which continue today. Capitalism has always reinvented itself through the process of "creative destruction." As the Fordist regime of production collapsed, deindustrialization, the rise of the NICs, and the microelectronics revolution led to the rise of post-Fordist, flexible production, with different manufacturing techniques and organizational forms. Many firms downsized, subcontracted many functions, adopted technologies such as just-in-time inventory systems, and entered into strategic partnerships. The new geographies of post-Fordist production include a handful of highly competitive regions scattered across the world.

Glossary

Assembly plants: factories in which labor and material inputs are combined to produce goods
Deindustrialization: the loss of manufacturing capacity and closure of plants
Economies of scale: the benefits firms accrue from producing in large quantities, notably by spreading fixed costs over a larger volume of output
Export processing zones: areas set up by governments with tax breaks and subsidies designed to lure firms to invest there
Flexible production: the type of manufacturing that arose in the late twentieth century that uses microelectronics to produce goods quickly and often in small quantities
Horizontal integration: the enlargement of market share in a given industry through the acquisition of other firms
Industrialization: the process of manufacturing job growth that frequently results in higher standards of living
International subcontracting: subcontracts that firms use with suppliers in other countries
Just-in-time manufacturing: a form of flexible production that allows firms to purchase inputs in small quantities, thus saving on storage costs
Manufacturing Belt: the industrial region of the United States stretching from southern New England to Chicago and Milwaukee
Multiestablishment enterprises: firms with facilities in more than one location
Offshore assembly: the use of assembly plants in other countries to make goods
Oligopoly: the concentration of production capacity in an industry in a small group of large firms
Outsourcing: the use of external suppliers, usually overseas, to provide inputs of goods and services
Supply chains: the networks of goods, parts, and knowledge that tie different places together in the production of goods
Taylorism: the process of organizing the labor process that maximizes efficiency by carefully designing each job
Transnational corporation (TNC): firms that operate in two or more countries
Value added: the profit added at each stage of the production process
Vertical disintegration: the process by which firms downsize their operations and increase subcontracts
Vertical integration: the growth of firms forward and backward along supply chains

Study Questions

1. What are the forces of production and the social relations of production?
2. Why are capital-labor relations a necessary starting point for studying economic geography?

3. What are the four major world regions of manufacturing?
4. Summarize the historical development of the US Manufacturing Belt.
5. When and why did the Manufacturing Belt begin to lose industry? Where, specifically, did it go?
6. What is flexible production?

Suggested Readings

Amin, A. (ed.) 2008. *Post-Fordism: A Reader*. Wiley.

Becattini, G., M. Bellandi, and L. De Propris. 2009. *A Handbook of Industrial Districts*. Edward Elgar.

Blackbourn, A., and R. Putnam. 2025. *The Industrial Geography of Canada*. Taylor & Francis.

Bole, D. 2021. "What is industrial culture anyway?" Theoretical framing of the concept in economic geography. *Geography Compass* 15(11): e12595.

Clark, J. 2022. Reframing deindustrialization. *International Journal of Urban Sciences* 26(1):29–35.

Dicken, P. 2015. *Global Shift: Mapping the Changing Contours of the Global Economy*, 7th ed. Guilford.

He, C. 2022. *Industrial Geography in Contemporary China*. Routledge.

Holl, A., R. Pardo, and R. Rama. 2010. Just-in-time manufacturing systems, subcontracting and geographic proximity. *Regional Studies* 44(5):519–33.

Inverardi-Ferri, C. 2023. Overtime: The cultural political economy of illicit labor in the electronics industry. *Economic Geography* 99(2):140–60.

Moretti, E. 2013. *The New Geography of Jobs*. Harper Business.

Pavlínek, P. 2020. Restructuring and internationalization of the European automotive industry. *Journal of Economic Geography* 20(2):509–41.

Peet, R. (ed.) 2024. *International Capitalism and Industrial Restructuring: A Critical Analysis*. Taylor & Francis.

Pike, A. 2022. Coping with deindustrialization in the global North and South. *International Journal of Urban Sciences* 26(1):1–22.

Schindler, S., T. Gillespie, N. Banks, M. Bayırbağ, H. Burte, J. Kanai, and N. Sami. 2020. Deindustrialization in cities of the Global South. *Area Development and Policy* 5(3):283–304.

Wood, S. (ed.) 2024. *The Transformation of Work? Skill, Flexibility and the Labour Process*. Taylor & Francis.

10 Services

Services, the tertiary sector of the economy, include a wide variety of industries and occupations, such as retail and wholesale trade, finance, real estate, law, advertising, education, and government. There is no single "service sector," a term that masks the huge heterogeneity of jobs involved.

Traditionally, services were viewed as part of **post-industrial society**, which viewed them as activities involved in the collection, analysis, and communication of information. Certainly, many services do involve information of one sort or another, such as clerical work, management, legal services, telecommunications, and the media. Postindustrial theorists argued that services were a qualitatively new form of economic activity, but this view is mistaken. First, many services do not involve information in any meaningful way, such as trash collectors, chefs, security guards, and janitors. More realistic approaches theorize services as another form of commodity production, similar to manufacturing and agriculture. Services do indeed exhibit differences from these other sectors, but they are nonetheless embedded in the broader social relations of capitalism.

All across the developed world, services have replaced manufacturing employment as the mainstay of the economy. This trend reflects the intense automation of manufacturing and the offshoring of industrial plants to the developing world. The percentage of the national labor forces employed in various types of services varies around the world (Figure 10.1, Table 10.1), but the more economically advanced countries have the highest proportions. In the United States, more than 85 percent of people work in services, whereas in poorer countries, most people work in agriculture.

This chapter defines services and notes how service industries differ from one another. Next, it summarizes the forces that drive the growth of the service economy. Third, it explores labor markets in services. Fourth, it describes the geographical dynamics of services, notably their tendency to agglomerate. Fifth, it offers brief summaries of several **producer services**, including finance, accounting, design, and legal services. Sixth, it examines the globalized, digitized world of services, including **electronic funds transfer**, **offshore banking**, and **back offices**. Seventh, it turns to the geography of **consumer services**. Eighth, it examines tourism, the world's largest industry by employment.

10.1 Defining Services

Broadly, services consist of the production of intangible outputs. They thus lie in sharp contrast to manufacturing, where the product can be "dropped on one's foot." But measuring the output of services is tricky. For example, what is the output of a lawyer, teacher, or doctor? Their output may be impossible to measure precisely, yet they are very real nonetheless. The federal government

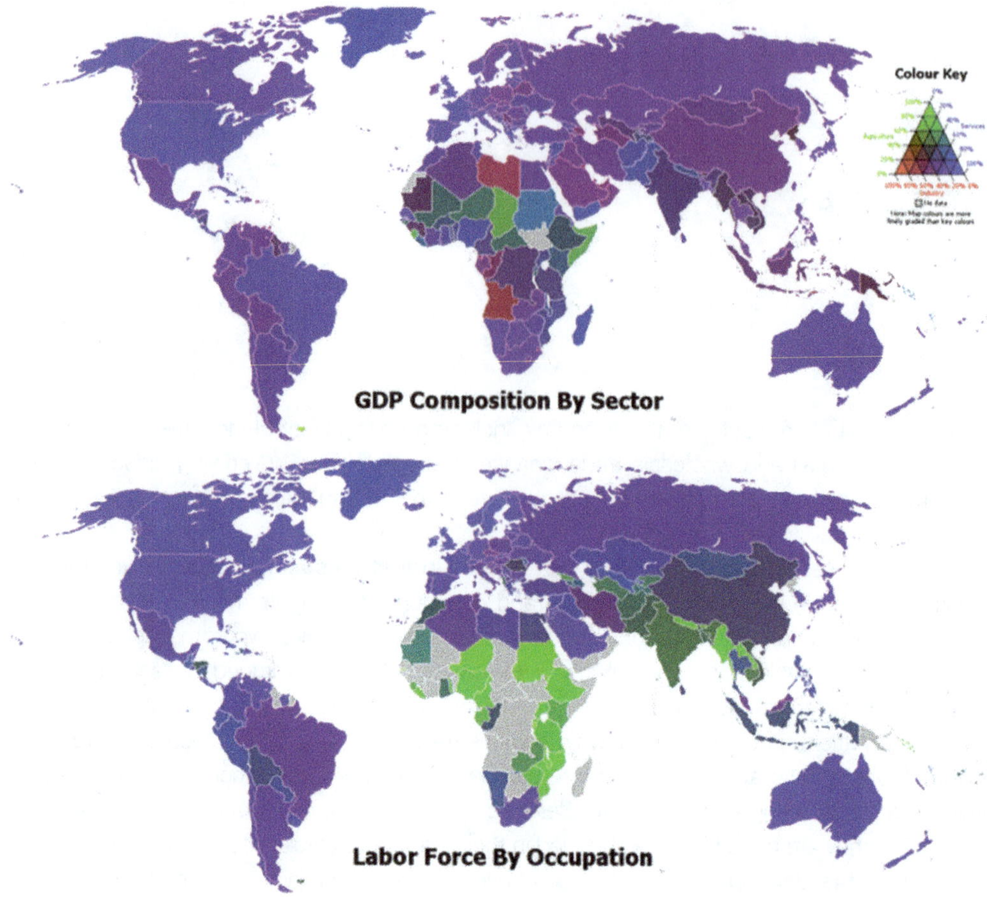

Figure 10.1 Global composition of GDP and labor force by agriculture, manufacturing, and services.

measures service output using corporate revenues as a proxy, but this measure is distorted by the relative prices of service outputs. Moreover, many services generate both tangible and intangible outputs. In a fast-food franchise, for example, the output is tangible, yet it is considered a service. The lines between services and manufacturing are often blurry.

The difficulty of measuring output in services has important analytical implications. For example, some observers argue that the slowdown in productivity growth reflects the rise of services. But if output in services cannot be accurately measured, how can one argue that output per employee in services is high or low, or rising or falling?

The major sectors of the service sector generally include the following:

a Producer services are sold primarily to corporations (intermediate demand) rather than private households (final demand). Many producer services firms serve both, such as attorneys who have both commercial clients and individuals. Producer services include two major groups:

- Finance, insurance, and real estate (**FIRE**) includes commercial and investment banking, insurance, and commercial and residential real estate;
- Business services include legal services, advertising, computer services, engineering and architecture, public relations, accounting, research and development, and consulting.

Table 10.1 **Composition of US Labor Force (Millions), 1960–2020**

	1960	*1970*	*1980*	*1990*	*2000*	*2010*	*2020*
Agriculture	5.0	4.8	3.9	3.6	3.5	3.3	2.2
Mining	0.9	0.7	0.6	1.0	0.7	0.5	0.8
Construction	2.1	2.9	3.6	4.3	5.1	6.6	11.0
Manufacturing	16.4	16.8	19.4	20.3	19.1	18.5	15.9
T.C.P.U.[a]	3.9	4.0	4.5	5.2	5.8	7.0	7.7
Wholesale trade	2.6	3.1	4.0	5.3	6.2	7.0	4.1
Retail trade	6.6	8.2	11.0	15.1	19.6	23.3	16.5
F.I.R.E.[b]	1.8	2.6	3.6	5.1	6.7	7.6	10.2
Other services:	5.3	7.3	11.4	17.5	27.3	39.9	53.9
Personal services	0.8	0.9	0.8	1.1	1.2	1.4	
Business services	0.6	1.4	2.5	5.1	9.7	15.5	
Health services	1.5	3.0	5.1	7.6	10.0	14.1	
Legal services	0.3	0.5	0.9	1.0	1.6		
Education	0.6	0.9	1.1	1.6	2.3	2.4	
Social services	0.5	1.1	1.7	2.8	2.7		
Engineering	2.5	3.3	3.8				
Government	5.8	8.2	12.3	16.1	18.0	20.4	23.1
Total	48.5	59.2	75.0	94.7	124.3	145.6	188.9

[a] Transportation, Communications, and Public Utilities
[b] Finance, Insurance, and Real Estate
Source: US Bureau of Labor Statistics.

b Transportation and communications include the media (television, radio, Internet), trucking, shipping, railroads, airlines, and local transportation (taxis, buses, Uber drivers).

c Wholesale and retail trade acts as intermediaries between producers and consumers.

d Consumer services subsume eating and drinking establishments, personal services (e.g., beauty salons), and repair and maintenance services. Entertainment, hotels, and motels comprise parts of tourism.

e Government or the public sector at the national, state, and local levels includes civil servants and everyone involved in the provision of public services (e.g., public education, health care, police, fire departments, etc.).

f The nonprofit sector includes charities, churches, museums, membership organizations, and private, nonprofit health care agencies.

Definitions of services are problematic. For example, do services refer to industries or occupations? The US government uses industries. But this approach has problems. Why is the secretary who works for a steel company part of manufacturing, while the secretary who works for a law firm is part of services? Yet when measured on the basis of their daily occupational activities, many workers in manufacturing are in fact service workers, including staff in headquarters, administration, clerical functions, janitors, and research workers who are often called **nondirect production workers**.

Using the standard definition, services generate the vast bulk of output and employment in economically developed countries of the world. More than 85 percent of the labor force of the United States is employed in services (Figure 10.2); similar proportions hold in Europe, Canada, and Japan. In addition, services comprise the overwhelming majority—often 90 percent—of all new jobs in these economies, meaning that they are not only heavily service-oriented but are becoming more so. Even in much of the developing world, services are a growing share of the labor force, including much of the "informal" (untaxed, unregulated) economy (Chapter 6).

10.2 Forces Behind the Growth of Services

There are several reasons why services have grown so rapidly, five of which are:

10.2a Rising Incomes

A common explanation for the rise of services is rising per capita incomes. The demand for many services is income-elastic, which means that increases in real personal income tend to generate proportionately larger increases in the demand for services, which is not the case for manufactured goods. Services with very high income elasticities include entertainment, health care, and transportation. Today, American households spend more on services than they do on durable and nondurable goods combined.

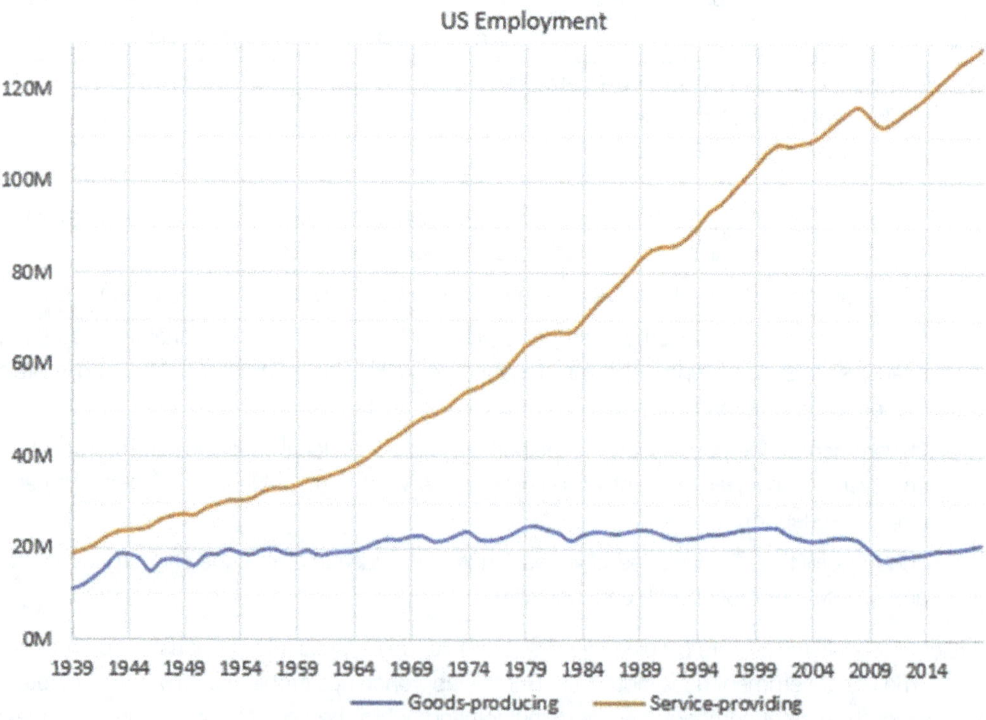

Figure 10.2 US employment in services and goods-producing sectors.

Rising incomes are accompanied by the increasing value of time, especially in households with two-income earners. As the value of time rises relative to other commodities, consumers minimize the time inputs needed for many tasks such as cooking and cleaning. This phenomenon explains the demand for dishwashers and automobiles, but it is especially important for the growth of services. The expansion of fast-food restaurants, for example, has little to do with the quality of food and much to do with attempts by consumers to reduce time spent cooking at home. The growth of repair services reflects increasingly sophisticated goods such as automobiles and an unwillingness to spend time (and lack of skills) fixing them. Thus, the increasing value of time has led to the externalization of household functions, so those that used to be done in-house become commodities purchased through the market.

However, this explanation is highly limited. First, it only explains consumer services. Equally important, real incomes in many countries have not grown much, but services continue to rise. Clearly, other factors are at work.

10.2b Rising Demand for Health Care and Education

The growing demand for health and educational services is an important part of the rise of the service economy. Health services employment and output increased steadily throughout Europe, North America, and Japan over the last century. In the United States, health care expenditures surpassed $4 trillion in 2018, roughly 15 percent of gross domestic product and much higher than any other industrialized country. The United States is the only developed country lacking a national health insurance system; today, about 9 percent of Americans are uninsured. Thus, the United States has the world's most inefficient system, spending far more per capita on health care than countries with single-payer systems. In many cities, health care is the largest part of the local economy, including, for example, Pittsburgh, Rochester, Minnesota, and Birmingham, Alabama.

One major reason why the demand for health care has increased is the changing demographic composition of the developed world. In wealthy countries, the most rapidly growing age groups today are the middle-aged and the elderly (Chapter 7), precisely those demographic segments that require high levels of medical care. As the baby boom enters its elderly years, the demand for health services will rise even higher. Higher life expectancies, soaring equipment, medical insurance and litigation costs, and pharmaceutical and research expenses also added to the growth of this sector.

Education is also part of the growth of services. Globalization and increasing demand for analytical skills at the workplace have driven the rising demand for educational services, a process reflected in higher enrollments in universities. A college degree has become a prerequisite for entrance to middle-class jobs. Whereas the bulk of graduating high school students in the United States did not attend college in the 1960s, almost 70 percent do so today.

10.2c An Increasingly Complex Division of Labor

The growth of services is also a product of the increasing complexity of the division of labor, as reflected in the rising proportion of nondirect production workers. All corporations must devote resources to dealing with the market and legal environment, including specialized, sophisticated clients, complex tax codes, numerous environmental and labor restrictions, international competition, mind-bogglingly complicated financial markets, and specialized real estate purchases and sales.

To survive and succeed in this environment, firms require staff to collect and process vast amounts of information and to make strategic decisions; clerical workers to process paperwork; researchers to analyze market demand and create new products; advertisers to market their output; and public relations, accountants, lawyers, and financial experts to assist in a complicated decision making environment. Similarly, the use of sophisticated machinery in factories and offices requires maintenance and repair personnel. Unlike income-based or demographic arguments, this approach explains the growth in producer services, the most rapidly growing sector of the economy of most developed countries.

This view embeds the growth of services in the broader dynamics of capitalism. For example, the **outsourcing** of services represents an extension of the division of labor. Services such as research and development, design, marketing, product testing, and sales are all essential steps in getting goods from producers to consumers. This means that the dramatic growth in business service employment reflects alterations in the way in which production is organized rather than the rise of a new type of knowledge economy.

10.2d The Public Sector

Another reason for the growth of services is the increasing size and role of the state or the public sector, and steady expansions in government employment. Governments contribute to the growth of services in two ways. First, public sector employment has increased steadily because the public wants the services that it provides. Governments provide services that are socially necessary but not profitable, such as public transportation and education, libraries, police, fire fighting, social workers, disaster relief, postal services, and public health. Today, the federal government is the largest single employer in the United States, employing more than 2.3 million people; local and municipal government employment is even larger.

The second way in which government contributes to the growth of services is indirectly, through laws and regulations such as tax codes, zoning ordinances, environmental restrictions, and anti-discrimination laws. To follow these mandates, firms must hire or subcontract with attorneys, accountants, consultants, and other specialists who assist them in negotiating the legal and regulatory environment.

10.2e Service Exports

A fifth reason for the growth of services is the growing service exports within and among countries. Some observers cling to the myth that services serve only local demand; that is, they are, by definition, nonbasic activities (Chapter 3) and are thus of secondary importance to manufacturing. This view reflects a long-standing bias against services in economic theory. In reality, in countries where the overwhelming majority of workers are employed in services, it would be odd if services were *not* traded among places. Many cities, regions, and countries derive a large portion of their revenues from selling services to clients located elsewhere. Many urban areas export services to other parts of the same nation. "Export" refers to the location of the source of revenues: if the demand is nonlocal, then sales of a service are exported and form part of the local basic sector. For example, cities like Las Vegas depend exclusively on exports of tourism and entertainment services to tourists, whether domestic or foreign. When a television company in one area sells advertising time to a client in another, it exports a service. New York City relies heavily upon exports of financial services domestically as well as globally, and exports legal and advertising services. The Washington, DC region exports federal government services to the rest of the country.

Services are also traded internationally, comprising roughly 20 percent of international trade. The United States is a net exporter of services, exporting $204 billion and importing $150 billion in 2021, leaving an annual trade surplus of roughly $54 billion. The United States also suffers major trade deficits in manufactured goods of more than $800 billion annually. As the United States has lost its comparative advantage in manufacturing, it has gained one in financial and business services. The data on global services trade are poor, but some estimates suggest that services comprise roughly one-third of total US export revenues. These sales overseas take many forms, including tourism, fees and royalties, sales of business services, and profits from bank loans; unfortunately, the US data in the Current Account include repatriated profits from manufacturing investments overseas as a "service," while in reality these comprise returns from capital investments abroad, many of which are in manufacturing. Service exports do not generate jobs and revenues in all places equally: cities such as New York, Los Angeles, and London, for example, which are critical to global capital markets, have benefited the most heavily.

10.3 Labor Markets in the Service Economy

In the world's developed countries, the overwhelming majority—more than 80 percent—of employment involves services. Roughly 90 percent of new jobs are in services. Services, however, include a very diverse array of industries and occupations. The labor markets these sectors create have characteristics in common. Among these are their relatively labor-intensive nature, income distribution, gender composition, lack of unions, and educational requirements.

10.3a Labor Intensity

In contrast to manufacturing, services tend to be labor-intensive, i.e., they use relatively more labor per unit of output. Wages and salaries for most service firms range from 70 to 90 percent of their total costs, compared to 5 to 40 percent in manufacturing, depending on the degree of labor- or capital-intensity. Some services, such as finance, can be very capital-intensive and generate enormous outputs with few workers; others, however, such as education and medical care, require numerous employees. Whereas manufacturing has been relatively easy to mechanize ever since the Industrial Revolution, in most services, it is much more difficult or costly to replace workers with machines, particularly for jobs that involve variations in tasks, judgment, or dexterity. Services do clearly exhibit technological change, such as computers in the office, artificial intelligence, or price scanners in grocery stores. In low-paying jobs, however, firms' incentives to replace workers with machines are usually low.

10.3b Income Distribution

Rising income inequality is a social concern worldwide. A common view is that manufacturing-based economies created an income distribution that was statistically "normal," that is, with a large middle class and relatively few rich or poor people. In contrast, services are often held to generate a bifurcated wage distribution split between well-paying, white-collar managerial/professional occupations, on the one hand, which require a university degree, and unskilled, low-paying jobs on the other, which require minimal education (e.g., retail trade, security guards, etc.). In contrast to the optimistic expectations of postindustrial theorists that services would generate unprecedented wealth and eliminate poverty, in reality, most new service jobs pay poorly, offer few benefits, and are part-time or temporary in duration. The growth of low-wage services has led to widespread concerns about the "McDonaldization" of the economy.

The most rapidly growing occupations in the United States over the last decade (Table 10.2) include professionals but also low-wage service workers, such as those in retail trade. The occupations with the greatest projected job growth include almost any work having to do with computers, as well as health care, such as home personal care aids and medical assistants. The largest number of new jobs will be in low-wage, unskilled positions such as food preparation, customer service, retail sales, clerks, and security guards. Further, the United States witnessed a steady growth of involuntary, part-time jobs that are usually filled by women and minorities. The growth of low-wage jobs signaled the rise of the "working poor," signifying that even employment does not pull people out of poverty. It has also swollen the number of homeless people caught between jobs that pay too little and housing that costs too much.

Rising in parallel with concerns about the distribution of income in services is the loss of relatively well-paying manufacturing jobs. On average, services tend to pay poorly compared to manufacturing: The average clerical job generates only 60 percent of the annual income of a blue-collar worker, and the average retail trade job only 50 percent as much. These trends have amplified fears about the declining middle class and the polarization of service-based economies into "haves" and "have nots."

However, there is scant statistical evidence to suggest that the distribution of incomes in services differs much from that in manufacturing. Services generate the jobs that allow for tens of millions of middle-class suburbanites to live comfortable lives. Others argue that the mounting income inequality reflects the increasingly regressive nature of taxation in the United States (where the affluent pay relatively little) as well as the growth of unearned income such as rents, royalties, and stock dividends, from which the wealthy derive the bulk of their earnings.

10.3c Gender Composition

A third dimension of labor markets in services is their gender composition. Manufacturing mostly employed males; while some women worked in factories, particularly textiles and garments, manufacturing-based economies usually have relatively low rates of female labor force participation. Most women who worked outside the home did so either for short periods prior to getting

Table 10.2 **US Most Rapidly Growing Occupations, 2006–2024 (Thousands)**

	2006	2024	% change
Home care aides	767	1,156	50.7
Home health aides	787	1,171	48.8
Customer service	2,202	2,747	24.8
Nurses	2,505	3,092	23.4
Teachers	1,672	2,054	22.8
Nurses' aides	1,447	1,711	18.2
Food preparation	2,503	2,955	18.1
Janitors	2,387	2,732	14.5
Office clerks	3,200	3,604	12.6
Retail sales	4,477	5,034	12.4

Source: US Bureau of Labor Statistics

Table 10.3 **Twenty Leading Occupations of Employed Women in the United States, 2022**

Occupation	% Women
Secretaries	96.7
Elementary School Teachers	96.7
Nurses	91.8
Home Health Aides	88.3
Cashiers	75.0
Administrative support	69.5
Retail sales	41.1
Customer service	70.1
Clerks	91.2
Accountants	60.8
Receptionists	93.9
Maids	89.7
Office clerks	83.8
Secondary school teachers	54.8
Waiters/waitresses	67.3
Financial managers	55.7
Teacher assistants	91.7
Preschool teachers	97.7
Social workers	76.1

Source: US Dept. of Labor

married or in a narrow range of occupations such as teaching or nursing (Table 10.3).

The growth of services has been accompanied by the steady rise in the number of women in the paid labor force throughout the developed world. The most rapid increase in women's labor force participation rates has been among married women with children. In the United States, women comprise 48 percent of all full-time employees. However, when women enter into services, it has often been in so-called pink-collar jobs, such as clerical and secretarial work, retail trade, health care (other than physicians), eating and drinking establishments, teaching, and childcare. Most of these jobs pay relatively poorly compared to jobs for men. This trend raised concerns about the feminization of poverty. Also, women's presence in high-paying occupations such as physicians, attorneys, or corporate management has historically been low, leading to concerns about the "glass ceiling" many women face at the workplace. Because they are clustered in low-paying jobs, women, on average, in the United States earn only 77 percent of men's incomes.

The growth of women in the paid labor force has several social implications. One view is that women's work serves as a strategic response to declining male incomes brought on by deindustrialization. The rise of the two-income family changed gender roles at home and led to pressures on men to do more housework. For dual-income families, childcare is a constraint to women's job opportunities and career advancement. For families with childcare, it often is a source of financial and emotional stress, complicating commuting patterns. For the poor, the lack of affordable childcare is a perpetual crisis. Unattended children, or "latchkey kids," spend long periods without adult supervision.

10.3d Low Degree of Unionization

Most services are not unionized. Since the Second World War, the percentage of workers belonging to unions has dropped steadily from 45 percent in 1950 to 12 percent today, mostly due to deindustrialization. Jobs in services are rarely unionized; the vast majority of workers in low-wage jobs, and in many skilled professions, do not belong to unions. Service employees do occasionally form unions, such as in health care (e.g., Union 1199), teaching (e.g., American Federation of Teachers), or the public sector (e.g., American Federation of State, County, and Municipal Employees), but these are exceptions.

10.3e Educational Inputs

Skilled managerial and professional services require considerably more education than jobs in manufacturing, particularly in numeracy, computer skills, and communications. To obtain the cognitive skills necessary to enter professional jobs, a college degree is usually required; a university education has become a prerequisite for obtaining middle-class lifestyles. Unemployment rates are typically much lower for well-educated workers compared to those without a high school degree. Moreover, the average income of workers with college degrees is considerably higher than that of those without. College degrees pay for themselves many times over if one considers lifetime earnings.

As a result, the demand for educational services has risen throughout the economically developed world, leading to a surge in college enrollments. In the United States, only 20 percent of high school graduates went on to universities after graduating in the 1950s; today, 70 percent do so (although some do not graduate). In the context of a knowledge-based economy, higher education is essential to national and local economic competitiveness. Within the United States, states that have developed high-quality university systems have enjoyed an advantage over those that neglect educational funding. While some non-college-educated people earn a fine living in professions such as construction, plumbing, and electrical, most workers without a university education, disproportionately consisting of the poor and minorities, are increasingly likely to face a lifetime of poverty.

10.4 Financial Services

Finance is a hugely important sector in contemporary economies. The development of capitalism historically included the formation of modern banking and money-based economies (Chapter 2).

Finance consists of several closely related sectors. Their primary function is to link borrowers and savers (i.e., individuals and institutions with excess capital and those who need capital). Banking revolves around the flow of money from savers to borrowers and the flow of assets (e.g., debt) from borrowers to savers. However, the markets involved, the size of firms, and the regulatory environment they face all vary markedly over time and space. Government regulations often strictly limit some types of firms from operating in some markets. Although deregulation has blurred the boundaries among them, the major sectors of finance include the following:

Commercial Banking

Commercial banks are the institutions involved in extending commercial loans (i.e., to firms, not individuals), including capital for business start-ups and commercial real estate. Commercial banks provide the capital that firms use to build office towers, hotels, convention centers, waterfront developments, and sports stadia. These institutions also offer personal loans, for example, to people who wish to purchase a boat or pay for a college education. They also provide retail banking services such as checking, savings accounts, and credit cards.

Investment Banking

Long legally differentiated from commercial banks, investment banks are primarily involved in buying and selling securities, such as stocks, bonds, futures, and derivatives. They also provide specialized expertise and capital in facilitating corporate mergers, takeovers, and leveraged buyouts. Both commercial and investment banks buy and sell foreign exchange.

Savings Banks

Also called thrifts or savings and loans (S&Ls), these firms were traditionally involved only in the market for mortgages, or loans to buy houses. They are the largest source of home loans and therefore play an important role in creating the residential geography of cities (including concerns over redlining, or denials of mortgages to low-income, predominantly minority communities). However, with the deregulation of finance in the 1980s, savings banks diversified, entering the commercial real estate market.

Insurance

Insurance is the commodification of risk. Insurance companies issue policies, funded by premiums paid by households and firms, that provide protection from unexpected loss. There are many types of insurance, including property (e.g., for houses), casualty, life, health, automobile, travel, and old age care.

10.4a The Regulation of Finance

Finance is highly regulated compared with most sectors. The reasons for this center on the importance of money and capital markets to current economies. Financial firms can affect the national money supply, which in turn shapes national interest, inflation, and exchange rates.

Almost all countries have a national bank to regulate their financial firms. In Britain, this role is played by the Bank of England; in Germany, the Deutsche Bank; in Japan, the Bank of Japan. In the United States, the Federal Reserve, set up in 1913, acts as a sort of national bank. The Federal Reserve consists of twelve Federal Reserve Banks in various parts of the country. It manages interest rates and the money supply by controlling the federal funds rate (the rate at which banks borrow and lend their excess reserves), the reserve ratios of commercial banks, and by buying and selling government securities such as Treasury Department bonds.

In the 1930s, as the banking system was devastated by bankruptcies caused by the Depression, the federal government introduced new regulations to stabilize the system, such as the separation between commercial and investment banking and laws prohibiting interstate banking, which were meant to protect small local banks from competition from large out-of-state ones, particularly those in New York. The result of these regulations was a highly fragmented, decentralized system of banking in the United States. Today, there are roughly 5,000 banks in the United States compared to 15 in Canada. The number of American banks is declining as mergers and acquisitions have reshaped the industry. Most banks are relatively small, with relatively few employees and assets, and serve local markets (i.e., they are nonbasic). The financial sector is dominated by a handful of large "money center" banks, commercial giants often headquartered in New York; only 2 percent of US banks own 50 percent of all bank assets. Thus, employees in finance are found in towns throughout the country; the largest numbers are concentrated in metropolitan areas such as New York, Boston, Los Angeles, and San Francisco.

The regulated system of banking worked well for fifty years (i.e., from the Depression of the 1930s to the enormous changes that swept the world economy in the late twentieth century). The international banking system was stabilized for much of this time by the Bretton Woods Agreement of 1947, in which the United States and its allies created the architecture for the global financial system. For decades, the number of banking bankruptcies was very low. In the 1980s, however, the banking system confronted a radically different environment. The Bretton Woods Agreement collapsed in 1973, and the shift to floating exchange rates created a new industry of currency trading. Rising debt in the developing world accelerated the globalization of banking as large banks extended loans to countries in Latin America and Asia. The microelectronics revolution created new ways of serving customers (e.g., automatic teller machines) and electronic funds transfer systems, which allowed money to be moved digitally.

10.4b Electronic Funds Transfer Systems

Financial markets were particularly heavily affected by the microelectronics and digital revolution. Telecommunications removed transmission costs for the movement of capital, much as deregulation eliminated regulatory barriers. Banks, insurance companies, and securities firms are highly information-intensive, so it is not surprising that they were at the forefront of the construction of a global network of fiber optic lines.

Specifically, banks and other firms adopted electronic funds transfer systems (EFTS), which constitute the nervous system of the global financial economy. EFTS allow banks to move capital around instantaneously, allowing them to take advantage of even tiny national differences in interest and exchange rates. EFTS take several forms. One is the Real Time Gross Settlement (RTGS) systems, which handle flows of funds among financial institutions and governments. The largest is the US Federal Reserve Bank's Fedwire system, which allows depository institutions to transfer funds to any other depository institution. In 2020, total Fedwire traffic amounted to $4 trillion per day. Other major US payment mechanisms include the Clearing House Interbank Payments System (CHIPS), which handles about $1 trillion in daily transactions. In Europe, the Society for Worldwide Interbank Financial Telecommunications (SWIFT) plays a similar role and processes €2.6 billion daily in transactions. In Japan, the Bank of Japan's Financial Network System fills this role.

Private firms also have systems for moving money. For example, Citigroup trades hundreds of billions of dollars daily in foreign exchange markets around the globe. Similar systems are run by MasterCard, Reuters, and many large stock exchanges. Yet others exchange cryptocurrencies.

Similarly, in the securities markets that trade in stocks and bonds, EFTS gave rise to 24-hour/day trading, linking stock markets together through computerized trading programs. Online trading also allows small investors to use the Internet for information, eroding the advantage held by specialists. Stock volumes on many exchanges rose exponentially as a result. The National Association of Securities Dealers Automated Quotations Stock Market (NASDAQ), the first fully automated stock trading system, is now the world's largest stock market. In Europe, the European

Association of Securities Dealers Automated Quotation System (EASDAQ) plays a similar role. Challenged by these electronic trading systems, many stock markets, such as those in Paris, Belgium, Spain, Vancouver, and Toronto, abolished their trading floors. The volatility of trading increased as computer trading programs allowed large swings in the price and volume of stocks.

Traveling through fiber networks at the speed of light as a series of zeros and ones, digital money washes over the world in enormous volumes. The world's currency markets, for example, trade roughly 6.6 trillion US dollars every day, which is much larger than the funds that pay for global trade in goods and services. Electronic money also fueled a shift from investing to speculation, particularly in currencies. Large amounts of foreign direct investments (FDI) have shifted to intangible investments such as stocks and bonds. Unlike FDI, which generates employment, accelerates technology transfer, and has material impacts on landscapes, financial investments create few jobs, are invisible to all but a few, and have unpredictable consequences.

EFTS had particularly important impacts on currency exchange markets. Since the shift to floating exchange rates in 1973, foreign currency trading (FOREX) has become the world's largest industry by volume. Currency exchange is driven by growing international trade, the removal of exchange controls, and the growth of nontraditional actors in finance, such as pension and mutual funds and insurance companies. The vast bulk (85 percent) of currency exchange transactions involves the US dollar. Typically, the trades involved follow the opening and closing hours of markets around the world. For example, the FOREX market opens in Asia while it is still nighttime in North America; funds then move west over fiber optic lines, say from Tokyo to Shanghai to Singapore to Bahrain to Frankfurt or London, then to New York, Los Angeles, and back across the Pacific Ocean.

10.4c Offshore Banking

Another aspect of the rise of a digitized global economy is the emergence of offshore banking centers, which are small, loosely regulated places. These originated in the Euromarket of the 1960s, when European countries enjoyed large trade surpluses with the United States, which left them with large quantities of dollars called Eurodollars. The Euromarket spread far beyond Europe and today includes all trades in financial assets outside of the country of issue (e.g., Eurobonds and Eurocurrencies), not just currencies. Unlike the regulated financial system in the United States, the Euromarket was essentially unregulated.

Digital globalization spawned a series of offshore banking centers (OFBs), which are usually but not always located on small islands. OFBs offer regular commercial services (i.e., loans), foreign currency trades, asset protection (insurance), investment consulting, and international tax planning. There are five major world clusters of OFBs, including the Caribbean (Cayman Islands, Bahamas, Panama); Europe (Isle of Man, Jersey, Luxembourg, Liechtenstein, San Marino); the Middle East (Cyprus, Bahrain, Abu Dhabi); Southeast Asia (Hong Kong, Singapore); and the South Pacific (Nauru, Vanuatu) (Figure 10.3). Such places are black holes in the global landscape of financial regulation, offering lax regulations and low taxes. Some of them have dubious reputations as centers of money laundering, i.e., as places in which illicit funds earned through international drug or weapons sales are converted into legitimate funds.

As the digital revolution allowed global capital to circulate more freely and rapidly, the technological barriers to moving money declined dramatically. Accordingly, spatial variations in the nature and degree of regulation rose in importance. Even small differences in regulations concerning taxes or repatriated profits may be sufficient to induce large quantities of capital to enter, or exit, particular places. Thus, many small states attempt to attract finance capital by deregulating as

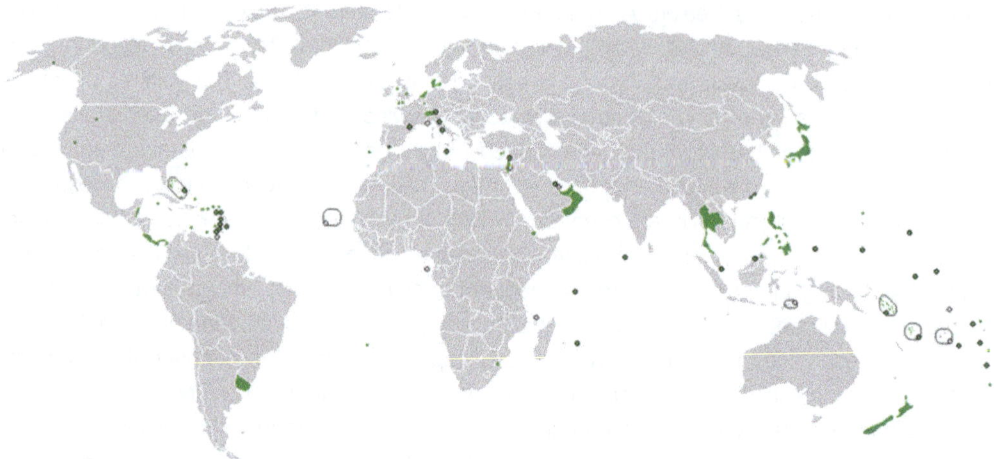

Figure 10.3 Distribution of the world's offshore financial centers.

much as possible, lifting controls over currency exchanges, investment, repatriated profits, and eliminating taxes in the hope that global money will select their locale. By allowing funds to cross national borders at will, offshore banking reworks the notion of national sovereignty, inducing nation-states to retain political sovereignty but to sacrifice financial sovereignty.

10.5 The Location of Producer Services

The geography of producer services reflects their need for large pools of highly skilled labor and the agglomeration economies offered by large cities. For these reasons, they tend to concentrate in large metropolitan regions. Two-thirds of business and professional services jobs in North America are located in large urban areas such as New York, Chicago, Los Angeles, and Toronto. Agglomeration economies allow the minimization of transaction costs between suppliers and subcontractors, pools of labor, transportation and communications infrastructure, and access to specialized information.

10.6 Back Office Relocations

Large service corporations often divide their operations into two major parts, headquarters and back offices. Headquarters use skilled professionals, including executives and managers, as well as those who perform research, public relations, finance, legal work, accounting, engineering, marketing, and product development. Back offices perform functions such as data entry, payroll and billing, bank checks, insurance claims, and subscriptions. Back offices rely on semiskilled labor, primarily women, and operate on a 24-hour-per-day basis. Unlike headquarters, they have few linkages to clients or suppliers. These facilities require extensive computer facilities, reliable sources of cheap electricity, and fiber optic networks.

In the past, back offices were located next to headquarters in downtown areas to allow close management supervision. However, with the microelectronics and telecommunications revolution, many firms began to uncouple their headquarters and back offices, moving them to cheaper locations in the suburbs, which offered pools of female labor, ample parking, and cheaper electricity.

In the twenty-first century, companies began to relocate on an international scale, especially to countries with low-wage workers. Many offshore back offices are in industries facing competitive pressures. For example, several New York–based companies relocated back-office facilities to Ireland. Likewise, the Caribbean has become an important destination for American back offices, which have moved to Jamaica, Barbados, and the Dominican Republic. In Asia, Manila, the Philippines, is a back-office center for British firms. Beyond simple data entry, offshore offices have attracted an increasing range of functions, including call centers, tax preparation, and radiologists who read digital medical records.

Another form of low-wage, low-value-added services involves call centers, including customer service, telemarketing, and technical support. These range in size from as few as five to several thousand employees. The workforce consists primarily of low-skilled women. There are roughly 100,000 call centers in the United States that employ about 3 percent of the labor force. The majority are located in urban or suburban locations.

Like other industries, call centers have become globalized. India has made itself into the call center capital of the world, with a large number of centers near its software capital, Bangalore. Indian workers are trained to speak American English and to be able to chat with customers about pop culture. Wages, which average $2,000 per year, are higher than most Indian salaries but are only 10 percent of what equivalent jobs pay in the United States.

10.7 Consumer Services

In addition to producer services, which are sold primarily to other firms, services include consumer services, which are sold to individuals and households. Consumer services subsume a wide variety of industries and occupations, including retail trade and personal services, eating and drinking establishments, tourism, sports and entertainment facilities, and health care and education.

In economic base theory (Chapter 3), consumer services are part of the nonbasic sector, the part of the economy that recycles revenues derived through the export base. However, often the boundary between basic and nonbasic industries is hard to draw; legal services, for example, may be consumed by both firms and households (e.g., divorces), and to some extent, they too are part of the consumer services sector. More recently, however, some consumer services have become basic sectors in their own right. Large cities with specialized stores (e.g., Macy's in New York) often lure nonlocal residents and thus form part of the economic base. Outlet malls in small towns perform a similar function. Tourism and conventions are likewise consumer services that make them part of the economic base.

Traditional geographic theorization of consumer services focused on the travel costs incurred in their consumption, such as Walter Christaller's famous but now-outdated Central Place Theory. Today, many large retail chains employ specialists to evaluate the market potential of stores, including the possible number of consumers, their economic and demographic characteristics, rents, and transportation routes. These analyses often utilize geographical information systems (GISs).

The distribution of consumer services closely reflects the affluence and disposable incomes of local customers. When incomes are high, consumer services do well, although the expenditures on goods and services do not rise equally (Chapter 11). The geography of consumer services is thus closely linked to local demographics, the tastes and preferences of different groups of people, their saving and spending habits, advertising and other sources of information that shape their preferences, and their age, gender, and ethnicity, all of which shape buying habits (Chapter 11). The most obvious example concerns suburbanization; as the urban periphery drew much of the middle class and its incomes, the urban geography of consumption was reshaped, leading to the growth of suburban malls of various sizes. Yet other forces that shape consumer services include work patterns, the rise of retail chains and franchises, government regulations, and the prices of imports.

More recently, as purchasing over the Internet has become more feasible, consumer services have migrated online (Chapter 11). The COVID pandemic accelerated this trend markedly. This phenomenon has contributed to the growth of giant firms such as Amazon, the world's largest online retailer, which sells a wide variety of goods and services.

10.8 Tourism

Tourism is usually defined as travel to a place outside of one's residence involving a stay for at least one night. Tourism is a huge consumer service with enormous international and local implications. There are many kinds of tourism, including personal as well as business travelers and conventions. The duration of tourist visits varies over a wide range of durations of stay. Domestic tourists usually outnumber international visitors, although both types comprise basic sector activities: when tourists spend money at a destination, the place exports a service.

The growth of global tourism reflects the rise in disposable income among the world's middle class. Mass tourism began in the nineteenth century, but grew particularly rapidly during the boom after the Second World War, when the working classes could afford vacations. The automobile and the wide-body jet airplane greatly reduced transport costs and times. Tourism reflects the income-elastic demand for leisure that accompanies rising disposable incomes.

Today, tourism is the world's largest industry by employment. It forms part of the economic base of many countries, including, for example, Italy, France, and Spain, as well as smaller island countries in the Caribbean, Hawaii, Florida, and cities such as New Orleans, Las Vegas, Los Angeles, and New York. Table 10.4 depicts the distribution of the world's 1.45 billion international tourist arrivals in 2024 (12 percent of the world's population): more than one-half of the world's tourists visited Europe. France, Italy, and Spain boast the largest national tourist markets in the world. To a lesser extent, Asia and North America are also significant tourist destinations.

Economically, the tourism industry comprises parts of several sectors, including international and local transportation, hotels and motels, eating and drinking establishments, entertainment, and retail trade. These sectors require enormous investments in water, transportation, communications, and other infrastructures.

Table 10.4 **International Tourist Arrivals, 2024**

Region	Visitors (millions)	% of World
Africa	69.9	4.8
Asia	360.7	24.8
Europe	744.3	51.1
Middle East	61.4	4.2
North America	101.4	6.9
Latin America	117.7	8.1
Total	1455.4	100.0

Source: World Tourism Organization

There is an enormous diversity in the types of tourism and tourist destinations, including low-impact ecotourism, highly urbanized urban areas, simple backpacking, luxury cruises, safaris, honeymoon retreats, package tours, health resorts, sex tourism, dude ranches, museums, ski resorts, and tropical playgrounds. The facilities that accommodate tourists are similarly varied, ranging from large hotel chains to small bed-and-breakfast establishments. Tourism is typically highly seasonal, fluctuating greatly throughout the year, with corresponding changes in prices for hotels and travel.

The volume of tourists to a given destination is determined by the information available to potential travelers; their disposable incomes and preferences for travel; currency exchange rates; available transportation and lodging and their costs; the cultural familiarity (e.g., language) or exotic appeal the destination has; and concerns about crime and terrorism. Sometimes, there are also political restrictions, such as those that long prohibited US tourists from visiting Cuba.

The impacts of tourism are numerous, but are usually confined to local sectors that deal with visitors, such as hotels, local transportation, and retail trade. In cases where a tourist operation imports many inputs, it can generate relatively small positive impacts, creating "tourist enclaves" with few multiplier impacts. Often, tourist-related jobs are unskilled and pay low wages. In fragile ecosystems, tourism can have detrimental effects on the local environment: skiing in the Alps or hiking in Nepal, for example, can generate garbage, and high-density tourist developments (e.g., Hawaii) create large amounts of waste.

10.9 Summary

Around the world, particularly in developed countries, the vast majority of the labor force is employed in services such as shops, offices, restaurants, hospitals, airlines, banks, universities, and hotels. A service is an economic activity not involve the transaction of a physical product, that is, the production and consumption of intangibles.

This chapter began by noting the difficulties in defining services and emphasized the diversity among services industries. Services encompass both highly skilled, well-paying positions, such as hedge fund manager or surgeon, and low-paying, unskilled ones, such as waitress, security guard, cashier, and janitor. The chapter discussed reasons underlying the massive structural shift from a manufacturing economy to services: the driving forces include changes in income and consumption patterns, the growth in public sector employment, health care expenditures, international trade in services, and the increasingly complex production environment created by a more refined division of labor.

Labor markets in services are noted for their labor intensity, lack of unions, large number of female workers, and, for professional services, higher educational requirements. Services do exhibit inequality in their income distribution, but it is not markedly different from that in manufacturing.

The geography of producer services is dominated by the pressures to agglomerate. Highly skilled, white-collar functions tend to concentrate in large cities, where they rely on face-to-face contact, while unskilled ones, such as back offices, have increasingly dispersed into rural areas and less developed countries.

Contrary to erroneous stereotypes that services are consumed where they are produced, there are, in fact, thriving trade networks in producer services. The chapter discussed offshore banking and the globalization of back offices. In finance, electronic funds transfer systems revolutionized movements of capital. Hypermobile, digitized money is a defining feature of contemporary globalization.

Finally, the chapter discussed consumer services, which tend to follow the disposable incomes of their clients. However, not all of them are nonbasic activities. Tourism, the world's largest industry in terms of employment, is very powerful in shaping national and local economies.

Glossary

Back offices: offices that provide data input services for firms Call centers
Consumer services: services consumed by individuals and households
Contingent labor: part-time and temporary labor, usually low-paid
Electronic funds transfer: digital networks designed to move large quantities of financial funds
FIRE: the finance, insurance and real estate sector
Income elasticity: the percentage change in supply or demand that occurs when incomes rise
Nondirect production workers: workers in manufacturing firms who are not immediately engaged in the production process, such as clerical staff and administration
Offshore banking: bank operations located overseas that allow firms to escape taxes
Outsourcing: the use of external suppliers, usually overseas, to provide inputs of goods and services
Post-industrial society: societies in which the economy is dominated by services rather than manufacturing
Producer services: knowledge-intensive services and expertise that are sold to firms
Services: economic activities that involve either the creation and analysis of information or the manipulation of goods, such as transportation and retail trade

Study Questions

1. What proportion of the labor force in North America, Japan, and Europe works in services?
2. Why is it difficult to measure the productivity of services?
3. What are the major industries that comprise "the" service sector?
4. What are five reasons services have grown so quickly?
5. What are five ways in which labor markets in services differ from those in manufacturing?
6. Where do producer services tend to locate, and why?
7. What are the components of the FIRE sector?
8. Why do US law firms do business overseas?
9. What are electronic funds transfer systems?
10. Where are the major areas of offshore banking?
11. How have many firms globalized their back offices?
12. Are all consumer services nonbasic?
13. Define tourism. What continent is the world's leader in this industry?

Suggested Readings

Althoff, L., F. Eckert, S. Ganapati, and C. Walsh. 2022. The geography of remote work. *Regional Science and Urban Economics* 93: 103770.

Bassens, D., L. Gutierrez, R. Hendrikse, D. Lambert, and M. Waiengnier. 2021. Unpacking the advanced producer services complex in world cities: Charting professional networks, localisation economies and markets. *Urban Studies* 58(6):1286–302.

Bryson, J., P. Daniels, and B. Warf. 2004. *Service Worlds: People, Organizations, Technologies*. Routledge.

Gibadullina, A. 2021. The birth and development of Anglophone financial geography: A historical analysis of geographical studies of money and finance. *Geoforum* 125:150–67.

Hall, C. (ed.) 2024. *The Wiley Blackwell Companion to Tourism*. John Wiley & Sons.

Hsieh, C.T., and E. Rossi-Hansberg. 2023. The industrial revolution in services. *Journal of Political Economy Macroeconomics* 1(1):3–42.

Knox-Hayes, J., and Wójcik, D. (eds.) 2020. *The Routledge Handbook of Financial Geography*. Routledge.

Martin, R., and J. Pollard (eds.) 2017. *Handbook on the Geographies of Money and Finance*. Edward Elgar.

McKercher, B. 2021. The impact of distance on tourism: A tourism geography law. In A. Lew (ed), *Tourism Spaces*. pp. 137–41. Routledge.

Narsiah, S. 2025. Global finance capitalism: Rebirth or denouement? *Human Geography* 18(3):449–56.

Neal, Z., B. Derudder, and P. Taylor. 2019. Should I stay or should I go: Predicting advanced producer services firm expansion and contraction. *International Regional Science Review* 42(2):207–29.

Nelson, V. 2021. *An Introduction to the Geography of Tourism*. Bloomsbury Publishing.

Öner, Ö., and J. Klaesson, 2017. Location of leisure: The new economic geography of leisure services. *Leisure Studies* 36(2):203–19.

Shaw, G., and A. Williams. 2004. *Tourism and Tourism Spaces*. Sage Publications.

Watson, P., and S. Deller. 2022. Tourism and economic resilience. *Tourism Economics* 28(5):1193–215.

Wójcik, D. 2021. Financial geography I: Exploring FinTech–maps and concepts. *Progress in Human Geography* 45(3):566–76.

Young, M., and F. Markham. 2020. Tourism, capital, and the commodification of place. *Progress in Human Geography* 44(2):276–96.

11 Digital Economic Geographies

One of the defining features of contemporary capitalism is digitization, the conversion of information to digital form. In the wake of the microelectronics revolution in the 1980s, which unfolded in conjunction with new rounds of globalization, digital technologies have become firmly interwoven in production processes, office work, consumption, and the daily lives of roughly 60 percent of the people on planet Earth.

11.1 Historical Perspectives on Telecommunications

The transmission of information is a critical part of the development of capitalist economies, on a par with the movement of people and goods. Information availability reduces uncertainty and helps individuals and firms make good decisions, and is thus a vital part of economic development. It follows that the history of capitalism has witnessed multiple waves of innovations in communications.

Telecommunications are not new. Starting in 1844, Samuel Morse's invention of the telegraph allowed communications to become detached from transportation. The telegraph made possible the transmission of information over long distances, including prices and shipments, and was vital to financial markets. Starting with the first transcontinental telegraph line in 1861, the telegraph was important in the American colonization of the West, where it helped to create a national market by allowing long-distance circulation of news, prices, stocks, and other information. In 1868, the first transatlantic telegraph line was completed and formed part of the time-space compression that accompanied the Industrial Revolution. A global network of telegraph lines became a nineteenth century version of the **Internet**.

The invention of the telephone in 1876 became essential to the growth of the American city-system. For example, it allowed corporations to spin off branch plants to smaller towns. Multi-establishment firms utilized the telephone to coordinate production schedules and shipments. In the 1920s, the telephone, like the automobile and the single-family home, became widespread among the growing middle class, with significant social effects on friendships and other ties. In the 1950s, the first international phone line was laid across the Atlantic Ocean. Today, despite the rise of the Internet, the telephone is still the most commonly used form of telecommunications.

The presence of telephones is a measure of a nation's communications infrastructure. Telephone penetration rates—the number per 1,000 people—are highest in the economically developed world. Africa has 20 percent of the world's people but only 1 percent of the world's telephones. All over the world, however, cell phones have rendered landlines obsolete. However, cell phone usage is spatially uneven, reflecting global patterns of wealth and poverty. Bypassing landlines in the developing world is an example of technological leapfrogging.

The microelectronics revolution was particularly important in the telecommunications industry. Advances in processing power led to dramatic increases in the ability of computerized systems to analyze and transmit data. No large corporation can produce, sell, and operate in multiple markets and coordinate the activities of thousands of employees within highly specialized divisions of labor without sophisticated systems of communications.

Today, fiber optic lines are the most widely used technology in the global telecommunications industry. Transnational corporations, financial institutions, and media firms use **fiber optics** either in the form of privately owned facilities or leased circuits from shared networks. In the 1980s, a network of fiber lines crossed the Atlantic Ocean; in the 1990s, a second one crossed the Pacific Ocean; and in the 2000s, a third network crossed the Indian Ocean (Figure 11.1). The world's network of fiber lines is essentially the nervous system of the global financial and services economy, linking countries, cities, suppliers, and clients around the world.

11.2 Telecommunications and Geography

The impacts of telecommunications on spatial relations have long been exaggerated, including claims that the Internet means "the end of geography." In the business literature, it is common to hear of a "borderless world," as if the Internet magically makes national boundaries disappear. These views rely on a simplistic technological determinism that oversimplifies the relations between technology, society, and space. In the nineteenth century, many thought the new technology of the telegraph would eliminate war; in the early twentieth century, others held the same for the telephone. Repeated proclamations that telecommunications would lead everyone to work at home through telecommuting would make cities obsolete have consistently been disproved by the ongoing vitality of urban areas. The reason for this is that telecommunications are a poor substitute for face-to-face encounters, the way in which most high-level corporate interactions occur, especially when the information involved is proprietary and or secretive in nature. While the

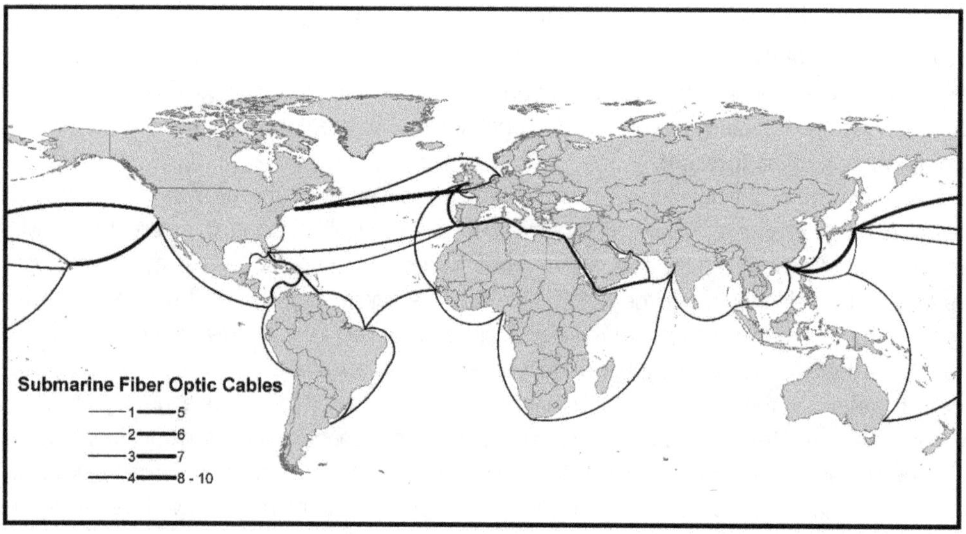

Figure 11.1 Map of the world's major fiber optic lines.

Internet allows some professionals to work from home, the agglomeration economies offered by urban cores are still enormous.

Due to these agglomerative pressures, most high-wage, white-collar, administrative functions remain clustered in downtown areas (Chapter 9). However, telecommunications are well-suited for the transmission of routinized, standardized data and have facilitated the dispersal of low-skilled jobs to low-wage regions. Examples include the spread of back offices and the growth of call centers overseas. In short, telecommunications enable the simultaneous concentration and deconcentration of economic activities. Thus, popular notions that telecommunications will make geography irrelevant are naive. Economic space, in short, will not disappear with the telecommunications revolution, although it is being thoroughly transformed.

Teleworking is often lauded as a means to reduce transportation costs and the demand for energy and to allow workers to balance their personal and professional lives. However, evidence indicates that workers who telecommute may drive fewer trips to work per week, but their trips are longer, and they may drive more during their free time, offsetting the savings from working at home. The growing number of home-office, temporary, and contract workers makes telework increasingly necessary; the COVID pandemic greatly amplified the demand to work at home. With workers demanding more flexible schedules, shorter commutes, and relief from traffic congestion, companies may find that telecommuting is useful for retaining or attracting high-quality workers. Videoconferencing and email are practical in a wide range of job categories and work situations. The growth of teleworking and telecommuting has put downward pressure on office rents in large downtown areas, indicating that it is challenging the agglomeration economies there.

Another set of impacts of telecommunications concerns transportation informatics, including improvements such as smart metering, electronic road pricing, synchronized traffic lights, automated toll payments and turnpikes, automated road maps, information for trip planning and navigation, travel advisory systems, electronic tourist guides, remote traffic monitoring and displays, and computerized traffic management and control systems, all of which are designed to minimize congestion and optimize traffic flow (particularly at peak hours). All of these enhance the efficiency, reliability, and safety of travel. Cell phones allow more productive use of time otherwise lost to congestion.

11.3 Geographies of the Internet

The Internet is a global system of interconnected computer networks, that is, a network of networks, which connects billions of computers across the planet. Its decentralized architecture allows any computer to communicate with almost any other using the Transmission Control Protocol and Internet Protocol (TCP/IP), the language that allows servers to exchange information with one another. In the twenty-first century, as cell phones have become widespread, the wireless Internet has become commonplace. The **World Wide Web** is part of the Internet that consists of documents written in the hypertext markup language (HTML), which allows them to be accessed by web browsers. The web contains countless hypertext documents, graphics, music, and videos.

Although **cyberspace** consists of the digitized "world in the wires," and is often conceived as a domain independent of the "real" or physical world, in reality, the virtual and analogs worlds shape each other continuously. The Internet delivers a number of services to homes, offices, and factories, including email, telephone calls, TV programs and other video images, text, and music. The Internet allowed for the rise of social media such as Facebook and Twitter. By now, digital reality and everyday life have become so thoroughly fused that it is difficult to distinguish them.

The Internet is used for so many purposes that life without it is simply inconceivable, including email, online shopping and banking, airline and hotel reservations, video games, chat rooms, Voice over Internet Protocol telephony, distance education, downloadable music and television shows, blogs, YouTube, telemedicine, to simply "Googling" information. Other applications include e-government (government applications to deliver information and services) and **e-commerce**. Cyberspace has become a necessity for vast swaths of the world's population. Thus, simple dichotomies like "offline" and "online" do not do justice to the multiple ways in which the "real" and virtual worlds are interpenetrated. Like the printing press and the automobile, the Internet has permanently altered the ways we work, interact, consume, and live.

11.3a Origins and Growth of the Internet

The Internet began in the 1960s as a project funded by the US Defense Department's Defense Agency Research Projects Administration (DARPA), which designed a computer network to withstand a nuclear attack (it was originally called DARPANET). Originally, it connected only a few major universities. In the 1980s, control of the Internet was transferred to the National Science Foundation (NSFNET), and in the 1990s, control was privatized when it was handed to a consortium of telecommunications corporations. In the 1990s, graphical interfaces developed at the CERN laboratory in Europe greatly simplified the use of the Internet and led to the formation of the World Wide Web.

The growth of the Internet has been phenomenal (Figure 11.2); indeed, it is one of the most rapidly diffusing technologies in world history. Internet usage climbed as it became clear that connected computers were more valuable than unconnected ones, as the price of equipment declined, and as cell phones made it possible to access the Internet from any location. In 2024, an estimated 5.6 billion people, or roughly 66 percent of the world's population, were connected to the Internet.

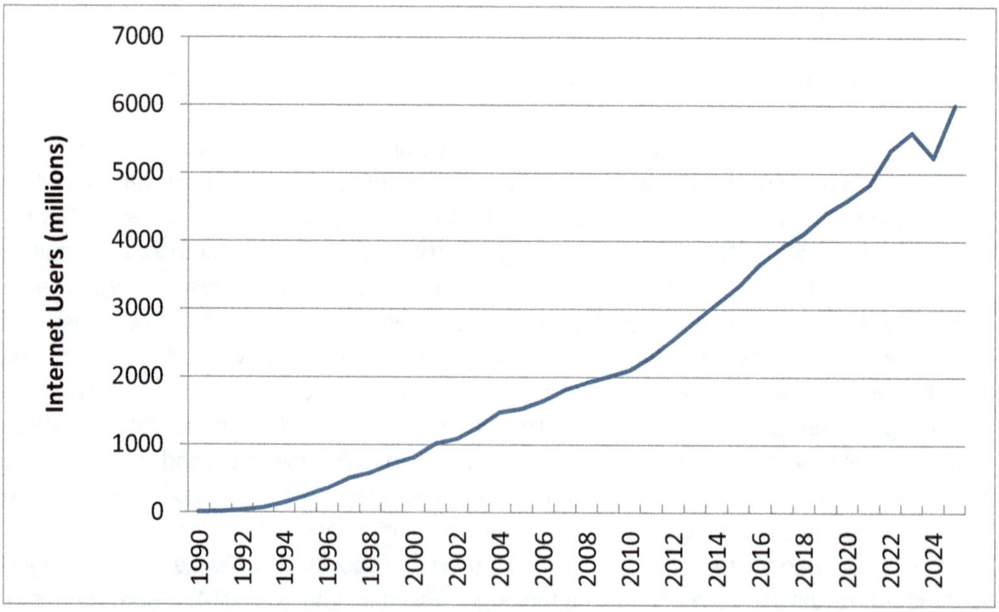

Figure 11.2 Growth of internet users worldwide.

However, there are large variations in Internet penetration rates (percentage of people with access) around the world. Roughly 93 percent of the US population in 2025 had Internet access. Internet use (penetration) rates vary considerably by country (Figure 11.3), forming a global digital divide that mirrors long-standing patterns of wealth and poverty. They are very high in the developed world and considerably lower in Africa, South Asia, Southeast Asia, and parts of Latin America. Thus, although two-thirds of the world uses the Internet, there are still vast numbers of people who are excluded from cyberspace and the benefits it offers.

11.3b Social and Spatial Discrepancies in Internet Access

Significant variations exist in terms of Internet access that are unevenly distributed by wealth, gender, and ethnicity. Access to computers linked to the Internet at home is highly correlated with income; wealthier households are far more likely to have a personal computer at home with a router and broadband than are poorer ones. In the United States, white households use networked computers more frequently than do African American or Latino households. The elderly likewise often find access to the Internet to be intimidating, although they are the fastest-growing demographic group of users.

Social and spatial differentials in access to cyberspace threaten to create a largely minority underclass deprived of the benefits of the Internet. Modern economies are increasingly divided between those who are comfortable and proficient with digital technology and those who neither understand nor trust it, disenfranchising the latter group from the possibility of citizenship in cyberspace. Despite the falling prices for computers, basic entry-level machines still cost more than many low-income households can afford. The recent COVID pandemic illustrated how poor people had to resort to going to places such as McDonald's to use the Internet there; for low-income households, assignments that can be done only online are a challenge. Most public libraries

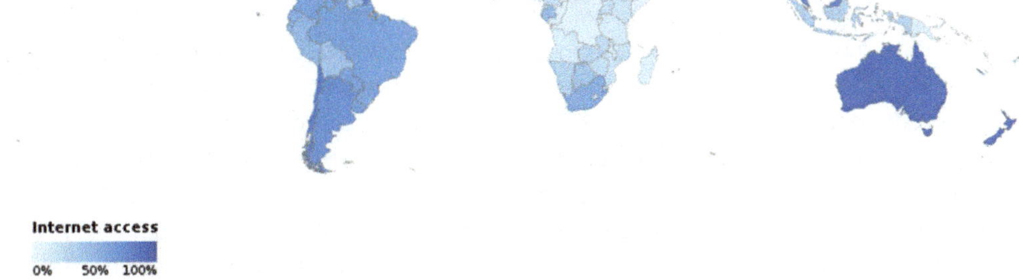

Figure 11.3 Internet penetration rates worldwide, 2017.

offer free access to the Internet, but they are unable to serve all of those without home access. Even within the most digitized of cities, there are pockets of "offline" poverty.

Internationally, access to the Internet is shaped by the reliability and affordability of national telecommunications systems, particularly fiber optics. Prices for Internet access at cafes vary; in countries with telecommunications monopolies, prices are much higher than in those with deregulated systems, and hence usage rates are lower. In some countries, gender norms inhibit women from using the Internet. Finally, many authoritarian governments are fearful that the Internet will undermine their monopoly over information and censor it, including restricting access for some groups.

11.3c Social Implications of the Internet

Like the printing press, the Internet has had enormous social, cultural, economic, and political impacts. Never have so many people been able to contact one another so easily, obtain news, file complaints, and save time as today. Seen this way, the division between offline and online does not do justice to the diverse ways in which the "real" and virtual worlds are interpenetrated. Cyberspace changes not only *what* we know about the world, but also *how* we experience it.

Many of the Internet's uses revolve around entertainment, personal communication, research, music, and online games. The Internet has given voice to long-marginalized groups with varied political interests and agendas. By facilitating the expression of views that otherwise may be difficult or impossible, the Internet allows for a dramatic expansion in the range of voices heard about many issues. Cyberspace allows for the construction of "communities without propinquity," groups of users who share common interests but not physical proximity.

Finally, there is also the "dark side" of the Internet, in which it is deployed for illegal or immoral purposes. Hackers can wreak havoc with corporate or government cybersecurity systems and steal large volumes of account information or unleash computer viruses and worms. The dark side also includes digital identity theft of drivers' licenses, passports, and Social Security cards; securities swindles; and scams. Credit card fraud is a mounting problem; 0.25 percent of Internet credit card transactions are fraudulent, compared to 0.08 percent for non-Internet transactions. The Internet has also allowed for a torrent of fake news and disinformation to circulate rapidly.

11.4 E-Commerce

The impact of cyberspace on businesses includes e-commerce, such as business-to-business transactions, as well as those that link firms to their customers. E-commerce was greatly accelerated by the introduction of smartphones. E-commerce includes online retailing, which has given rise to firms such as Amazon. Information technology lowers the transaction costs among corporations, which helps to increase productivity. Many firms sought to accelerate information flows within the firm and to lower costs by reducing intermediaries and distribution costs.

The Internet's economic effects extend into every economic sector. Firms use it to monitor supply and demand, keep an eye on inventories, and manage supply chains and logistics. Farmers can use the Internet to control irrigation systems remotely; farm equipment with global positioning systems (GPS) can use water and fertilizer minimally. Telecenters in the developing world allow farmers to acquire information about crop prices and to learn best practices, acquire land titles, and bypass corruption and intermediaries. In manufacturing, the Internet has given new powers to robots using artificial intelligence. Producer services, too, have been markedly altered, such as with the movement of many clerical functions, call centers, tax preparation, and radiology jobs to the developing world. Internet banking has reduced the need for labor, lowering costs, improving efficiency, and convenience. Digital currencies such as Bitcoin are affecting money supplies and currency markets; in El Salvador, Bitcoin is the official currency. Crowdfunding has allowed many firms to become start-ups that otherwise would have been starved for investment funds.

The Internet has also changed transportation. Smartphones and laptops allow people to work while traveling or at home. GPS has made navigation much easier. Smart traffic systems have made roads faster and safer. Firms such as Uber and Lyft have created enormous competition for taxis, an example of creative destruction. Travelers can use the Internet to obtain information about their destinations, book airline and hotel tickets, and reserve cars.

Energy production and consumption have also been affected by the Internet. Parallel and distributed computing has improved the efficiency of energy suppliers in the face of constraints such as security and environmental limitations. Smart grids—computerized controls and sensor networks—allow for more efficient ways of managing the supply and demand for electricity. Smart homes use Internet-enabled devices to control temperatures, humidity, and security. **Smart cities**, which integrate information technology in numerous ways, including smart traffic systems and smart homes, to promote sustainable development and reduce carbon use.

One important dimension of e-commerce concerns **electronic data interchange** (EDI) systems, which are used in business-to-business (B2B) interactions. EDI uses include advertising, online product catalogues, the sharing of sales and inventory data, submissions of purchase orders, contracts, invoices, payments, delivery schedules, product updates, and labor recruitment. E-commerce reduces delays, marketing and delivery costs, and improves customer service. E-tailing or electronic retailing means online shopping for goods and services: in 2021, 70 percent of all websites were commercial (i.e., "dot com") sites. Shopping on the Internet requires only access, a credit card, or Paypal and a parcel delivery service. It allows effortless comparison shopping.

The most successful example of e-commerce is Amazon.com, started by Seattle entrepreneur Jeff Bezos in 1994. Amazon is the largest Internet-based company and retailer in the world. It accounts for 51 percent of the world's online retail sales. Amazon began as an online bookstore, and still sells half of the world's books, but quickly expanded to encompass a large variety of products and services including appliances, video games, office products, household, garden, and kitchen supplies, apparel, arts and crafts, luggage, drones, food, music, jewelry, furniture, cosmetics, software, and magazine subscriptions. Amazon offers an app store, digital games, digital comics, and downloadable music. Amazon Drive is the world's largest provider of cloud services. Amazon Wireless offers telephone service. Amazon Tickets offers sales to concerts and performances. Amazon Destinations finds vacation spots near customers' homes. Handmade by Amazon links producers of craft items with customers. Amazon Prime offers free two-day shipping (in some places, same-day service) and also offers access to Amazon Video, which provides instant streaming of movies and television shows, some of which are produced at Amazon Studios. Amazon Fire offers streaming video to television sets, and in 2014, the company entered the smartphone market with the Fire Phone.

However, e-commerce still only comprises 14 percent of total global retail sales, perhaps because it lacks the emotional content of shopping. Customers using the Internet tend to be relatively wealthy and well educated. Internet-based sales of stocks (e.g., E*Trade) now comprise 26 percent of all trades. One particularly successful application has been in the travel reservation and ticketing business, where web-based purchases of hotel rooms and airline seats (e.g., through services such as Travelocity) have caused a steady drop in the number of travel agents, another example of creative destruction. Other services offer Internet searches of databases and classified ads; Google is the giant of online searches, comprising more than 90 percent of searches worldwide. Streaming television shows led to the rise of companies such as Netflix and its competitors. Internet advertising reaches specialized markets rather than mass audiences. E-advertising comprises two-thirds of total global advertising expenditures (including streaming television).

Another dimension of e-commerce is Internet gambling, which includes betting services, particularly regarding sports events, and online games and slot machines in which gamblers may use their credit cards. The geography of legal gambling is highly uneven; the existence of online gambling challenges communities in which gambling is illegal. Offshore gambling centers have grown quickly, particularly in the Caribbean, where Antigua licensed its first Internet casino in 1994. By 2020, an estimated 10,000 online casinos attracted roughly 120 million users. Asia was the largest source of clients and revenues.

11.5 Summary

Communications and information technology are having enormous effects across the planet. In many ways, digitization is as significant a dimension of contemporary capitalism as is globalization. Capitalism has witnessed several telecommunications technologies dating back to the telegraph in the nineteenth century, each of which unleashed a round of time-space compression. Far from eliminating the importance of geography, however, information systems have changed the meaning of space: as communications times and costs declined, other factors grew in significance.

Today, the core of this transformation is the Internet, which connected more than 5.3 billion people worldwide in 2022, about 66 percent of the planet. The chapter explored the origins, growth, and size of the Internet; noted the uneven access to it that people around the world and within the United States have; and summarized some of its impacts, including the growth of cybercommunities. Telework allows many workers to work at home, with long-term implications for the structure of cities. It pointed out that the social divisions that exist offline are replicated online. It also explained the nature and impacts of e-commerce, including e-tailing.

Glossary

Cryptocurrency: currency that exists only digitally and is not under the regulation of the state
Cyberspace: the world in the wires, the Internet
Digital divides: social and spatial inequalities in access to the Internet
E-commerce: buying and selling conducted over the Internet
Electronic data interchange: networks that allow enormous volumes of data to flow at the speed of light
Fiber optics: glass wires that transmit large amounts of information at the speed of light
Internet: the network of digital networks that forms cyberspace and has had enormous impacts on the world
Smart cities: cities that deploy digital technologies to improve their energy systems, transportation, and housing networks, emphasizing environmental sustainability
Telework: using the Internet to work from home or other nonwork space
World Wide Web: the part of the Internet that uses user-friendly web pages

Study Questions

1 How does the global distribution of telephones mirror colonialism?

2 What role does fiber optics play in global telecommunications?

3 What factors shape access to the Internet?

4 What is e-commerce?

5 What is Electronic Data Interchange?

Suggested Readings

Autio, E., R. Mudambi, and Y. Yoo. 2021. Digitalization and globalization in a turbulent world: Centrifugal and centripetal forces. *Global Strategy Journal* 11(1):3–16.

Capello, R., C. Lenzi, and E. Panzera. 2024. The nexus between the digital service economy and intraregional wage inequalities. *Economic Geography* 100(3):246–73.

Forman, C., A. Goldfarb, and S. Greenstein. 2018. How geography shapes-and is shaped by-the internet. In G. Clark, M. Feldman, M. Gertler, and D. Wozcik (eds.), *The New Oxford Handbook of Economic Geography*. pp. 269–85. Oxford University Press.

Howson, K., F. Ferrari, F. Ustek-Spilda, N. Salem, H. Johnston, S. Katta, R. Heeks, and M. Graham. 2022. Driving the digital value network: Economic geographies of global platform capitalism. *Global Networks* 22(4):631–48.

Kenney, M., and J. Zysman. 2020. The platform economy: Restructuring the space of capitalist accumulation. *Cambridge Journal of Regions, Economy and Society* 13(1):55–76.

Lehdonvirta, V., O. Kässi, I. Hjorth, H. Barnard, and M. Graham, 2019. The global platform economy: A new offshoring institution enabling emerging-economy microproviders. *Journal of Management* 45(2):567–99.

Shaban, A. 2024. *Digital Geographies—Urbanisation, Economy, and Modelling: A Machine-Generated Literature Review*. Springer Nature Singapore.

Standage, T. 1998. *The Victorian Internet*. Walker and Company.

Vale, M., D. Ferreira, and N. Rodrigues. 2024. *Geographies of the Platform Economy: Critical Perspectives*. Springer.

Warf, B. (ed.) 2018. *Encyclopedia of the Internet*. Sage.

Warf, B. (ed.) 2020. *Geographies of the Internet*. Routledge.

Yang, C., and T. An. 2020. The Internet reshapes China's economic geography: Micromechanisms and macro effects. *China Political Economy* 3(2):341–65.

12 Geographies of Consumption

In contrast to production, which has been studied extensively, consumption has often been ignored in economic geography. Only recently has consumption been analyzed in social, not individual, terms. Yet consumption is vitally important in understanding how economies are structured geographically, how they are organized, and their effects. Consumption and production cannot be easily separated because they are closely interwoven: people work to consume and consume to live and work. Thus, consumption and production are two sides of one process. Consumption lies at the intersection of production, class, gender, markets and states, and individuals and society. It is simultaneously an economic, cultural, and psychological process.

Economically, consumption is hugely important, constituting the majority of economic activity in most countries. In the United States, about 55 percent of the GDP consists of household consumption; business investments are only 6 percent, foreign trade is 23 percent, and government spending is 17 percent. Consumer confidence plays an important role in shaping the course of the economy. How consumers spend their incomes has serious implications for the demand for goods and services; for example, in the United States, spending on food and clothing has declined as a percentage of disposable income, while spending on education and medical care has soared. American consumers do not propel the United States economy; by purchasing imports, they shape the economies of other countries. Thus, oscillations in American consumer spending have global effects. The same is true for consumers in other wealthy countries, although no other economy rivals the United States in its size of consumption and enormous trade deficits.

12.1 The Historical Context of Consumption

Consumption is a social process with a history and geography that are shaped by how people have consumed over time. Most of our knowledge of premodern consumption is based on the spending habits of elites, who had the largest purchasing power. In medieval Europe, most goods were produced and consumed locally; trade was largely confined to luxury goods such as silks, spices, porcelain, and jewelry that were affordable only to the very rich. The rise of the colonial world economy in the sixteenth century (Chapter 2) created a class of people with sufficiently large incomes to spend and made available products from diverse places around the globe, such as sugar from the Caribbean, cod from New England, and spices from Asia.

The rise of the modern form of **mass consumption—consumerism**—unfolded with the Industrial Revolution in the late eighteenth century. Industrialization and mechanized production dramatically brought down the prices of many goods and lifted millions of people out of poverty. Food and clothing became increasingly affordable to the growing middle class. Mass production was accompanied by mass consumption, and rising incomes turned workers into consumers. In the **commodification** of consciousness, the self is defined through want, not social obligations. Consumption was an integral part of the rise of modernity: if tradition is largely concerned with the regulation of desire, modernity tends to unleash it. This process had huge implications for class consciousness, allowing consumption to override the politics of work. Consumption also shaped economic landscapes: in the 1830s, the first department stores appeared in Paris. As the middle class grew in the nineteenth and twentieth centuries, department stores became major centers of consumption.

Capitalism depends on innovation, in consumption as well as production. Thus, mass consumption also saw the birth of advertising as a means of creating desire, often converting luxuries into necessities. As capitalism unleashed successive waves of new products, the cycles of consumption accelerated over time (Figure 12.1).

Consumerism entails much more than the simple purchase of goods, but the entire experience of shopping. Consumption was abetted by early catalogues, mail orders, and the postal system. As workers struggled for higher wages and lower working hours, including the eight-hour working day and the five-day working week, consumption became equated with leisure. Everyday life changed accordingly. For example, child rearing came to include the purchase of goods for children. Buying came to be naturalized as the key to happiness, status, success, relief from boredom, and a positive self-image.

Consumerism began in Britain and the United States, the world's two wealthiest countries in the nineteenth century, and spread rapidly to Western Europe. After the Second World War, consumerism diffused to large parts of the world during the postwar boom. The spread of this consumerism was aided by the enormous influence of the United States in the world economy; American-style capitalism excels at advertising and consumer goods. Transnational corporations, increased international trade, and television were all important in this process. Consumerism's popularity was very uneven geographically: typically, its growth was slowest in societies with deeply entrenched religious objections. Over time, consumerism came to be the model for other kinds of social relations, such as education, religion, and family life, reducing everyone to a buyer or seller.

The expansion of consumerism was also linked to the growth of credit. Prior to the Second World War, people had to save for expensive items. During the Great Depression of the 1930s, General Motors introduced financing for cars, an early form of credit. It was credit cards, however, that increased the ease of consuming, essentially eliminating the old Protestant idea

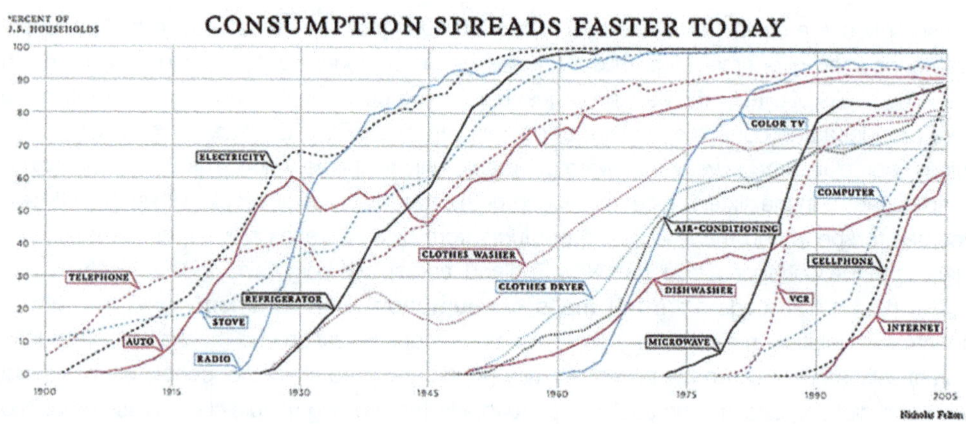

Figure 12.1 Accelerating cycles of consumption.

of **delayed gratification**. Credit cards were sold as a sign of social sophistication and a carefree lifestyle. Following the deregulation of finance, banks aggressively targeted social groups that they had previously ignored, such as college students and the elderly. In addition, many consumers faced growing problems arising from deindustrialization, globalization, and stagnant incomes (Chapters 4 and 9). As spending rose but incomes did not, the United States savings rate declined, and many households found themselves with debt they could not afford. In the United States, 260 million Americans held 2.3 billion cards in 2020, with a total of roughly $1 trillion in debt. Average household consumer debt levels have risen explosively. The popularity of credit cards, including among small businesses, created a wave of personal bankruptcies; more people go bankrupt each year than go to college.

12.2 Conceptual Perspectives on Consumption

Consumption has been studied from several theoretical angles, including the sociological, neoclassical economic, and Marxist views. There are strengths and weaknesses to each of these.

12.2a Sociological Views of Consumption

Sociologists long examined consumption in relation to individuals' and households' living standards, class, and social status. Max Weber argued that delayed gratification was at the heart of the Protestant ethic. American sociologist Thorsten Veblen examined the ostentatious consumption of the late nineteenth-century robber barons and coined the famous phrase "**conspicuous consumption**" to indicate consumption as a display of wealth and status, including, for example, large mansions. Many sociologists studied the relations between consumption and identity, emphasizing shopping as something much more than simply the purchase of goods and services but as a set of signals used to define identities and status. Ironically, in societies such as the United States, many people find their "unique" individuality through the purchase of mass-produced commodities.

Sociologists have noted how consumption varies greatly by social class, age, ethnicity, and gender, which is of great importance to marketers. Consumption is highly differentiated by class and income: low-income families are apt to consume more of the cheaper foods, such as potatoes, and less of the expensive foods. Consumption varies by age: typically, consumption levels are relatively constant as people age (although the types of goods and services they buy vary), but incomes fluctuate with the life cycle, leading to different patterns of spending and saving over time. As people age, they tend to spend more on health care, a major force behind the growth of that sector (Chapter 10). Consumption is also highly gendered: men and women tend to buy very different items. For example, men are more likely to purchase alcohol, motorcycles, and cigars, while women are the primary buyers of household foods, clothes, and cosmetics. Markets for magazines, movies, shoes, and cars are also gendered. Because women are often responsible for buying household necessities, they are the bulk of consumers, and advertisers tend to appeal to them directly. Similarly, different ethnic groups have different preferences for foods, clothing, magazines, television shows, automobiles, and other goods and services. It is worth emphasizing that consumption habits change constantly, due to changing age and ethnic demographics, tastes and preferences (e.g., concerns over health), and incomes.

Sociologists have also analyzed the role of advertising, which arose in the nineteenth century in the wake of mass production and consumption. Advertising is a means of linking consciousness, desire, and identity to the commodity. Consumerist societies such as the United States are saturated by advertising, which shapes people's roles and views of the world and themselves. The average American child has seen one million television ads by the age of 20. Advertising invites us to meet social and psychological needs through material ends, creating the belief that commodities will give us desired status, sexual appeal, physical prowess, lifestyle, and freedom.

Extreme, widespread consumerism has led to an epidemic of "affluenza," the obsessive pursuit of more, a never-ending search for pleasure, and materialism run haywire. Because commodities alone cannot make us happy, many people find themselves in a state of permanent discontent: never has so much meant so little to so many. To sustain this lifestyle, many Americans suffer a "time famine" in which they work too much and sleep too little. Studies of satisfaction levels over time, which peaked in the United States in the 1950s, reveal that money does not necessarily buy happiness. Even as consumption has increased, so have depression and suicide.

Consumption also has serious political and civic implications. Consumerism turns citizens into consumers and collapses social obligations into the self. As a result, acknowledgment of social obligations is diminished when the sense of citizenship extends no further than shopping. The effects of this process are demonstrated in long-term declines in voting, volunteering, and public disinvestment in public areas.

12.2b Neoclassical Economic Views

Another perspective on consumption comes from neoclassical economics, which focuses on consumer demand more than the dynamics of production in the analysis of social relations. Neoclassical economics arose in the 1870s and uncoupled economics from political economy. Neoclassical economics assumes consumer demand is the motor that drives the economy.

This type of economic analysis, which dominates the United States, Canada, and much of Europe, begins with the individual decision maker. A hypothetical individual consumer, ***Homo economicus***, is used as the model of decision making. This person is held to be all-knowing about his or her opportunities and their respective costs, benefits, and consequences. The level of satisfaction that individuals achieve by purchasing a specific combination of goods is reflected in indifference or utility curves that map the tradeoffs each person makes to achieve optimal satisfaction. Because consumption levels are always limited by income, each person maximizes their utility or happiness by allocating their income among different goods.

When the individual utility-maximizing decisions are aggregated, a picture of collective demand emerges. Consumer spending reflects the status of the labor market and real incomes, unemployment rates, prices, proclivities toward saving and spending, and the forces that shape utility curves, such as the demographic composition of a society. Consumer tastes and preferences constantly change in the face of fads, fashions, and concerns over health and safety. For example, in the United States, beer and wine are displacing hard liquor, and chicken has surpassed beef consumption. Consumption patterns also reflect the income- and price-elasticities of demand for different commodities, that is, how sensitive consumers are to changes in incomes and prices. In the neoclassical economic view, markets are always optimally efficient.

The neoclassical economic view is consistent within its own terms of reference, but it suffers from several problems. First, it does not embed consumption in any historical or social context. It is silent about the symbolic and social dimensions of consumption. These failures arise because neoclassical economics does not view consumption as a *social* act that is part of broader relations of class, gender, ethnicity, and power, but as a purely individual one. Thus, it does not offer any account of the origins of utility curves or how or why they assume their particular form; they are simply taken as given. In reality, individual choices are always part of a broader social fabric, including public policies. For example, the simple act of buying butter includes the effects of federal government subsidies to dairy farmers, health laws, and the use of artificial chemicals. To the extent that they are visible at all, social categories in neoclassical economics are defined by consumption: for example, class refers to income and socioeconomic status, not the division of labor. The so-called hidden hand of self-interest assumes individual choices are independent, not socially produced, and that they do not affect each other; none of this is true in reality.

Finally, it should be noted that even when individuals act rationally by maximizing their utility, they can create collective irrationalities or market failures. These occur when the market price does not reflect the social costs of a good, including traffic jams. Driving a car creates serious environmental externalities. If each person at a concert stands up for a better view, everyone is forced to do so and is therefore worse off.

12.2c Marxist Views of Consumption

Another view of consumption arises from Marxism, which asserts that to understand it in its social context is to go beyond its outer appearances and reveal the social relations that lie beneath them. The historical expansion of capitalism has involved the steady widening and deepening of commodity relations, including the search for new markets and the transformation of goods and services that were formerly acquired outside of the market into commodities (e.g., housing, childcare, transportation, education). Karl Marx argued, therefore, that commodities are not simply *things*, but the embodiments of social relations. To view commodities apart from their social origins is to engage in the error of **commodity fetishism**, succumbing to the opaqueness through which market relations hide relations between workers and capital. Viewed this way, commodities are not simply goods on the shelf or advertised on television; they are combinations of labor power, nature, and ideology.

Marxism drew upon classical political economy to differentiate the **use values** of commodities—their qualitative, subjective dimensions—from their **exchange values**, the quantitative prices they command on the market. For example, the use value of a banana is its taste and the relief from hunger it offers; its exchange value is the price for which it can be bought.

Labor, too, is a commodity whose use value to employers is less than its exchange value in wages, which is where surplus value and profit originate. Class is thus defined by relations to production, not consumption (i.e., income). Marxism suggests that the extraction of surplus value by employers over time inevitably leads to **underconsumption** by the working class, which generates unsold inventories, lower prices, and periodic crises. Employers cannot pay their workers the full value of their output, or they would receive no profit. Thus, capitalists must pay workers too little because they produce too much. For Marxists, this issue is the most severe internal contradiction that capitalism faces.

12.3 Geographies of Consumption

Geographers, too, offer insights into consumption, notably how commodities are sutured to their spatial origins. For example, global patterns of consumption mirror those of wealth and poverty (Figure 12.2): wealthy countries consume much more per capita than poor countries.

Because incomes and consumer preferences vary considerably around the world, patterns of consumption differ markedly among countries. Goods such as meat are consumed at different levels across the world: most meat is consumed in wealthy countries. Notably, meat production is also a leading cause of the gases that create climate change, notably methane. Sometimes

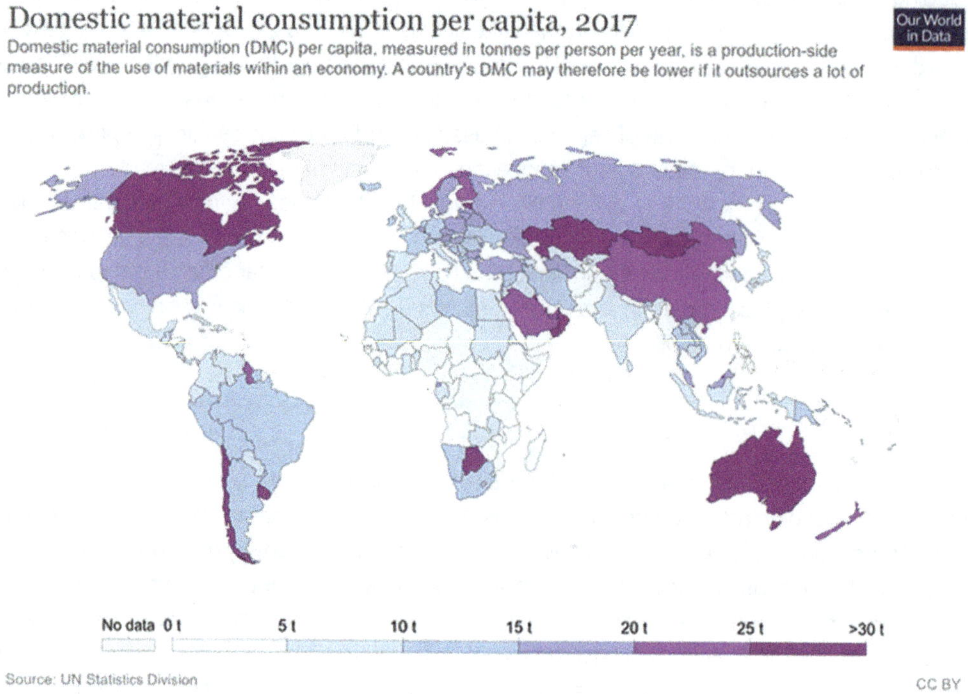

Figure 12.2 Material consumption per capita.

religious prohibitions shape consumption, such as Jews and Arabs avoiding pork or Hindus avoiding beef. Laws may prohibit the consumption of some goods (e.g., cannabis) in some countries but not others. Political ideologies and movements can shape consumption as well: in the nineteenth century, abolitionists avoided buying sugar because it was produced by slaves; recent boycotts of grapes to protest the exploitation of farm workers are another example.

Some geographers have examined consumption at the most intimate spatial scale, the body and individual experience. The body is the site of intentionality, what gives the mind a location in time and space, and a basis of identity. While bodies appear "natural," they are in fact social constructions; bodies are gendered, racialized, aged, and abled/disabled. The geography of the body locates it within social relations, situating the self in time and space. For example, geographers have looked at food and its origins and cultural meanings in different geographic contexts. Eating is the most direct relation between a body and the environment, and food consumption plays a large role in shaping bodies, such as the epidemic of obesity plaguing the United States and much of the world today.

Similarly, geographers have examined the shopping mall not simply as an economic phenomenon, but as a cultural phenomenon filled with meanings. The shopping mall is perhaps the only remaining pedestrian space in which to congregate. For example, the West Edmonton Mall in Canada, the largest megamall in the world, has 600 stores, 18,000 employees, and generates 1 percent of all of Canada's retail trade. Inside, it contains a golf course, a skating rink, a fantasyland hotel, and four submarines. Similarly, the Mall of America in Bloomington, Minnesota, has 520 stores, chapels, a roller coaster, an aquarium, and a rainforest. In this environment, fantasy, fun, and the commodity are merged into a seamless whole, illustrating how "the economy" is inseparable from culture. Malls, of course, have suffered a serious decline in the face of the massive switch to online shopping.

Economic geographers have studied consumption in the context of the global economy, notably the way in which commodities are produced, distributed, and consumed. These can be studied by focusing on **commodity chains** (also called value chains), a way of uniting production and consumption. Commodity chains are networks of labor and the stages of production that give rise to a commodity; they extend from the raw material to various steps in processing and delivery, and end with the acts of purchase and consumption. For example, coffee commodity chains begin with growers, often impoverished farmers, and extend to the processing plants (e.g., roasting), exporters, traders, wholesalers, retailers, and, finally, the consumer (Figure 12.3). Similarly, meat commodity chains begin with ranchers and feedlots, then slaughterhouses and to packinghouses, cold storage facilities, grocery stores, and finally to consumers. Gold commodity chains begin with miners working under horrific conditions in countries such as South Africa, to diamond cutters (typically in Europe), to retailers and consumers in the developed world. By obscuring the earlier steps and the labor that went into the production of gold jewelry, commodity fetishism allows consumers to celebrate the romantic aura of gold without considering the conditions of its production. At each stage in the chain, commodities are transformed, and value is added. One company may control several stages in a commodity chain depending on the degree of vertical integration or disintegration. Because these activities are carried out in spatially separated locations, commodity chains are geographical as well as economic and cultural phenomena.

Commodity chains are thus a means of illustrating the varying steps in the production process over time and space, and help us to overcome the artificial separation between consumption and production. Chains enable us to see commodities as more than just things, but as outcomes of processes operating at several spatial scales. This approach allows us to trace the impacts of consumption decisions back through the production and distribution process, enlarging our scale of analysis from the local to the global.

As capitalism has become steadily more globalized, commodity chains have become longer, often reaching halfway around the globe. This perspective offers insights into the ways in which globalization has created a flood of cheap imports that has raised consumer spending in societies such as the United States to new heights. By uniting consumption with production, chain analysis

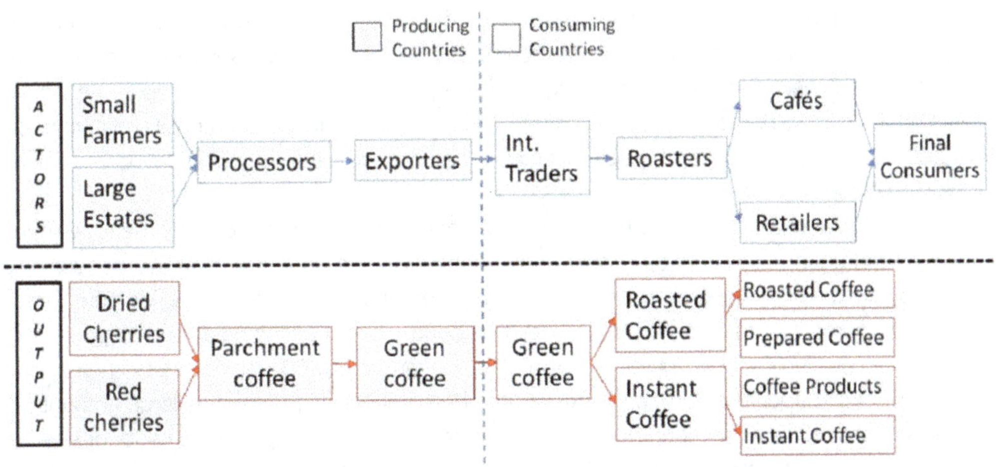

Figure 12.3 The coffee commodity chain.

points to the sacrifices made by low-wage labor working in sweatshops in the developing world in order to provide consumers with cheap goods. This approach illustrates consumption as an economic, cultural, psychological, and environmental act that firmly links the world's most abstract space, the global economy, with the most personal, the individual subject and body.

12.4 Environmental Dimensions of Consumption

In addition to being an economic and cultural process, consumption is also an environmental process. Every act of consumption imposes changes on the environment, using resources and energy and generating waste. Because commodity chains are often very long, the environmental impacts may occur thousands of miles away.

Traditional views of environmental destruction typically focused on the poor: with origins in Malthusianism (Chapter 7) and the high birth rates found in many developing countries, the world's impoverished populations have been blamed for clearing forests for farms, soil erosion, and other environmental problems. While there is no doubt that rapid population growth contributes to the degradation of the environment, blaming the poor ignores the even more destructive role played by overconsumption in the developed world, which causes most of the planet's environmental problems.

When standards of living, and thus consumption, rise, a society puts increasing pressure on the environment by using more resources and producing more waste. For example, in 2020, total US consumption of raw materials was fifteen times the volume of that in 1900, but its population was only three times as high. Mass consumption across the globe has been a disaster for the world's ecosystems, which have been sacrificed to support our lifestyle. Examples include the deforestation of tropical regions to produce plywood and beef. Entire landscapes are destroyed by the tremendous demand of the developed world, and increasingly, the developing world, for minerals, food, wood, water, and energy to produce houses, hamburgers, and clothes. Global demand for seafood is emptying the world's oceans: 75 percent of fish stocks are overfished, leading to empty oceans. Every year, humans dump millions of tons of plastic into the oceans, which destroys marine ecosystems and is the oceans' second greatest threat after overfishing. Across the planet, enormous levels of air and water pollution, deforestation, soil erosion, and habitat destruction are generating the largest wave of biodiversity loss in the planet's history.

Transportation using the automobile is perhaps the most wasteful of all consumption activities. In the United States, urban sprawl and long commutes have locked us into an extremely energy-intensive lifestyle. Today, there are more cars than drivers. Transportation consumes 40 percent of the American total energy budget; with 5 percent of the world's people, the United States consumes 25 percent of its total energy. Moreover, the high levels of fossil fuel consumption are the primary driver of anthropogenic global warming, which is wreaking havoc on regions across the globe.

A fundamental characteristic of the world economy is uneven spatial development, which is reflected in the geographies of purchasing power and resource consumption. This pattern was shaped by centuries of colonialism and the ongoing reproduction of uneven development today. Roughly 20 percent of the world, that is, the developed world, consumes about 80 percent of its vital materials, including 40 percent of the global meat supply, 60 percent of its energy, and 80 percent of its paper and vehicles. In return, the developed world generates large amounts of waste. As consumerism has spread to parts of the developing world, global demand for energy, meat, and other materials has soared. Most scientists are doubtful that the world can generate

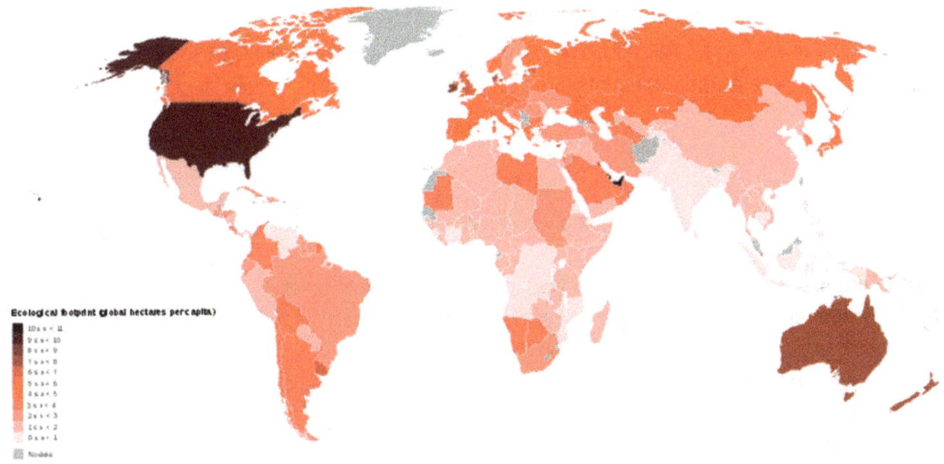

Figure 12.4 Ecological footprints of individuals in the world's countries.

the resources necessary to support the planet's population at any level approximating that of the United States today.

One way to understand the uneven ecological impacts of consumption is the **ecological footprint**, which portrays the quantity of resources necessary to support an individual's lifestyle. For example, over his or her lifetime, one US citizen has an impact roughly two hundred times that of a child in sub-Saharan Africa. The economically developed world has vastly larger footprints than do societies in Asia, Africa, or Latin America (Figure 12.4).

12.5 Summary

Consumption is a complicated process with economic, cultural, psychological, political, and environmental dimensions. Mass consumption and the culture of consumerism arose historically during the Industrial Revolution, which gave rise to a working class with disposable incomes. Consumerism reached its apex in societies such as the United States and has diffused to much of the rest of the world. However, despite higher incomes and consumption levels, the populations of societies such as the United States are not necessarily happier than they were a century ago.

Sociological views of consumption portray it as a social act in which individual choices reflect their social status. In this view, demand does not simply arise in individual consumers, but is generated by pressures such as advertising. In neoclassical economics, demand is modeled using utility curves that describe the behavior of *homo economicus*. In Marxism, the labor theory of value leads to the conclusion that employers must extract more value from workers than they can consume, giving rise to a chronic problem of overproduction and underconsumption.

Geographically, consumption unfolds at multiple spatial scales. The body, for example, reflects the networks of consumption in which people are engaged. Shoppers engage in consumption in environments consciously constructed to lure people to spend as much as possible. Commodity chains are a useful way of tying consumption with production, noting the various stages through which commodities pass from raw material to final product.

Consumption inevitably entails impacts on the environment. While many people blame the developing world for most of the world's environmental ills, in reality, it is the mass consumption lifestyle of the developed world that creates problems such as habitat destruction and global warming. A relatively small proportion of the world, 20 percent, consumes as much as 80 percent of the world's resources, a reflection of uneven spatial development. As consumerism has spread throughout much of the world, the consequences for ecosystems have increased proportionately. Ecological footprints are a useful way of portraying the environmental costs of consumption.

Glossary

Commodification: the process by which capitalism turns things and processes into commodities to be sold on the market for profit

Commodity chains: a good or service produced for sale on the market for a profit

Commodity fetishism: the view of commodities as things rather than outcomes of social processes, which hides the labor that goes into their making

Conspicuous consumption: sumptious consumption by the rich to display social status

Consumerism: a culture centered on the purchase of commodities

Delayed gratification: the process of forgoing consumption in order to save funds for future investments

Ecological footprint: the impacts on the environment of specific social processes such as consumption or burning fossil fuels

Exchange value: the price of a commodity on the market

***Homo economicus*:** a model of human behavior popular among economists that reduces humans to consumers

Mass consumption: the purchase of vast amounts of goods and services by a large population with some spending power

Underconsumption: low levels of consumption necessitated by low incomes and purchasing power

Use value: the qualitative benefits of consuming a good in terms of the satisfaction that it delivers to the consumer

Utility maximization: the process of making oneself as happy as possible by consuming specific bundles of goods and services that reflect an individual's preferences and income

Study Questions

1. When and where did consumerism originate historically?
2. Why don't rising levels of consumption make everyone happier?
3. What are the major theoretical perspectives on consumption?
4. What factors shape the world's geographies of consumption?
5. How is the body a geographical locus of consumption?
6. Are most of the world's environmental problems generated in the developing world?
7. What is an ecological footprint?

Suggested Readings

Aspers, P., and E. Le Grand. 2024. Consumption and place: The phenomenology of relational economic geography. *Geografiska Annaler: Series B, Human Geography* 104(3):1–17.

Bissell, D., 2020. Affective platform urbanism: Changing habits of digital on-demand consumption. *Geoforum* 115(3):102–10.

Boström, M. 2020. The social life of mass and excess consumption. *Environmental Sociology* 6(3):268–78.

Dawson, J., A. Findlay, and L. Sparks. 2020. *The Retailing Reader*. Routledge.
De Graaf, J., D. Wann, and T. Naylor. 2001. *Affluenza: The All-Consuming Epidemic*. Berrett-Koehler.
González, S. 2020. Contested marketplaces: Retail spaces at the global urban margins. *Progress in Human Geography* 44(5):877–97.
Kucuk, S. 2016. Consumerism in the digital age. *Journal of Consumer Affairs* 50(3):515–38.
Mansvelt, J. 2005. *Geographies of Consumption*. Sage.
Paterson, M. 2023. *Consumption and Everyday Life*. Routledge.
Patterson, M., and N. Hoalst-Pullen. 2023. *The Geography of Beer*. Oxford University Press.
Redclift, M. 2020. *Wasted: Counting the Costs of Global Consumption*. Routledge.
Shannon, J. 2021. Dollar stores, retailer redlining, and the metropolitan geographies of precarious consumption. *Annals of the American Association of Geographers* 111(4):1200–18.
Shao, R., B. Derudder, and F. Witlox. 2022. The geography of e-shopping in China: On the role of physical and virtual accessibility. *Journal of Retailing and Consumer Services* 64:102753.
Shove, E., F. Trentmann, and R. Wilk (eds.) 2020. *Time, Consumption and Everyday Life: Practice, Materiality and Culture*. Routledge.
Singleton, C., L. Wright, M. McDonald, I. Archer, C. Bell, G. McLoughlin, B. Houghtaling, K. Stowers, and E. Steeves. 2023. Structural racism and geographic access to food retailers in the United States: A scoping review. *Health & Place* 83:103089.
Stearns, P. 2001. *Consumerism in World History: The Global Transformation of Desire*. Routledge.
Wang, S., and P. Du. 2020. *Retail Geography*. Routledge.

Index

www.ingramcontent.com/pod-product-compliance
Lightning Source LLC
LaVergne TN
LVHW062227200926
848484LV00006B/1488
9781538168318